P9-EJZ-273

"*Macrowikinomics* takes the art of mass collaboration and breaks it down to a science—complete with tools and strategies for rebuilding our institutions in a time of profound change."
—**Lazaro Campos, CEO, Swift**

"Once again, Don Tapscott and Anthony Williams provide essential insight into the deep and rapid changes that are reshaping our world. Their five principles for a world of networked intelligence explain exactly how companies, governments, and NGOs can make lasting change in a complex world."
—**Aron Cramer, President and CEO, Business for Social Responsibility**

"The *Macrowikinomics* assertion that 'there has never been a more exciting time to be human' is spot on. The new engine of innovation driven by collaboration, openness, stewardship, and the power of the social web gives all of us an opportunity to drive even more rapid, meaningful change across global institutions."
—**Michael Dell, President and CEO, Dell Inc.**

"*Macrowikinomics* connects the big picture of business, culture, and society with what is really going on in the trenches of the new digital world. Tapscott and Williams affirm their authority as the leading thinkers by offering facts in the style of demographers and by storytelling in the manner of acute social observers."
—**Brian Fetherstonhaugh, Chairman and CEO, OgilvyOne**

"In *Macrowikinomics*, Tapscott and Williams define a new era of networked intelligence that is changing the way we approach business, government, education and health care. When I read this book I was inspired by the ideas on entrepreneurship and innovation."
—**Dr. Jim Goodnight, CEO, SAS Institute**

"*Macrowikinomics* sets out, comprehensively and holistically, what it will take to reset the postcrisis world, harnessing the power of the Net and the Net-Generation."
—**Ian Hudson, President EMEA, Dupont**

"The idea that global leaders need to stop tinkering with old, tired models, and instead embrace transformative change is key to our future. The challenges facing our planet are greater than we think and the disruption of legacy models, as *Macrowikinomics* portrays, can unleash innovation."
—**Kevin Johnson, CEO, Juniper Networks**

"A magnificent work! Tapscott and Williams guide us through the labyrinth—how everything is being redesigned by collaboration on the Internet. *Macrowikinomics* follows the threads of openness and collective intelligence as they reweave the fabric of our institutions."
—**Kevin Kimberlin, Chairman, Spencer Trask & Co**

"An extraordinary book . . . should be required reading for every university President—and all who want to challenge themselves about how to create positive change for institutions and organizations."

—Linda Lorimer, Vice President, Yale University

"A provocative and powerful series of ideas and case studies that challenge every leader to seize the opportunity to make a difference. The fresh perspectives in *Macrowikinomics* should inspire all of us to collaborate in new ways to meet the challenges we are facing today."

—Craig Mulhauser, CEO, Celestica

"The opportunity for businesses today is to become networks—with a culture of collaborative innovation, stewardship, and integrity. *Macrowikinomics* is a very good guide to get there."

—Filippo Passerini, President, Global Business Services and CIO, Procter & Gamble

"Tapscott and Williams have crafted a blueprint to a better world for our children. *Macrowikinomics* defines the agenda for the systemic change we need and ignore at our peril."

—Kal Patel, Executive Vice President, Best Buy

"A majestic book—breathtaking in its breadth, coverage, and richness of detail. Don and Anthony have taken the trends of collaboration, openness, sharing, integrity, and interdependence, and placed them in pragmatic real-life contexts across a plethora of spheres covering government and citizenship, health care, education, transportation, media, and banking."

—J. P. Rangaswami, Chief Scientist, BT Group PLC

"It's true that business cannot succeed in a world that is failing. The new multipolar and highly interdependent world is requiring us to reboot and rebuild. There is no better guidebook than *Macrowikinomics*."

—Carlos Rodriguez-Pastor Percivale, CEO, Intergroup Financial Services Corporation, Lima, Peru

"*Macrowikinomics* is the leadership guidebook for the emerging networked age. Maps an exciting future. Not starry eyed and naïve, it frames the potential dark side and pitfalls, and calls for all of us to step up our game as leaders. A must read for those who want to control their own destiny."

—Noel Tichy, Professor and Director, Global Citizenship Initiative, University of Michigan and Co-author of *Judgment*

"*Macrowikinomics* underscores the critical importance of transparency, participation, and collaboration among business, government, and citizens in addressing global challenges like corruption."

—Nancy Zucker Boswell, President and CEO, Transparency International-USA

MACROWIKINOMICS

ALSO FROM DON TAPSCOTT AND ANTHONY D. WILLIAMS

Wikinomics: How Mass Collaboration Changes Everything

OTHER BOOKS BY DON TAPSCOTT

Grown Up Digital: How the Net Generation Is Changing Your World

The Naked Corporation: How the Age of Transparency Will Revolutionize Business (with David Ticoll)

Digital Capital: Harnessing the Power of Business Webs (with David Ticoll and Alex Lowy)

Creating Value in the Network Economy

Blueprint to the Digital Economy: Creating Wealth in the Era of E-Business (with David Ticoll and Alex Lowy)

Growing Up Digital: The Rise of the Net Generation

Who Knows: Safeguarding Your Privacy in a Networked World (with Ann Cavoukian)

The Digital Economy: Promise and Peril in the Age of Networked Intelligence

Paradigm Shift: The New Promise of Information Technology (with Art Caston)

Planning for Integrated Office Systems: A Strategic Approach (with Del Henderson and Morley Greenberg)

Office Automation: A User-Driven Method (with Del Henderson and Morley Greenberg)

MACROWIKINOMICS

Rebooting Business and the World

Don Tapscott
and Anthony D. Williams

Portfolio / Penguin

Dedicated to the entrepreneurs and social innovators everywhere who are rebooting business and the world.

PORTFOLIO PENGUIN
Published by the Penguin Group
Penguin Group (USA) Inc., 375 Hudson Street, New York, New York 10014, U.S.A. • Penguin Group (Canada), 90 Eglinton Avenue East, Suite 700, Toronto, Ontario, Canada M4P 2Y3 (a division of Pearson Penguin Canada Inc.) • Penguin Books Ltd, 80 Strand, London WC2R 0RL, England • Penguin Ireland, 25 St. Stephen's Green, Dublin 2, Ireland (a division of Penguin Books Ltd) • Penguin Books Australia Ltd, 250 Camberwell Road, Camberwell, Victoria 3124, Australia (a division of Pearson Australia Group Pty Ltd) • Penguin Books India Pvt Ltd, 11 Community Centre, Panchsheel Park, New Delhi — 110 017, India • Penguin Group (NZ), 67 Apollo Drive, Rosedale, North Shore 0632, New Zealand (a division of Pearson New Zealand Ltd) • Penguin Books (South Africa) (Pty) Ltd, 24 Sturdee Avenue, Rosebank, Johannesburg 2196, South Africa

Penguin Books Ltd, Registered Offices:
80 Strand, London WC2R 0RL, England

First published in 2010 by Portfolio Penguin,
a member of Penguin Group (USA) Inc.

10 9 8 7 6 5 4 3 2 1

LIBRARY OF CONGRESS CATALOGING IN PUBLICATION DATA
Tapscott, Don, 1947–
Macrowikinomics : rebooting business and the world / Don Tapscott and Anthony D. Williams.
p. cm.
Includes bibliographical references and index.
ISBN 978-1-59184-356-6
ISBN 978-1-59184-372-6 (export edition)
1. Information technology—Economic aspects. 2. Information technology—Social aspects. 3. Online social networks—Economic aspects. 4. Business networks—Social aspects. 5. Technological innovations—Social aspects. I. Williams, Anthony D., 1974– II. Title.
HC79.I55T369 2010
303.48'33—dc22 2010023338

Printed in the United States of America
Set in Janson with Daily News
Designed by Daniel Lagin

CONTENTS

I

FROM WIKINOMICS TO MACROWIKINOMICS

1. REBOOTING THE WORLD

On Sunday, January 17, a full five days after a devastating earthquake struck Haiti, a text message sent from a cell phone in Port-au-Prince was translated from Creole into English and posted on an interactive crisis mapping site that was being closely monitored by emergency responders. The text was a cry for help from a survivor and it appeared to have been sent from beneath the rubble of one of Haiti's largest supermarkets. By that time, the odds of finding survivors had diminished sharply and many in the emergency relief community were giving up hope. The situation on the ground was dire indeed: without access to food or water some tens of thousands had already perished beneath the immense piles of concrete strewn across the city. But the text message posted online suggested a miracle: could the person who sent it still be alive? Was it possible they made it through the excruciatingly long wait for help? An American search-and-rescue team raced to the scene to find out. Many hours later, after having cut through several feet of concrete, the rescuers had a horrible realization: the body being pulled from the rubble was that of a child. The small, frail frame of a seven-year-old girl emerged from the supermarket wreckage, deeply shaken and barely alive. The little girl, overwhelmed with relief and emotion, recounted her terrifying experience to her astonished family. She had managed to survive on a small ration of leathery fruit snacks, and a whole lot of hope.

It was a glimmer of light in an otherwise tragic story. Indeed, few people will soon forget the horrendous damage inflicted by the magnitude 7.0 earthquake that struck near Port-au-Prince on January 12, 2010, causing more human misery and economic damage than any earthquake on record. In a mere forty-five seconds of seismic contortions, an astonishing 15 percent of the nation's population—1.5 million people—was rendered homeless. Tens of thousands were dead, and hundreds of thousands more were injured. Any semblance of the usual infrastructure emergency crews depend on (roads, hospitals, water, sanitation, electrical power, and communication networks) was obliterated. Vast regions of the 250-year-old city utterly toppled.

The ruthless and indiscriminate wrath of nature's forces, however, was just a prelude to the real misery. Circumstances on the ground made life astonishingly difficult for first responders. The sea- and airports were congested and there were too few trucks to transport supplies and no safe place to store them. No one—not the army, the government, or the aid community—had a clear picture of the full scale of the catastrophe unfolding around them. There was confusion about precisely which supplies had been received, and in what quantities. There was also a lack of coordination among aid agencies and other entities about which people and areas to prioritize and how to overcome this logistical nightmare. This initial lack of coordination, in turn, left Haiti's earthquake victims (already among the poorest people in the world) utterly destitute, without food, water, or clothing, separated from their loved ones, and many in desperate need of medical attention. Yet, out of the rubble, and in the face of tremendous suffering, came a powerful story of how an ad hoc team of volunteers from around the world came together to concoct an information management solution that far surpassed anything the official crisis response team had mustered, including the world's largest emergency relief organizations, the U.S. State Department, and even the U.S. Army.

At the heart of the volunteer effort was a small Kenyan-born organization called Ushahidi whose crisis-mapping site allows users to submit eyewitness accounts or other relevant information in a disaster situation via e-mail, text, or Twitter—and then visualize the frequency and distribution of these events on a map. Ory Okolloh, a prominent Kenyan lawyer and blogger, first came up with the idea in 2008 when violence erupted in the aftermath of Kenya's disputed election. After hearing many disturbing reports of rape, looting, and murder from friends and family across the country, she suspected that the government and the official news agencies were grossly underreporting the violence. The proof came when her own vivid reporting on her blog Kenyan Pundit triggered a flood of e-mails and texts from hundreds of Kenyans who had witnessed or experienced violence firsthand. The volume of reports soon overwhelmed Okolloh's ability to authenticate and document them using her blog, so she sketched out the basic parameters of an Internet mapping solution, and with the help of some fellow Kenyan technology whizzes, built the Ushahidi platform over a long weekend. Within hours of its launch, the site was collecting user-generated cell phone reports of riots, stranded refugees, rapes, and deaths and plotting them on a map, using the information supplied by informants. For the first time, interested parties could see at a glance which areas of the country were experiencing trouble. Indeed, the site collected more testimony with greater speed and broader reach than the media or the local officials, except in Ushahidi's case there was a big difference: Okolloh didn't have government grants, official mandates, formal command structures, or elaborate communication protocols; just a loose group of committed individuals under effective grassroots leadership harnessing rudimentary open-source technologies to help those in need.

When disaster struck Haiti two years later, Ushahidi's director of crisis mapping, Patrick Meier, sprang into action. Meier had been enjoying a quiet evening watching the news at his home in Boston. It was 7:00 p.m. when he first learned about the earthquake. By 7:20, he'd contacted a colleague in Atlanta. By 7:40, the two were setting up a dedicated site for Haiti on the Ushahidi platform. By 8:00, they were gathering intelligence from everywhere, in a global effort to crowdsource assistance for Haiti.

Since the majority of incoming text messages were in Creole, they needed a translation service. And since most reports lacked sufficient location details, they needed a way to quickly identify the GPS coordinates so that incidents could be mapped as accurately as possible. So Meier reached out to dozens of Haitian communities for help, including the large diaspora in Boston. Soon hundreds of volunteers around the world were using Ushahidi-Haiti to translate, categorize, and geo-locate urgent life-and-death text messages in real-time. Many of the volunteers spent weeks on end on their laptops in a dimly lit school basement in Boston that Meier converted into a makeshift situation room. Although located some 1,640 miles from the scene, the volunteer crisis mappers used Skype to relay critical information about the location of potential survivors to search-and-rescue teams on the ground in Port-au-Prince. They responded to requests from the World Food Program and the aircraft carrier USS *Carl Vinson* in the middle of the night. And to better link calls with specific GPS coordinates, they even got direct access to DigitalGlobe's high-resolution satellite imagery and to the U.S. Army's video footage from military drones. By the time Meier's group had honed their process, text messages were being translated into English and posted online just minutes after they left a mobile phone in Haiti. And as a result of their dedication, Ushahidi's crisis mappers found themselves center stage in an urgent effort to save lives during one of the largest relief operations in history.

"If a relief worker from the Red Cross has a field office in the neighborhood of Delmas," says Meier, "they could subscribe to Ushahidi to receive information on all reports originating from their immediate vicinity by specifying a radius." Not only were responders able to specify their geographic area of interest, but they could also select the type of alert, say collapsed buildings, medical emergencies, food shortages, or looting. Now as the focus shifts from crisis relief to rebuilding in the years to come, Meier thinks Ushahidi's crisis-mapping tools could just as readily be used by Haitians to hold crisis-relief organizations, private contractors, and the local government accountable for higher standards than have been the norm during the many years of failed efforts to lift the impoverished Caribbean nation out of poverty. Indeed, this everyone-as-informant mapping heralds some pretty profound changes as the wiki world revolutionizes the work of humanitarians, journalists, and soldiers who provide aid and assistance in some of the most unforgiving circumstances imaginable.

In the old crisis management paradigm, big institutions and aid workers parachute into a crisis, assess the situation, and dispense aid with the limited information they have. Most aid organizations don't have good systems for sharing information, and certainly don't like ceding turf or marching to the beat of another organization's drum. The resulting fragmentation leads to poor decision making, redundancy, and confusion, and often to wasted money and wasted opportunities.[1] To make matters worse, the end recipients of disaster relief are almost always treated as helpless victims and passive consumers of other people's charity. This makes for perversely compelling television drama (so-called disaster porn temporarily boosted CNN's ratings by 95 percent), but it fails miserably in delivering results. Indeed, a report produced by the International Federation of Red Cross and Red Crescent Societies following the international community's response to the 2004 Indian Ocean tsunami highlighted the need for better coordination as well as victim participation in future disaster relief efforts to help ensure that the needs and interests of disaster victims are not sidestepped in the rush to implement solutions.[2]

The new paradigm for humanitarian efforts turns much of the conventional wisdom upside down. Rather than sit idly waiting for help, victims supply on-the-ground data using cell phones or whatever communication channels are available to them. Rather than simply donate money, a self-organized network of volunteers triages this data, translating and authenticating text messages and plotting incidents on interactive mapping displays that help aid workers target their response. And rather than just forge ahead with narrow institutional priorities, new communication channels like Ushahidi enable the whole emergency relief ecosystem to operate like a coherent entity. Sure, a lot could go wrong with this distributed model. People could get the address wrong or exaggerate their situation. But as data accumulates, interactive crisis maps can quickly reveal emerging patterns and vital information in an emergency situation: How many miles inland did the tsunami kill? Which roadways are passable and where are the closest temporary emergency wards? Are the incidents of violence and looting broadly dispersed or concentrated around certain neighborhoods?

Given an open platform and a complement of simple tools, it turns out ordinary people can create effective new information services that are speedier and more resilient than traditional bureaucratic channels. Yes, one can argue that tales of heroic volunteer efforts are not particularly unusual. Disaster situations do tend to bring out humanity's finer traits. What is remarkable is that the Ushahidi-Haiti project might have taken a government agency with loads of money a year or more to execute. Yet, thanks to social innovators like Okolloh and Meier, the crisis-mapping community rallied to pull it together in a matter of days with absolutely no cost to the taxpayer.

Indeed, the story of how Ushahidi got started, and where it's gone since, reveals a great deal about a powerful new form of economic and social innovation that is sweeping across all sectors—one where people with drive, passion, and expertise take advantage of new Web-based tools to get more involved in making the world more prosperous, just, and sustainable. Okolloh created Ushahidi in crisis; it never crossed her mind to patent or monopolize it. Knowing that computers are out of reach for most Kenyans, Okolloh made sure Ushahidi could work on cell phones.[3] And in the absence of venture capital backing, Okolloh used open-source software and allowed others to reuse her tools for new projects. To date, the versatile platform has been used in Africa to report medicine shortages; in Gaza to track incidents of violence; and in India and Mexico to monitor elections. *The Washington Post* even partnered with Ushahidi in 2010 to map road blockages and the location of available snow blowers during the infamous Snowmageddon, D.C.'s largest snowfall in nearly a century.

With every new application, Ushahidi is quietly empowering millions of ordinary individuals to play a larger role in everything from democratic decision making to crisis management to protecting public health. In doing so, Ushahidi highlights a profound contrast between a set of deeply troubled and stalled institutions that revolve around industrial age thinking and hierarchical organizational designs versus a new set of bottom-up institutions that are being built on principles such as openness, collaboration, and the sharing of data and intellectual property. This new model of collaboration and social production goes way beyond disaster relief efforts to affect the modus operandi of virtually every institution in society, including government, education, health care, science, finance, and international diplomacy.

These industrial age institutions brought us mass production of goods, mass media like newspapers, radio, and television, mass education and learning for everyone, mass marketing and mass democracy and government in which elected officials produced and distributed laws and services. As a mode of production the industrial economy was infinitely superior to what came before it (the agrarian craft society), dramatically advancing wealth, prosperity, and the standard of living for many. But this was a centralized, one-way, one-size-fits-all mass model controlled by the powerful owners of production and society.

Now because of the new Web the old industrial models are all being turned on their heads. There is now a new engine of innovation and wealth creation and a powerful new force that radically drops collaboration costs and as such enables communities to collaborate on shared concerns, endeavors, and challenges. Greater openness in innovation and science, for example, is creating more economic opportunity for citizens and businesses that learn how to tap into global innovation webs. In the fight against climate change, ordinary people are forging a mass movement to bring greater consumer awareness and a sense of community to making ordinary household and business decisions that can reduce our carbon footprints. In education, leading universities are breaking down their ivory towers and building a global network for higher learning—a rich tapestry of world-class educational resources that every aspiring student on the planet can use and return to throughout his or her lifetime. Innovators across the public sector are harnessing the Web to generate more productive and equitable services, bolster public trust and legitimacy, and unlock new possibilities to co-innovate solutions to local, national, and global challenges. Put it all together and it becomes increasingly clear that we can rethink and rebuild many industries and sectors of society on a profoundly new, open, networked model. Indeed, for the first time in history, people everywhere can participate fully in achieving this new future.

In our previous book, *Wikinomics* (Portfolio 2006), we called this new force "mass collaboration" and argued that it was reaching a tipping point where social networking was becoming a new mode of social production that would forever change the way products and services are designed, manufactured, and marketed on a global basis. But, in the four years since penning the idea, it's clear that wikinomics has gone beyond a business or a technology trend to become a more encompassing societal shift. It's a bit like going from micro- to macroeconomics. In which case, wikinomics, defined as the art and science of mass collaboration in business, becomes macrowikinomics: the application of wikinomics and its core principles to society and all of its institutions. Just as millions have contributed to Wikipedia—and thousands still make ongoing contributions to large-scale collaborations like Linux and the human genome project—there is now a historic opportunity to marshal human skill, ingenuity, and intelligence on a mass scale to reevaluate and reposition many of our institutions for the coming decades and for future generations. After all, the potential for new models of collaboration does not end with the production of software, media, entertainment, and culture. Why not open-source government, education, science, the production of energy, and even health care? As we will discover in later chapters, these are not idle fantasies, but real opportunities that the new world of macrowikinomics makes possible.

A TURNING POINT IN HISTORY

When the economy crashed in 2008, it cost American taxpayers trillions of dollars. Faced with a historic market meltdown, the worst recession in three generations, plus government guarantees that exceed the cost of every war the United States has ever fought, American taxpayers are understandably furious. It is pretty much the same story around the world. Many people are reviving calls for updated regulations, more government intervention, and even the breakup or nationalization of the big banks. In the meantime, the lingering effects of the financial meltdown threaten to engulf not just companies but entire countries in a sovereign debt crisis. In early 2010, Greece appeared unable to make good on its debt payments to worldwide bondholders. The fear about state defaults quickly spread to Spain, Portugal, and Ireland in a fiscal domino effect that jeopardizes all sixteen nations sharing the euro as a common currency. Governments everywhere are awash with unprecedented and potentially unsustainable debt. The United States looms largest, with Congress contemplating a budget that by 2020 would nearly double America's national debt, to $22 trillion—twice the size of the U.S. economy.[4] Clearly we need to rethink the old approaches to governing the global economy. But rebuilding public finances and restoring long-term confidence in the financial services industry in the United States and other nations will require more than government intervention and new rules; it's becoming clearer that what's needed is a new modus operandi based on new business principles like transparency, integrity, and collaboration.

The financial system is not the only institution that's in desperate need of a makeover. The sparkling possibilities described above contrast sharply with the precipitous decline of the industrial economy as a whole. Many of the institutions that have served us well for decades—even centuries—seem frozen and unable to move forward. Sure, the industrial economy brought us three centuries of unprecedented productivity, knowledge accumulation, and innovation that resulted in undreamt-of wealth and prosperity. But that prosperity has come at a cost to society and the planet. And it is clear that the wealth and security enjoyed in advanced economies may not be sustainable as billions of citizens in emerging markets aspire to join the global middle class. If we continue on a business-as-usual path, today's global instability will surely increase. Indeed, we believe the world has reached a critical turning point: reboot all the old models, approaches, and structures or risk institutional paralysis or even collapse. It's a question of stagnation versus renewal. Atrophy versus renaissance. Society has at its disposal the most powerful platform ever for bringing together the people, skills, and knowledge we need to solve many of the issues plaguing the world. And if this book shows anything, it's that good things happen when we, as individuals and as organizations, seize the opportunity to contribute our ideas, our passion, and our creativity. The question is whether the world is ready to truly embrace the social and economic innovations that this collaboration could unleash.

This may sound like a fairly radical perspective, but it's one that mainstream voices increasingly endorse. None other than *Time* magazine recently saw fit to ring the alarm bells over the trouble it sees on the horizon. In a special feature on the ten ideas that will define the next ten years, Christopher Hayes asserts that it is twilight time for the elites. He says, "In the past decade, nearly every pillar institution in American society—whether it's General Motors, Congress, Wall Street, Major League Baseball, the Catholic Church or the mainstream media—has revealed itself to be corrupt, incompetent or both. And at the root of these failures are the people who run these institutions, the bright and industrious minds who occupy the commanding heights of our meritocratic order. In exchange for their power, status and remuneration, they are supposed to make sure everything operates smoothly. But after a cascade of scandals and catastrophes, that implicit social contract lies in ruins, replaced by mass skepticism, contempt and disillusionment."[5] So what are we to do in the wake of this broad-based implosion? This new decade, Hayes suggests, will have to be about reforming our institutions to reconstitute a more reliable and democratic form of authority.

In fact, the evidence is growing daily that a powder keg of sorts is in the making. Fifteen million people between the ages of fifteen and twenty-four were out of work in North America and Europe as of this writing. There was 25 percent youth unemployment in France and Italy. In Spain, 45 percent of young people were jobless. At these levels, we are talking about structural unemployment for the whole generation and the picture ain't getting any rosier. Youth jobs are likely to be the last to come back after the recession and research shows that prolonged periods of unemployment can have lasting effects on one's career prospects as skills and education quickly get dated. The risk that a large cohort could fall behind has policy makers questioning how they're going to prevent a "lost youth generation" from becoming one of the recession's long-term casualties. Is there a way forward? Don't look to the government or the big corporations for the answers. The Kauffman Foundation's analysis of recent U.S. Census Bureau data shows that companies less than five years old create nearly two thirds of net new jobs in America.[6] In other words, if the U.S. economy is going to have a sustained recovery in jobs, it will be up to entrepreneurs to lead the way.

Indeed, as the global economic crisis forced companies in every sector to cut their costs to stay alive, many began to see that hunkering down, while necessary, was insufficient for long-term success. From manufacturing to retail, smart managers have begun to initiate long overdue changes to their structures and strategies. It seems there is growing consensus that we are finally entering a very different economy. Economist Robert Reich asks, "What will it look like? Nobody knows. All we know is the current economy can't 'recover' because it can't go back to where it was before the crash." Instead he suggests, "we should be asking when and how the new economy will begin."

The upheaval is now spreading to other sectors—from the universities, health care, and science to energy, transportation, and government. Many old media empires are crumbling. The continuing collapse of many newspapers in the United States is a storm warning of more to come. As of this writing in May 2010, the Tribune Company, owner of the *Los Angeles Times* and *Chicago Tribune*, is bankrupt, as is the owner of *The Philadelphia Inquirer*. The *Rocky Mountain News* and *Seattle Post-Intelligencer* are gone, and the *San Francisco Chronicle* and others are in deep trouble. *The New York Times*' debt has been downgraded to junk status and lovers of the "paper of record" (like us) are on a deathwatch, wondering how many more months it can service its debt. Magazines are in trouble too, and your favorite has probably never had such a dearth of advertising.

Basically, the Internet has destroyed the business model for print. Compared to the massive physical plant of, say, *The New York Times*, online newspaper *The Huffington Post* has almost zero "printing" and distribution costs. *The New York Times* employs over a thousand people in its editorial department alone. HuffPo employs sixty and a volunteer roster of thousands of writers. As a result the site is thriving with 20 million readers. Journalism will surely survive, just not in its present form. Whether core journalistic values like objectivity, quality, and truth will prevail remains to be seen. And how will journalists make a living?

Around the world, health care systems are under severe stress, but none more so than in the United States. In 1960, the United States spent only 5.2 percent of GDP on health care. By 2009 that number had risen to 17.3 percent, which means America now spends more on health care than it does on food.[7] The underlying causes are complex and include factors ranging from the expansion of medical possibilities, thanks to new technologies and research, to demographic and lifestyle factors such as aging populations and poor diets. Such spending levels might be justified if the system delivered superior results, but many indicators point to the contrary. Despite being the world's largest spender, the United States ranks forty-ninth out of 224 countries in health care performance, with disproportionately lower life expectancies and higher levels of infant mortality than other rich nations like Germany and Japan.[8] Meanwhile some 46 million Americans with inadequate or nonexistent coverage fear an encounter with the health care system as much as, if not more than, they fear getting sick in the first place. Even those with "good coverage" are often poorly protected. A study by Harvard researchers found that medical problems caused an astonishing 62 percent of all personal bankruptcies filed in the United States in 2007. But even more surprising was that 78 percent of those filers had medical insurance at the start of their illness.[9]

Factor in high levels of administrative inefficiency, and it is clear that current models of health care are unsustainable. If the current growth trajectory continues, health care costs will nearly double to $4.5 trillion in less than a decade, eroding carefully orchestrated tax and social security systems and plunging the nation further and further into debt.[10] While politicians tinker with funding models, physicians and patients increasingly recognize that only a deeper transformation in the way society promotes physical well-being and cares for the sick will be sufficient to safeguard the system. As Dr. Michael Evans, a physician at St. Michael's Hospital in Toronto, put it, "today's health care institutions are like the old media: centralized, one way, immutable and controlled by the people who created and delivered it. Patients are passive recipients." In the new model, patients become more like partners—they self-organize, contribute to the total sum of knowledge, share information, support each other, and become active in managing their own health.

However, with much more serious problems looming, fretting about the future of U.S. health care could seem like a luxury. Water, or more precisely the lack of fresh water, is shaping up to be a catastrophe for humanity. About 2.8 billion (or 44 percent) of the world's population lives in regions where fresh water resources are under severe stress. This troubling figure is set to rise to 3.9 billion by 2030.[11] As yet, nobody has determined exactly how the world's need for fresh water will be met. The challenge is so immense that we will need some of our smartest minds focused on finding solutions before water shortages erupt into conflicts.

In a time of growing capacity and wealth, our world is a very unequal place. Some countries—like China and India—have been immensely successful at lifting a large percentage of their populations out of poverty in a short time period. But vast regions of the world have not shared in this prosperity. Around the world, ten children die of hunger every minute. Nearly one quarter of the global population struggles to eke out an existence on less than two dollars a day. Is it acceptable that such a large share of humanity has been entirely bypassed by the modern world? And how long can this imbalance persist?

The demise of the Cold War was supposed to bring an era of lasting peace. Yet the world is seemingly less secure than ever, despite the $1.46 trillion spent on defense on an annual basis.[12] The Faustian nuclear standoff between the superpowers of yesterday's bipolar world has been replaced by a volatile powder keg where rogue nations and rogue groups inch closer to having their own personal weapons of mass destruction. Rather than an improbable rain of missiles we are confronted by the real possibility of a backpack or courier package delivering the toxins or explosive power to destroy a city.

Then there is the most daunting of challenges: weaning the world off its dangerous addiction to fossil fuels and building a new green energy economy that can sustain human civilization for centuries to come. At the current pace of development we are still decades away from truly mass-market deployment of any clean energy solutions, let alone all of them. "You easily spend 10 years maturing a new technology from the lab to a first commercial plant. And that's just the beginning," says Peter Voser, the CEO of Royal Dutch Shell. "It usually takes another 25 years for this new energy type to conquer 1 per cent of the global market."[13] Biofuels are reaching that mark about now. Wind could do so by 2015, twenty-five years after the first large-scale wind parks were built in Denmark!

It's true that investment is set to increase. China has recently ramped up its investments in green energy technology to the tune of US$9 billion per month according to some accounts.[14] Al Gore is challenging the United

States to produce every kilowatt of electricity through wind, sun, and other climate-friendly energy sources within ten years, an audacious goal he hopes the Obama administration will embrace. Even people in the green power industry are calling it "ambitious." But one thing is increasingly certain: it will take a new way of thinking about the challenges and opportunities associated with the green energy economy—including an unprecedented level of transparency, collaboration, and technology sharing—to bring changes like these about.

It's easy to see these issues as isolated, but in fact they are highly interrelated. Excruciating levels of poverty provide a fertile ground for extremism.[15] The failed states in which many of the world's poorest live provide a safe haven in which terrorists set up base camps and hijack oil tankers to fund their nefarious deeds. Speaking of oil, the world's unrelenting addiction to its diminishing supplies is sowing the seeds of deeper global instability in the years to come, not to mention the environmental ruin inflicted by catastrophes like the Gulf of Mexico oil spill. And if that's not already bad enough, runaway climate change could displace hundreds of millions of people, creating a permanent state of emergency that will make the tragic events in Haiti seem like a mere warm-up exercise for the international community.

But is the international community up for the job? Arguably not, according to Klaus Schwab, founder of the World Economic Forum, which recently launched an ambitious Global Redesign Initiative to develop new institutions for global problem solving. "Our existing global institutions require extensive rewiring, and a fundamental shift in values and political culture is vital if we are to foster the global cooperation necessary to confront contemporary challenges in an effective, inclusive and sustainable way," he says. He's right. Decades of economic development, integration of product and service markets, cross-border travel, and new technologies enabling virtual interaction have created a world that is much more complex and bottom-up than top-down. People increasingly perceive their interdependence and seek ways to express it outside of formal national political structures. When the United Nations was founded in 1945, for example, there were only a few dozen nongovernmental organizations (NGOs) in the entire world. And they sure didn't have a seat at the table. Today, an estimated 100,000 such organizations operate internationally in virtually every field of human endeavor. The upshot, argues Schwab, is that the world's citizens "have become more aware that global problems require global trusteeship and that efforts to solve problems solely through traditional negotiating processes, characterized by the defense of national interests, are inadequate in the face of critical global challenges." Large international organizations like the United

Nations and the World Bank now recognize that NGOs can powerfully influence the marketplace and the public sector—either as high-profile challengers or as partners in finding new solutions.

Such inflection points have come before. John Gerard Ruggie, director of the Center for Business and Government at Harvard University, warns that history can repeat itself. "The lesson that capitalist countries needed to combine the efficiency of markets with the broader values of community . . . did not come to them easily. It took the calamitous collapse of the Victorian era of globalization—into worldwide war, followed by extreme left wing revolution in Russia, extreme right wing revolution in Italy and Germany, militarism in Japan, the Great Depression, unprecedented financial volatility and the shriveling up of world trade." Ruggie, who served as a special adviser to former UN secretary-general Kofi Annan, thinks the new era of globalization requires a new social contract. Unregulated free markets, he warns, could spawn another series of cataclysmic events if adequate social and environmental protections are not somehow embedded in the global economy. In other words, the efficiency of markets must be combined with the values of community to sustain a viable global society. The institutions that established the historic social bargains that underpinned post–World War II prosperity (i.e., national governments, business associations, and organized labor) are no longer the right ones to help rebuild the global economy or fashion a new form of sustainable governance.

A TIME FOR RENEWAL AND TRANSFORMATION, NOT FOR TINKERING

In his opening address to the 2010 meeting of the World Economic Forum in Davos, French president Nicolas Sarkozy remarked on the financial crisis that brought the world to the very edge of economic apocalypse. "This is not just a global financial crisis," he said, "it is a crisis of globalization." He called on world leaders to correct the systemic imbalances that led to the triumph of markets over democracy and justice. "In the future, there will be a much greater demand for income to better reflect social utility and merit," he said. "There will be a much greater demand for justice. There will be a much greater demand for protection. And no one can escape this. Either we change of our own accord, or change will be imposed on us by economic, social and political crises. Either we are capable of responding to the demand for protection, justice and fairness through cooperation, regulation and governance, or we will have isolation and protectionism."

Like many heads of state, President Sarkozy certainly means well, but he doesn't really show the way forward. He calls for more "international cooperation" and points to the G20 as a source of solutions and new models of global governance. He proposes taxes on financial speculation to help fund the fight against poverty. He demands that the world move quickly to adopt a more robust, binding global agreement on climate change.

All of these things, while necessary, are only the beginning of what is to be done. Like most heads of state, Sarkozy tends to see the same institutions that produced the current mess as the source of solutions and stability in the future. He argues for a change in values, but still takes most of the old assumptions about how the world works for granted. For example, he doesn't consider the fact that markets may have triumphed precisely because our models of government and democracy are broken. He doesn't call for a complete rethink of the top-down approach to global problem solving. He merely calls for the same old elite club of decision makers, except this time with a few additional members at the table. He doesn't seem to recognize that the new models of social innovation and wealth creation that offer genuine promise are fundamentally incompatible with his outmoded vision of the role of the nation-state in a global economy. Sarkozy proposes traditional instruments like taxes and legal agreements, but they won't be enough. Many of the time-urgent situations we face in this century won't be solved without a more dynamic way to marshal and fully exploit the collective ingenuity of citizens and businesses around the world.

Sarkozy is hardly alone. Most world leaders—indeed, most leaders of business and government anywhere—harbor the same old tired set of assumptions about how to solve the world's problems. And more often than not, they seem focused on tinkering with old models rather than moving to something new and viable. Consider the dysfunctional financial services industry. Conventional policy wisdom demands more regulation over financial markets. But no one stops to ask whether the current models of regulatory oversight and enforcement are truly equipped for the job. Can a patchwork of national financial regulators—all operating in silos with a skeleton crew of underpaid and overworked employees—really be expected to exert meaningful control over a global financial system that operates at light speed and employs some of the smartest and most highly paid people on the planet? Isn't it time for a new model of regulation: one that uses the Web to disclose pertinent information and enables a worldwide network of experts—including the thousands of analysts already employed by government regulators today—to pool their tips, risk models, and analyses in a wiki-like fashion? Sure, it would mean setting aside aspects of national sovereignty and requiring companies to disclose more information in more usable formats than they do today. Rather than cling to the status quo, we'll argue in chapter 3 that these are the kinds of changes that we should be debating.

Climate change presents similar challenges to reigning orthodoxy about how to address global problems. The December 2009 meeting of world leaders in Copenhagen was once heralded as a defining moment for humanity and a chance to prove definitively that international cooperation can and will prevail against the challenges facing the planet. Despite years of preparation and many heads of state in attendance, Copenhagen produced a twelve-paragraph "accord" with weak targets, no details, and no binding commitments. The failure to secure a meaningful deal in Copenhagen has many questioning whether a political deal is possible at all. "The forces trying to tackle climate change are in disarray, wandering in small groups around the battlefield like a beaten army," said one senior British diplomat.[16]

Many politicians and pundits aren't even in the right card game. They want to legislate climate change out of existence with a "cap and trade" system or a tax on carbon when evidence suggests that, over the long term, nothing short of a complete reindustrialization of the planet is in order. Getting the economic incentives right is an important start. But among other things, we need to rethink transportation, adopt new manufacturing and shipping practices, pull off a dramatic shift toward greener products and lifestyles, and retool our energy system, all while devoting enormous intellectual and financial resources to protecting the world's most vulnerable peoples and locations from the effects of rising sea levels and other consequences. Surely a little bit of political tinkering will be insufficient to achieve all of this.

In short, many of our institutions are stalled, lacking vitality, leadership, and dynamism. It's like every last ounce of oxygen has been squeezed out, leaving a mess of deflated expectations and chronically underutilized resources. This apparent paralysis, in turn, begs some pretty fundamental questions: if the knowledge, leadership, and capability required to solve the really tough problems can't be found in the corporate headquarters and national capitals around the globe, will it be found at all? And if so, where will the new insights and leadership come from? Indeed, if our problems are not solvable by fine-tuning existing institutions, what new models and structures should replace them? Are you, wearing your various hats as an employee, manager, learner, teacher, entrepreneur, voter, consumer, community member, or citizen of the world, prepared to take on a larger role in reinventing our beleaguered institutions? What must be done to reboot business and the world and how can you participate?

These are just some of the tough questions we tackle in this book. In these pages you won't read of a single superficial tweak or tinker to old failing institutions. Rather, we'll meet individuals, companies, and organizations that are forging new models of problem solving in their sectors and industries—models that rely less on central control and more on getting a self-organizing critical mass of people and organizations working to initiate small experiments and social innovations that can mushroom into pervasive changes in societal behavior. Put simply, these individuals and organizations have learned how to tap into the world's decentralized sources of knowledge and capability using an approach that mobilizes not just the world's largest companies and nations, but a whole ecosystem of citizens and organizations around the globe. As citizens, and as leaders within our organizations, we need to look beyond the borders of nations and think about society in broad, global terms. If our problems are global in scale, then we need to come together as global citizens to solve them. A system erected around the primacy of national and corporate self-interests just isn't going to cut it for this century.

CREATIVE FRICTION AT THE FRINGES OF NEW AND OLD

The good news is that while many institutions are in various stages of decline, for each we can now see the clear contours of fresh thinking, new approaches, and rebirth. Indeed, while this crisis of globalization paints a bleak picture, there is a more intriguing and optimistic story to be told—one that does much more than catalogue a long list of institutional failures; one that offers new solutions for outmoded institutions in business, government, and society. Billions of people can play active roles in their workplaces, communities, national democracies, and in global forums and institutions, too. At the same time, the new world of wikinomics gives organizations an opportunity to tap into new sources of insight and value. Closed, hierarchical corporations that once innovated in secret can now tap, and contribute to, a much larger global talent pool—one that opens up the world of knowledge workers to every organization seeking a uniquely qualified mind to solve their problem. Scientists can accelerate research by open-sourcing their data and methods to offer every budding and experienced researcher in the world an opportunity to participate in the discovery process. Doctors can collaborate with self-organizing patient communities where people with similar medical conditions share insights, provide mutual support, and contribute to medical research. As this book shows, the possibilities are literally endless.

To be sure, collaborative innovation can have downsides—including tough adjustments for industries whose business models were based on scarcities that no longer exist. Open-source software benefits some companies—particularly users—while hurting those who depend on selling proprietary wares. Health care sites can provide a rich and reliable resource for people seeking information or mutual support, but they can also sow confusion and misinformation. The Web provides an enormous asset for fostering reasoned democratic debate, but so-called online town halls frequently deteriorate into senseless banter and echo chambers where the like-minded rail against their ideological opponents.

The growing prominence of collaborative endeavors also raises a number of tough questions about the roles and responsibilities of different actors in society. Can we really rely on the self-organized masses to deliver critical services such as compiling life-and-death information in a crisis? What happens if the funding dries up or people lose interest and move on to something else? Who will take responsibility if something goes wrong, or claim success when things go right? And who's ultimately accountable when everyone is on everyone else's turf?

In the old paradigm, there were clear roles and responsibilities. In the new world of wikinomics, the lines between sectors and institutions are blurring. Nonprofits increasingly act like entrepreneurial start-ups. Businesses are taking on some of the functions of government. Governments are caught in a network of powers and counterinfluences of which they are just a part. And though most people recognize that problems get solved more quickly when governments, businesses, nonprofits, and citizens work together, there is still a dearth of understanding about how to make partnerships across sectors work at the pace of wikinomics.

These are just some of the genuine concerns that we will return to throughout the book. Indeed, in each sector—whether education, media, health care, energy, finance, or government—there is a mixture of promise and peril ahead. On balance, however, wikinomics offers profound social benefits, including the opportunity to broaden access to science and knowledge, impose greater transparency on financial markets, accelerate the invention and adoption of green technologies, and help make today's leaders in business and government accountable for delivering outcomes that enhance well-being around the world. Collaborative communities not only transcend the boundaries of time and space, they can reach across the usual disciplinary and organizational silos that inhibit cooperation, learning, and progress. In doing all these things, mass collaboration provides an attractive alternative to the hierarchical, command-and-control management systems that are failing many of our key institutions.

Of course, there is the risk that old institutions will crumble before new ones emerge in their place, leading to troubling dislocations. But the creative friction that emerges from the old and new models of innovation is healthy and constructive; it demonstrates that our institutions are not static and that our dynamic and diverse societies move on, even when policies—and policy makers—lag behind. What will be of paramount importance now is laying a social and political foundation that recognizes the emergence of a new economic model whose potential has not even begun to be exhausted. Indeed, if we are serious about rebooting business and the world, then we must be able not only to talk innovation, but to do innovation and do it fast. Every stakeholder involved—not only companies operating in diverse sectors such as transportation, media, health care, and energy, but also universities, scientific institutions, and governments—must summon the courage and creativity to reinvent themselves, using technology and collaboration as an enabler, a catalyst, and a driver of change with the ultimate goal of providing better outcomes for citizens and users. This is not about tinkering at the edges; this is about devising, living, and experiencing a new model of innovation that is fit for the twenty-first century.

Changes of this magnitude have occurred before. In fact, human societies have always been punctuated by periods of great change that not only cause people to think and behave differently, but also give rise to new social orders and institutions. In many instances these changes are driven by disruptive technologies that penetrate societies to fundamentally change their culture and economy. But today's Internet is the most powerful platform yet for facilitating and accelerating new creative disruptions. People, knowledge, objects, devices, and intelligent agents are converging in many-to-many networks where new innovations and social trends spread with viral intensity. Organizations that have scrambled to come up with responses to new phenomena like peer-to-peer downloading, free Internet telephony, or the blogosphere should expect much more of the same—at an increasing rate—in the future.

For individuals, this is an exciting new era—an era in which we can participate in production and add value to large-scale cultural, political, and economic systems in ways that were previously impossible. For large and small companies alike, new models of mass collaboration provide myriad ways to harness external knowledge, resources, and talent for greater competitiveness and growth. For governments and society as a whole, evidence is mounting that we can harness the explosion of knowledge, collaboration, and business innovation to lead richer, fuller lives and spur prosperity and social development for all.

Needless to say, these transitions will not be easy, and not everyone will come out a winner. According to IBM's 2010 Global CEO survey, eight in ten CEOs expect their environment to grow significantly more complex, and fewer than half believe they know how to deal with it successfully.[17] This is not surprising. Whenever such a shift occurs, there are always realignments of competitive advantage and new measures of success and value. To succeed in this new world, it will not be enough—indeed, it will be counterproductive—simply to carry on with the current stimuli, policies, management strategies, and curricular approaches. So let's use the opportunity that the digital revolution presents to rethink and rebuild all of the old approaches and institutions that are failing. Many promising solutions to issues ranging from the current health care crisis to climate change already exist at the fringes of established institutions and in the collaborative spaces of the Web. Organizations that tap this new force can claim leadership roles in a world where collaborative innovation across borders, disciplines, and cultures is becoming a societal imperative. Those that resist, or fail to get on board, will find themselves ever more isolated—cut off from the networks that are sharing, adapting, and updating knowledge to solve problems, big and small.

2. FIVE PRINCIPLES FOR THE AGE OF NETWORKED INTELLIGENCE

Around the turn of the century, in the midst of the dot-com frenzy, many bold predictions were issued about how the Internet was going to revolutionize the economy and society overnight. You've probably heard some of these before: most physical retail would disappear; television was toast; we would all soon work from telecottages in the south of France as masses of white-collar workers fled the oppressive confines of their dull gray cubicles. That didn't quite happen. But then history tells us that the impacts of technology-driven revolutions are often long delayed. We expect too much in the short term, but underestimate or completely miss the longer-term implications for society. That is not surprising. Culture is sticky. Old habits and old ways of working die hard. Vested interests fight the transformations unleashed by revolutionary forces so that the anticipated changes take decades, and sometimes many generations, to unfold.

When Gutenberg introduced Western Europe's first press and printing process in 1440, for example, the world was a very different place. The sum of mankind's knowledge was contained in oral traditions or inscribed in rare and fragile manuscripts that were closely guarded by the elites of feudal society.[1] Feudalism's rigid social structures and relatively static natural economy draped a blanket of inertia over estates, communities, and towns. Most products were made and consumed locally, thanks to a large indentured labor force that was neither paid for its efforts, nor in control of the lands in which it toiled. The elites, who jealously guarded access to knowledge, also controlled society's institutions. Unless you were royalty or belonged to the clergy, there was no prospect of formal education and there was no such thing as reading. There was no voting and no say in political decision making. There was no freedom or economic opportunity. There was no concept of progress. Apart from family, there was merely servitude, and there was survival.

In those days, wars among medieval kingdoms were fought with cannons, large cavalries, and brute men carrying swords. But it was arguably Gutenberg's printing press that first pierced the armor of the old established orders and gave rise to new social classes, new institutions, and new principles by which to guide them. His printing technology made possible the economical mass distribution of written materials—not just books, but also government edicts and records, maps, scientific and engineering documents, how-to guides, pamphlets, posters, banners, and flyers. Scientific knowledge, ideas, and artistic and cultural expression could be produced, marketed, criticized, revised, and preserved on an ever-expanding scale.

Many of the people who produced and distributed these new publications joined an emerging merchant class. They took on new roles as patrons of scientific research, art, theater, music, and education. They created the soul of a new age—the concept of self and the right of the individual to "ownership" of his (not yet *her*) own faith; and the right to consider ideas, laws, and religious edicts in a critical and free manner. As the ideas of the Enlightenment took hold, society began to create, accumulate, and harness knowledge in new ways.[2] Engineers, mechanics, chemists, physicians, and natural philosophers formed circles in which access to knowledge was the primary objective. For the first time in history, knowledge about the natural world became increasingly nonproprietary. Scientific advances were shared freely within informal scholarly communities and with the public at large. The world opened up in new ways too. As European trade advanced, new knowledge was acquired from the Arab world, Africa, Asia, and the Western Hemisphere, which in turn spurred an ever-greater spirit of adventure and discovery. Open scientific communities created a new medium for intercultural exchange and places like Córdoba (Spain) provided a major cosmopolitan meeting place that helped bridge Europe and the Arab world.[3]

As with today's Internet, the printing press caused calamity, confusion, and disruption in many aspects of society. Just as the printed word could be a vehicle for enlightenment, it could also spew hatred and misinformation. The nascent publishing industry met with resistance as vested interests fought change. But nothing could stop the deep changes prompted by the printing press, and ultimately every institution in society was challenged. Just as the clergy could no longer control science or medicine, the monarchs could no longer dominate every aspect of political life or grip the reins of a burgeoning capitalistic economy. Power was leaching down through social hierarchies, freed by the dissemination of information and knowledge. Eventually a rising class of informed and powerful businessmen, professional soldiers, and intelligentsia in countries like France and in the British colonies demanded a new kind of economy and new forms of rule free of the old power of the Church and feudal nobility. Through armed insurrection, they did indeed lead their populations in "changing the very warp and woof of history."[4]

THE AGE OF NETWORKED INTELLIGENCE

The men and women of the Renaissance period likely had little inkling of the truly titanic changes that were unfolding around them, let alone the ability to predict where it would all end up. Nevertheless, many of their legacies remain with us today: the birth of modern nation-states; the expansion of political rights and freedoms; the rise of the university, the media, and the industrial corporation; and a Cambrian explosion of science, medicine, knowledge, and cultural expression. Clearly, the printing press can't account for all of this. European imperialism, the assembly of large standing armies, and other powerful inventions such as the mariner's compass, gunpowder, and the steam engine were influential in shaping the modern world. But it is equally true that one can't begin to understand how today's world came into existence without comprehending how a modest innovation in movable type helped broaden the distribution of power and knowledge in society.

Thanks largely to the Internet we are crossing a similar chasm today. Long-standing monopolies and power imbalances are once again being challenged as more people from more regions of the world now connect, collaborate, and compete on the global stage. Young digital natives everywhere are questioning the historic traditions of venerable institutions such as the university, the newspaper, the medical establishment, and the entire apparatus of representative government. Ailing industrial sectors such as energy and transportation are in the midst of profound change as the digital age unearths new opportunities to accelerate research and collaborate around sustainable alternatives. Old truths about how to motivate and retain human capital are proving inadequate in the face of new innovation models where organizations source ideas and the skills to implement them from a vast global talent pool.

In the long run we will look back at this period as a time when the world began a historic transition from industrial capitalism to a new kind of economy based on new principles and new ways of thinking and behaving. And while there are certainly many similarities between what is happening today and what happened over five hundred years ago, there are also some profound differences. Printing gave humanity the written word. The Web makes everyone a publisher. Printing enabled the distribution of knowledge. The Web provides a platform for networking human minds. Printing allowed people to know. The Web enables people to collaborate and to learn collectively. Printing played a key role in the rise of the industrial revolution and the creation of capitalism. The Web is now enabling new models for creating wealth and prosperity on a global basis. However, the biggest difference between the printing press and the Internet is that what took four centuries to unfold then is occurring in as little as four decades today.

The long-term effects of these shifts on all our institutions are hard to forecast with accuracy. As media analyst and author Clay Shirky put it in his essay about the impending collapse of the newspaper industry: "That is what real revolutions are like. The old stuff gets broken faster than the new stuff is put in its place." He makes the case that even the revolutionaries can't predict what will happen. "Ancient social bargains, once disrupted, can neither be mended nor quickly replaced, since any such bargain takes decades to solidify," he says.[5] Shirky is right, but it's worth remembering that the future is not something to predict, it is something to achieve. And through the haze and all the turmoil, the contours of fresh new forms of human organization, as revolutionary as those of centuries ago, are beginning to reveal themselves. Thanks to a new global medium for collaboration and an unprecedented level of social connectivity, people in business, government, and society at large have powerful new tools for reinventing our institutions around a new set of organizing principles for the twenty-first century.

Having said that, let's be clear. Wikinomics, *by itself*, is not a panacea or a complete recipe to fix all of the world's ailments. It is not a *wholesale replacement* for good government, the corporation, professional journalism, our health care systems, or our institutions of higher learning. Nor is it an argument to replace the dynamism of capitalism with some new form of online collectivism or central planning by committee. Financial markets and corporations will remain the underlying engines of innovation, prosperity, and job creation. Governments will still collect taxes, provide social security, and enact new laws on their population's behalf. Universities will continue to be an oasis for learning, advanced research, and free thinking, as well as a place where young people go to "grow up." Health care will still largely consist of highly trained experts treating sick patients with the latest medical techniques. Web-savvy citizens will still seek out reliable news, information, and advice from trained journalists and credentialed experts.

In each instance, however, we are seeing how the new age of networked intelligence renders conventional approaches to value creation insufficient, and in some cases, completely inappropriate. Collaborative innovation, for example, is killing the old, hardwired "plan and push" mentality taught in business schools. Citizens who co-create policies and services are exposing the one-size-fits-all model of government as an anachronism. Twenty-first-century universities increasingly look and feel more like a global network than an ivory tower. And, just as journalists now coexist with a much broader ecosystem of knowledge producers, self-organizing patient communities and a greater emphasis on education and preventative medicine are beginning to augment conventional health care.

Organizations can succeed and even thrive in this new environment by embracing the five principles of wikinomics: collaboration, openness, sharing, integrity, and interdependence. Indeed, three years of research and hundreds of interviews have convinced us that these are not only key business principles; they are principles for achieving a world that is secure, prosperous, just, and sustainable. In the remainder of the chapter, we explain why.

1. Collaboration

Human hierarchies are power structures, defining rank, importance, status, and accountabilities, and as such have worked well as a means of organizing the way most organizations have operated over the centuries. But an increasingly complex and interdependent global economic environment is now exposing hierarchical institutions as being deeply limited and perhaps even a liability as we enter an era in which the upper bounds of human ingenuity will be tested. From a CEO's perspective, an increasingly self-organized and networked world makes the old models of top-down corporate management and industrial planning increasingly feeble—a bit like erecting a straw house to provide shelter from an oncoming hurricane. Science and technology now evolve at such a great speed that even the largest companies can't possibly research all the fundamental disciplines that contribute to their products. Nor can they can control an end-to-end production process or seek to retain the most talented people inside their corporate boundaries. Instead, smart companies increasingly collaborate globally to get things done. Whether designing a new product or launching a global marketing campaign, mass Internet-enabled collaboration has become a powerful modus operandi for business and many other institutions, as we will see shortly.

As we explained in *Wikinomics*, collaboration used to conjure up images of people working together happily and productively. Google CEO Eric Schmidt said it best: "When you say 'collaboration,' the average forty-five-year-old thinks they know what you're talking about—teams sitting down, having a nice conversation with nice objectives and a nice attitude. That's what collaboration means to most people." But for Google and many other companies and organizations, collaboration is now a profoundly new approach to orchestrating capability to innovate, create goods and services, and solve problems. Social networking is becoming social production, where self-organizing groups of peers can design and produce everything from software to motorcycles. Sometimes these collaborations occur on an astronomical scale involving thousands, even millions, of people. Organizations that learn how to participate in these networks can access a greater diversity of thought and talent than they could ever hope to marshall internally. Indeed, Schmidt argues that collaborative innovation is now an essential skill, as important as budgeting, R&D, and planning.

Google may seem like an obvious champion, but since *Wikinomics* was published in 2006 this new model of collaborative innovation has rippled across every sector of the economy. Even the embattled and much-maligned auto industry is on the verge of profound change thanks to a new generation of automotive innovators. In chapter 4, we'll meet an Iraq War veteran who launched a new breed of car company to rekindle innovation in an industry that has long suffered from a declining infusion of new thought and ingenuity. His company, Local Motors, is "staffed" by 4,500 competing designers and supplied by microfactories around the country. His customers aren't buying mass-produced models; they design, buy, service, and even recycle their vehicle at a Local Motors factory in their region. It's a radical new model that combines hyperlocalism, customer engagement, and online collaboration to build cars that meet the needs and tastes of the local markets.

While enterprises are predictably in the vanguard of collaboration innovation, something similar is happening in society at large. In our chapter on science we'll meet an innovative young astrophysicist named Kevin Schawinski who enlisted almost three hundred thousand amateur astronomers to help him identify and analyze new galaxies. It was a highly unorthodox move in that it challenged the deeply held assumption that ordinary people can't participate meaningfully in scientific research. But conventional wisdom turned out to be wrong, and now thanks to a massive Internet-enabled collaboration called Galaxy Zoo, Schawinski and his colleagues are investigating possibilities that most astronomers only dream of exploring.

The common thread in the examples above is the growing realization that the collective knowledge, capability, and resources embodied within broad horizontal networks of participants can accomplish much more than one organization or one individual can acting alone. Of course, hierarchies will not disappear from the economy in the foreseeable future. Nor are we likely to see large top-down bureaucracies erased from the societal landscape either. But new forms of bottom-up collaboration now rival the hierarchical organization in its capacity to create information-based products and services and, in some cases, to solve the critical challenges facing the world. Whether analyzing the human genome or designing a smart energy grid, organizations that make their boundaries porous to external ideas and human capital outperform those that rely solely on their internal resources and capabilities and outmoded ways of working.

2. Openness

The term "open" is rich with meaning and positive connotations. Among other things, openness is associated with candor, transparency, freedom, flexibility, expansiveness, engagement, and access. "Open," however, is not an adjective often used to describe the inner workings of today's economy or many other institutions in society either. When it came to sharing information with the world, the *default position* of most organizations was secrecy and opacity. Organizations were closed, in part, because they were able. Companies kept important information to themselves, especially with regard to flaws, errors, or weaknesses. Most government agencies would rather shield important information from the public and only grant access when forced to by regulation or statute. From the patient's perspective, medical institutions are frustratingly opaque. Even scientists must fight against deep institutional incentives to keep their most important discoveries secret and patent potential applications.

Recently, because of the digital age, smart organizations have been rethinking openness, and this is beginning to challenge conventional business wisdom and transform a number of important functions and institutions. Why? In large part because organizations have no choice. The world is becoming more transparent: from the customers with unprecedented information about the true value of products and services to employees with access to previously unthinkable knowledge about their firm's strategy, management, and challenges.

To be sure, this new ubiquity of information, about everything and everyone, can be unsettling. Consider the rude awakening facing the CEO whose company is engulfed in scandal thanks to the secret memo leaked by a whistleblower. But for a growing number of organizations, openness is not simply an obligation to report information to an external party like a regulator or an institutional investor; it's a new competitive force and an essential precondition for building productive relationships with potential collaborators. Take the public sector, where most of the innovators we've met agree that one of the best, and fastest, routes to fostering change in government is to open up the mountains of public data buried in inaccessible databases and file drawers to broader public input and innovation. Citizens of the

European Union, for example, can now visit the European Environment Agency's (EEA) Eye on Earth portal to access real-time information about environmental quality (including air and water quality readings) in the EU's twenty-seven member countries. Access to this level of detail would have been impossible until recently, but thanks to online mapping technologies users can browse the visual imaging interfaces and drill down for detailed, neighborhood-level data about ozone levels, nitrogen dioxide, particle matter, and carbon emissions. Citizens can even contribute their own data and observations about the environment around them, including firsthand experiences of climate change or potential explanations for environmental degradation in specific areas. In turn, this new openness benefits policy makers, says Jacqueline McGlade, the EEA's executive director. "We can now bring complex strands of information together into a single, simple-to-use and easy-to-understand application," she says. "And, as more people understand what's happening in their area, more will contribute to solving environmental problems."

Of course, the mind-boggling opacity that helped produce the global financial crisis reminds us that we have much further to go. But evidence suggests that society, on the whole, is becoming more open. Fifty years ago, few countries routinely released information about their economies—indeed, many treated such information as a state secret. Now scores of countries post detailed economic statistics on the International Monetary Fund's Web site. A half-century ago, no country had laws specifically requiring government officials to provide information to their citizens. Now, nearly seventy countries do, and the number is still growing. Until as recently as the late 1990s, environmental regulation consisted largely of governments telling corporations which production processes to use. Now regulation is increasingly about telling them they simply have to report what they're polluting, and making that information public. Even the world's most ardent freedom haters—including the despotic regimes in countries like Burma and Iran—cannot restrain the nascent forces of openness that are percolating in their societies. As the Iranian youth mobilization for freedom so vividly demonstrated, an explosive combination of youthful demographics and the spread of the Internet is helping oppressed peoples everywhere release the authoritarian stranglehold over their social and economic destinies.

3. Sharing

If openness is about the communication of pertinent information to stakeholders of firms, governments, and other organizations, sharing is about the releasing or handing over of assets—by placing them in "the commons" for others to use or by sharing them with interested users under agreements that may generate license revenue. Of course, conventional wisdom says organizations should control and protect all proprietary resources and innovations—especially intellectual property—through patents, copyrights, and trademarks. According to this scenario, if someone infringes your IP, you get the lawyers out to do battle. But increasingly and seemingly paradoxically, firms in electronics, biotechnology, and other fields find that maintaining and defending a strict proprietary system of intellectual property often cripples their ability to create value. So, smart firms treat intellectual property like a mutual fund—they manage a balanced portfolio of IP assets; some protected and some shared.

In 2007, for example, Novartis, the Basel, Switzerland–based drug maker, did something almost unheard of in the high-stakes, highly competitive world of global drug manufacturers. After investing millions of dollars over three years to unlock the genetic basis of type 2 diabetes—a disease which poses one of the most common and costly public health challenges in the industrialized world, and offers potential windfalls to any company that can contain or remedy it—Novartis posted its raw research data on type 2 diabetes on the Internet. That data is now available for free to any outside scientist or company, including Novartis's competitors. Has the company suddenly lost its mind? Not necessarily. "These discoveries are but a first step," says Mark Fishman, president of the Novartis Institutes for BioMedical Research. "To translate this study's provocative identification of diabetes-related genes into the invention of new medicines will require a global effort."[6]

Global effort, indeed. By placing its data in the public domain, Novartis is actually being very smart; the company hopes to leverage the talents and insights of a global research community that stretches well beyond its 100,000 employees. In other words, the type 2 diabetes research Novartis has already conducted contains far more leads than any one lab could possibly follow up on alone. The company stands to benefit not by hoarding that early research behind a wall of patent protection, but by opening up the data to the eyes of the world and inviting thousands of researchers outside of the company to join a global search for solutions.

It's worth noting that Novartis didn't reveal everything. For example, the company didn't disclose its own notes or commentary on the data, which it spent three years compiling. This means Novartis retains a substantial lead time on other companies that attempt to exploit the data. But Novartis's decision to release its type 2 diabetes research to the world is a nice illustration of the principle of sharing. Rather than keeping everything secret, Novartis shares some intellectual property in order to increase demand, foster relationships, and stimulate progress in other areas where they will see profits.

The power of sharing is not limited to patents and copyrights, nor is it confined to business. Educational institutions are sharing their courseware. Government agencies are swapping their software code. Internet users share computing power, bandwidth, and content, while scientists share data and knowledge. Indeed, there is a case to be made that, in light of the escalating pressures on "common pool" resources such as air, fresh water, and ocean fisheries, we need a sustainability commons—a place where businesses, nonprofits, universities, and governments can share and collaborate around green innovations. "We need to do a lot fast and being really good at sharing resources is critical," says Dave Witzel of the Environmental Defense Fund. "Increasingly that means sharing within businesses, across businesses, across business sectors, between the private sector and the public sector, and across national boundaries." For Witzel, this includes not just patents related to green technologies, but what he calls the human bit: the smart, dedicated people contributing, using, filtering, and improving our shared resources. "We need channels that allow us to discuss what we are doing and learning, ways to find each other, venues for small, private conversations as well as large, inclusive discussions," he says. "We also need imaginative incentives that encourage widespread participation." If all goes as planned, emerging marketplaces such as the GreenXchange described in chapter 5 will allow us to invent, scale, and disseminate new sustainable technologies at a vastly accelerated rate. At the same time, these shared assets will create new business opportunities, giving companies an open platform on which to build innovative and profitable solutions to the environmental challenges we face.

Perhaps the most significant impact of the new culture of sharing, however, will be on the way we redefine public space and public goods in an increasingly crowded and interdependent world. Like a park in a village, we need new global parks in the emerging global village. Indeed, when it comes to managing our shared cultural and ecological heritage, scientific assets like the human genome, or even essential platforms like the Internet itself, it's time to take a fresh look and assess whether the conventional approaches are working as well as they could. After all, competition through free enterprise and open markets may remain at the heart of a dynamic economy, but we can't rely on competition and the pursuit of short-term economic gain alone to promote innovation and economic well-being. Vibrant markets rest on robust common foundations: a shared infrastructure of rules, institutions, knowledge, standards, and technologies provided by a mix of public and private sector initiative.

4. Integrity

Years ago corporate social responsibility advocates coined the optimistic adage, "you do well by doing good." They were trying to make a business case for good corporate behavior. Few were persuaded. The main reason for the lack of success in winning support for corporate responsibility was that the "doing well by doing good" adage was not true. Many companies did well by being bad. Creative accounting, unfair labor practices, corporate secrecy, monopolistic behaviors, externalizing costs to society, and shady environmental behaviors could help beef up the bottom line. Not to mention that corporate executives themselves could "do well" by paying astronomical bonuses, even while their companies were struggling.

But all that is changing. Because the previously discussed principles are driving success, increasingly the "doing well by doing good" adage is becoming true. Companies that do bad things tend to fail. There is strong evidence that companies and other organizations are being forced to act with integrity, not just by regulators and institutional shareholders, but also because of the forces of this complex networked age, in particular, transparency.

The subprime lending scandals and excessive bonuses on Wall Street, for example, have profoundly undermined trust and confidence in business leaders and reinforced the need for more effective governance, better transparency, and greater integrity in corporate management. In the minds of the broader public, corporations are no longer solely accountable for financial results; they're expected to be exemplary corporate citizens and good stewards of the environment. But perhaps more important, a growing number of business leaders now believe that good values make good business sense too. As Nike's CEO Mark Parker put it, "We believe in what we call return on investments squared, which brings a return on investment not only for the business but also for the environment and the communities we work in."

In politics, citizens are increasingly fed up with political gamesmanship, insider influence, and perceived mismanagement in public office. Remarkably, only 13 percent of Brits think their elected officials tell the truth, a dismal showing compared to doctors and teachers, who were rated as truthful by 92 and 88 percent of the population respectively.[7] In America, the cloud of cynicism hanging over government is just as palpable. When President Obama declared in his State of the Union address that his administration had cut taxes for 95 percent of Americans, just 21 percent of voters nationwide believed him.[8]

Political integrity violations are now more easily exposed and ultimately more damaging than in the past. If a politician appears to be hiding information, citizens use social media sites to publicly unearth dishonesty, highlight failed promises, or reveal unsavory relationships with influential lobbyists. The danger is that public apathy will grow as people increasingly perceive democratic institutions as remote and disengaged from popular experience, feelings, and aspirations. Even nonprofit organizations—which are accustomed to digging up dirt on their adversaries in business and government—are coming under increased scrutiny and heeding growing calls for greater integrity in the way they campaign for issues or manage the funds contributed by donors.[9]

The bottom line is that in an age of transparency all organizations need integrity as part of their DNA—not just to secure a healthy business environment, but for their own sustainability and competitive advantage. Society will be increasingly alert to individuals and organizations that cultivate an aura of responsibility, when their business practices don't measure up in reality. In everything from motivating employees, to negotiating with partners, disclosing financial information, or explaining the environmental impacts of a new factory, companies and other organizations must tell the truth, be considerate of the interests of others, and be willing to be held accountable for delivering on their commitments.

These three values—honesty, consideration, and accountability—together with transparency are the foundation of trust and integrity. Or put another way: trust in business and society is the expectation that another party will be honest, considerate, accountable, and open. When organizations build a broad foundation of trust their networks reciprocate with cooperative behavior. For example, customers decide whether to give companies their money and communities decide whether to let a company locate in their area. Voters decide whether to put up a lawn sign during an election and employees decide whether to share innovative ideas with their employer or defect to a competitor. The better an organization's relationships with its stakeholders, the better its access to resources. A lack of trust, on the other hand, generates conflict, friction, and inefficiencies, while consuming management time and resources with defensive activities.

Keeping up with society's evolving expectations can be tough, but integrity shouldn't be a millstone. Leaders in business and government with a combination of vision, energy, and communication skills can help convince other leaders to share the risk and responsibility in meeting the challenges of the twenty-first century. The promise is that public and private enterprises will become more integrated with the societies they serve, more attuned to social and environmental concerns, and better equipped to develop pragmatic and profitable solutions for advancing the common good.

5. Interdependence

If the financial system meltdown shows us anything, it is that we live in an interconnected world. In an age when everything and everyone is interconnected through networks of glass and air, no one, no business, organization, government agency, country, or society, is an island. The meltdown started in the mortgage offices of America, then crashed down on Wall Street, and spread almost instantly to London, Asia, and the rest of the world. Regulators, if not bankers, are getting the message. "Changes in financial rules and accounting standards . . . must be coordinated globally in the effort to help avoid a recurrence of the economic crisis," said Federal Reserve governor Elizabeth Duke to an accounting industry audience on the first anniversary of the collapse of Lehman Brothers. "Accounting standard setters, regulators and policy makers around the world are discussing and proposing preventative measures. Now the challenge lies in integrating those changes smoothly and seamlessly," she said.[10] The Securities and Exchange Commission has also been pushing for a single, global set of rules for financial services. Said James Kroeker, chief economist at the SEC: "The financial crisis that erupted last year has highlighted for us the importance of global solutions to complex issues."[11]

Sometimes the pejorative word to describe interdependence is "globalization," but talk to your typical anti-WTO protester and they'll tell you that globalization needs to be achieved more fairly, *not* that we should or can turn back the hands of time. While globalization may not be new (prominent scholars point out that globalization has been a feature of human civilization for centuries, expanding and contracting, intensifying and declining, throughout history), it has never been so vast, so intense and all-encompassing as it is today. Global interdependence has become a defining feature of our time as people, money, technology, products, services, culture, and ideas relentlessly cross borders in a vast network of transactions and social exchanges.

The intensification of climate change shows that environmental disasters do not respect national borders, just as the flow of drugs, diseases, and weapons moves readily from continent to continent despite the vast sums spent on border protection. In effect, the world is shrinking and we are becoming ever more aware of the multileveled ways our fortunes are overlapping. This phenomenon of interdependence is only intensifying as the Internet not only renders distant events more proximate and immediate, but also reinforces our consciousness of this interconnectedness.

Growing interdependence has profound ramifications for the way organizations act and make decisions. The more we become connected and interdependent as societies, the more we want and need to know about the affairs of others and how the actions of a distant third party could potentially affect our individual or aggregate welfare. Similarly, the interdependence of actions and events means we have no option other than to try to encourage and enforce mutual cooperation through a new division of labor among the four key pillars of society: business, government, the civic sector, and a new pillar enabled by the Internet—the individual citizen. There is simply no room for unilateralism in a world in which trust, transparency, and collaboration will be essential to ensuring near- and long-term stability.

Interdependence plays out in other ways too, particularly in relation to the environment and our obligations to future generations. For most of the last century, Western society has acted as though the ecological resources that sustain our existence would be forever available in unlimited quantities. But as scientific analysis reveals the devastating environmental toll of rampant consumption and waste, it has become increasingly apparent that we must become much better stewards of the planet, to ensure that what we enjoy today will be available for future generations. This basic notion of sustainability should be baked into every economic and political decision.

Unfortunately, the problems that lead to the massive waste and inefficiency of industrial capitalism are so deep-rooted that we often don't even recognize them. Like knowledge, brand, reputation, and other intangibles, our natural environment forms part of the invisible fabric of value creation. The dilemma is that this value creation is currently hidden in, indeed excluded from, our economic assumptions and measurements, largely because growth and progress are measured in money, and money does not give us information about the health of ecological or social systems. So, for example, we count the economic value created when trees are converted into timber and other products, but we don't account for the loss of species habitat or consider the important role trees play in absorbing carbon, producing oxygen, and regulating the climate. Building the "cost" of these vital ecological services into everything from the pricing of assets to our measures of productivity and growth is one of many options for redressing the imbalances.

The longer unsustainable practices continue, the greater the threats to the essential foundations of our economy. Firms are dependent on their employees to work productively; on customers to buy their products; on investors to buy and hold their stock; and on government to provide human, legal, and physical infrastructures for economic development. All of these people and institutions, in turn, are collectively dependent on the shared resources and services provided by the natural environment. If a financial crisis makes the point, a potential global meltdown of our environmental ecosystems does so a thousand times stronger: on an increasingly interdependent planet, no organization can succeed in a world that is failing.

PLANTING SEEDS

Just as the printing press expanded freedom and fostered new forms of personal expression that led to new social orders and institutions, the Internet is finally reweaving the fabric of society as millions and eventually billions of people connect and collaborate around any conceivable shared interest. For better or worse, there is a new fabric of connectivity in society that is leading to a deep change in our institutions. There is no guarantee that the changes unleashed by this social connectivity will always lead to good. But if guided by the principles of wikinomics, organizations that harness this new force can spur social and economic innovations that will alter society for the better.

In the chapters that follow, we explore the contours of this new age with a series of investigations into how macrowikinomics is changing everything from innovation and financial services to science and the university, from energy and transportation to the nature of government and democracy. The evidence will show that the seeds have been planted for pervasive and enduring changes in the way we orchestrate human thought and capability to achieve superior outcomes in these diverse domains. Across all sectors, a growing pool of innovators is showing the way forward. But ultimately it will be up to you the reader to help grow these seeds of creativity into systemic changes in the way our institutions operate. Everyone has a stake, and increasingly everyone can participate. Whether your passion is developing innovative products and services or new social innovations that can help solve problems in society, the new world of possibilities is beckoning.

II

RETHINKING THE FUNDAMENTALS

3. OPENING UP THE FINANCIAL SERVICES INDUSTRY

In early 2008, as many bankers were celebrating their fifth year of fabulous profits, one of our colleagues sat down for lunch with the CEO of a major investment bank. The banker was, to put it mildly, very happy. "We're minting money—just minting money," he said. Bank profits were shooting up at a time when nonfinancial companies in the United States were reporting little real profit growth. So how did the banking CEO account for this amazing performance? "We're exploiting discontinuities in the mortgage market," he said.

A translation is in order. A discontinuity, the dictionary says, refers to a break or gap in the logical sequence or cohesion of things. By the year 2008, in the financial world, the term became a deeply cynical euphemism summing up the reckless, sometimes unscrupulous behavior that led to the global financial and economic crisis of 2008–9. Mortgage brokers convinced people to buy houses they couldn't afford, on the premise that housing prices would forever go up, and then, with the help of investment bankers, handed the risk of default to unsuspecting institutions and investors. This was a scheme that greatly enriched the brokers and bankers, at least for a while, because of one key feature: it was so obscure that virtually no one could understand what was going on, and at what real risk.

When the game ended, and the market toppled, it cost American taxpayers billions of dollars. Faced with a historic market crash, the worst recession in three generations, plus government guarantees that exceed the cost of every war the country has ever fought, American taxpayers are understandably furious. Many people are reviving calls for updated regulations, more government intervention, and even the breakup of the big banks.[1] But most of the proposals are pretty much tinkering within the existing framework.

Restoring long-term confidence in the financial services industry in the United States and other industrialized nations will require more than government intervention and new rules. What's needed is a new modus operandi, one where all of the key players (banks, insurers, investment brokers, rating agencies, and regulators) embrace transparency and other wikinomics principles in order to develop credible practices and policies that will ensure the crisis doesn't reoccur. If the investment community and the broader public understand the true causes of the financial disaster, perhaps they can protect themselves against financial adventurers in the future.

In short, it is time to shine the light on the opaque products and activities of the financial sector that have threatened the entire economy, and the Web provides a platform to do this. A digital response involving collaboration on a mass scale may be the best way to properly evaluate and assess the value and risk of new financial instruments as they are produced. New models based on openness, transparency, and participation are already changing many parts of the industry from venture capital to mutual funds and even lending, so why not apply the same thinking to the obscure mathematical models that value the risk and expected returns of the most complex instruments: expose them to the scrutiny of the thousands of experts who have the knowledge to vet the underlying assumptions?

Arguably, this is the perfect time for fresh and even radical thinking. When it comes to evaluating risk, this interconnected digital crowd comprises people who are financially sophisticated and can provide the innovative insight that will clarify the questionable dealings in the financial services business. A more open and collaborative approach would restore trust in banks, kick-start venture capital, unfreeze the paralysis of lending markets, and lay a foundation for a financial service industry that continues to underpin the growth and prosperity of the world's economies.

WHERE THINGS WENT WRONG

If the financial crisis taught us one thing it is that the financial services industry is absolutely at the core of our economic structures. When the system fails the repercussions are felt in virtually every economy and industry across the globe. At the most basic level, financial institutions facilitate the flow of money through the economy—taking deposits and making loans, taking savings and making investments. But these retail functions have long been eclipsed by an ever more sophisticated constellation of banks, investment funds, brokers, and insurance companies that create diverse and intricate services that touch virtually every aspect of today's economy. Financial institutions provide an engine for innovation by funding new ideas and companies. They aggregate and manage risk so that individual investors don't lose everything by making a bad decision and people and organizations are not wiped out by unforeseen circumstances. They facilitate the trade in various goods ranging from grain to crude oil, helping transfer wealth from consumers to producers and establishing prices in the process. They also enable day-to-day commerce for billions of people so that the mechanics of purchasing something or paying a bill are an afterthought. Innovation in the industry brought us ATMs, online banking, better mortgage rates, higher yielding investments, and new insurance products to protect ourselves from unexpected events. So how does an industry so vital to the economy go so badly wrong?

If you're a banker or a mortgage broker, it worked like this: you give a mortgage for a $300,000 house to a gardener in California who's earning $24,000 per year who can't afford the payments and is therefore almost certain to default. No matter: you have a nickname for loans to people who can't possibly pay—"liar loans," or NINJA loans, meaning No Income, No Job, No Assets. Yet in a deregulated marketplace, you can hide this inconvenient fact by offering the NINJA customer a nice line of credit, so that they could make their mortgage payments and avoid going into default, at least for a while. Then Wall Street's investment bankers step in with what sounds like an ingenious method for getting these risky mortgages off the bank's books: you bundle the mortgages up with a bunch of other similar mortgages into a CDO—a collateralized debt obligation—that will be sold to investors for the promised cash flows or even for the potential profit on the underlying assets, the mortgaged houses. However, a lack of transparency about the underlying assets and associated risks makes these financial products hard to understand and evaluate, even for the financial advisers who retail them.

Now comes the key moment in this story of discontinuities: Using the bank's good name, you simply unload the risk to others. To do this, you have to dress up this financial dud—which after all is made up of mortgages people can't pay—to be sold to investors. You go to a rating agency and get it to rate at least part of the CDO as an AAA investment product. The ratings agency will do it partly because you are paying the bills. Now with the AAA rating, you can make it look even "safer" by guaranteeing your product, perhaps by an insurance company, before selling it to an investor like a big pension fund. Investors will never be able to value the underlying assets, even if they manage to plow through your three-hundred-page prospectus. It will be particularly hard to figure out who's responsible in the event the financial product tanks.

When you sell that CDO, you take a small percentage of the cash flow promised to your investors as part of the deal. If you add up that small percentage over thirty years, it's a sizable amount of money. You know that a lot of people will probably default on these loans for houses they can't afford, but you still record it on your income statement as revenue earned this year. So now you can bonus yourself up front. You've just transformed an utterly foolish business into a highly profitable one, for a while anyway.

To be sure, the description above is oversimplified, as many entities (not just banks) got in on the "minting of money" and should carry some blame for creating the meltdown. Moreover, the actions of any one player could be a combination of good faith, perfidy, ignorance, and mass delusion. In fact, one could argue that bankers, investors, and mortgage lenders were simply doing what public policy makers had asked of them. Gord Nixon, the CEO of RBC, Canada's largest bank, says you can trace the "discontinuities" back to loose money and government policies in the United States that encouraged home ownership by low-income individuals.[2] Like its Canadian counterparts, Nixon's bank survived the financial crisis largely unscathed. In fact it has benefited, having grown in relative terms to become the tenth largest bank in the world.[3] But RBC is not staking a claim on brilliance. To Nixon, the Canadian banks avoided this mess "primarily because the structure of the mortgage market is very different." In the United States, 75 percent of all mortgages are securitized, that is, bundled together and packaged into offerings in turn sold to others. In Canada, 75 percent of all mortgages are on the balance sheets of the banks, so banks are very cautious about whom they give mortgages to because if the borrower defaults they'll be on the hook.[4] "But ironically," Nixon points out, "even though lending standards are much higher, the level of home ownership in Canada is higher than in the U.S. A lot of loose money doesn't lead to higher home ownership levels."

Naturally, the record profits were a factor in promoting reckless behavior, one that made many smart people turn a blind eye to the potential downsides of risky lending strategies and even riskier investment crazes. "Everyone was making money," said Nixon. "Mortgage brokers made money because real estate sales were booming. Investment bankers made money by taking these mortgages, bundling them and selling them to investors. And traditional banks made money through fees and loans which hit record levels." When the housing market collapsed, it affected not just the people who defaulted on their payments, but the whole chain of people and institutions that were invested in the subprime sector. Homeowners started losing their homes. Insurance companies went bankrupt. Rating agencies were discredited. Pension funds and other investors saw their investments become worthless. By then many of the major banks and investment funds had U.S.

mortgages on their books that they had not been able to securitize. Without enough capital to cover their losses, many banks were facing insolvency—only a few months after they were toasting many years of record profits. Then as we all know, the U.S. government rescued the banks with hundreds of billions of dollars in taxpayers' money, which according to one estimate added up to more than $2,300 for each man, woman, and child in America.[5] Most economists agree that there were no plausible alternatives to the government-funded bank bailouts (although there is plenty of disagreement over whether the TARP funds were administered appropriately in the immediate aftermath of the crisis). Allowing the large financial institutions to fail would have unleashed unimaginable consequences for the global economy. The bottom line is that the rescue package saved the big banks and transferred the pain from the investor to the taxpayer. But the more lasting consequence is that people's trust in the soundness of the overall system evaporated along with the billions of taxpayer dollars used to douse the fire.

WHY NOT OPEN UP FINANCIAL MODELS?

Financial scandals are a predictable part of the Wall Street landscape.[6] They're typically followed by waves of fury—and demands for reparations from the banks, new taxes, criminal prosecutions, and as with the current crisis, breakups of the big banks and even nationalization of the industry. Many pundits have said the cause of the financial collapse was that regulation did not keep pace with innovation in the industry itself. Yet regulators are faced with a new problem, one that stems from the fast-paced, digital, global world of innovation. Just as soon as they come up with new solutions to solve yesterday's problems, the financial services industry dreams up a new set of services—or threatens to go elsewhere. When that doesn't work, they lobby hard. In one meeting on July 31, 2009, nearly two years into the crisis, Secretary of the Treasury Timothy Geithner bemoaned the slow progress of regulatory change. "The reality is that there are pretty powerful vested interests fighting this," he said.[7]

So regulation is only part of the solution to ensure that the marketplace is fair and that deposits in banks are safe. Yes, we need to continually update rules to encourage financial practices that are in line with the public good. We obviously need to enforce existing rules, which is one place where the system broke down. Regulators also need to take care of the basics, including proper funding to pay for the people and the technology needed to do their jobs. But in today's global financial environment, the current patchwork of regulation is not enough. "We have a perfectly networked financial industry but a much less networked regulatory community," says Peter Gruetter, former secretary-general at the Swiss Federal Department of Finance. "At this point collaboration among various regulators involves a lot of physical meetings. What we need is the use of better communication and networked technologies to help facilitate the collaboration process."

It is one thing to coordinate banking reforms internationally through policy networks and multilateral forums like the G20. But why not go further to include real-time collaboration among regulatory agencies around the world? This would include a robust information platform that would allow various regulators to swap intelligence and better sense and respond to potential threats to the stability of the global financial system. Common standards for sharing data, like XBRL, will go a long way in helping make such a platform possible.

Beyond this, we think collaboration can bring about much deeper changes to the whole operating model of the financial services industry itself. Indeed, regulators, bankers, and everyone could stand to benefit if the industry was to embrace the kinds of innovations made possible by wikinomics. More specifically, we think collaboration on the Web combined with increased transparency would have the power to reshape the industry by engaging a global community of financial experts in monitoring the soundness of the financial system.

To see how this might work, let's take an example. Why not open up financial modeling and make pertinent assumptions and data transparent to all interested parties? Financial modeling is one of the most important tools in a banker's arsenal. It allows analysts to estimate the value of goods ranging from a company's stock to a barrel of oil six months into the future. The valuations, and associated risk estimates, are behind almost everything financial, from mortgage lending to more complex investment products like derivatives and options. Over time, both the products, and the underlying financial models, have gotten increasingly complex, and therefore opaque.[8] Indeed, the financial services business has developed an arcane discipline of applying complex mathematics to the underlying financial data to determine the price of financial service products and calculating the associated risk. These models, often closely held secrets, are supposed to help companies make safe investments, protect them from risk, and give them a competitive advantage. In hindsight they weren't even close. In 2010 *The Economist* reported that the predictions made in 2007 for losses to be expected in 2010 were off by orders of magnitude. For one CDO rating (BBB) the industry estimated a default rate of less than 0.5 percent, which turned out to be well over 50 percent three years later.[9]

In the spirit of pointing out that "the Emperor has no clothes," may we suggest that the current approach to valuation and risk management does not work? These models not only failed to deliver competitive advantage, they also contributed significantly to the collapse of the entire industry, and harmed many other parts of the economy. We believe that the proprietary nature of these risk management models is the Achilles' heel of this risky business and that a radical new transparency provides a necessary antidote. Professor John C. Hull, a thought leader in derivatives at the University of Toronto's Rotman School of Management and author of the industry bible, *Options, Futures and Other Derivatives*, said, "Markets are frozen because investors have no real idea of what they are buying [or] have bought. Huge amounts of aggregation and the absence of the low level data," he argues, "make the true values of these assets and pending losses very difficult to determine. With no way for most investors to sort out good from bad, they have chosen to abandon whole sectors of the market." He concludes that "An open mathematical algorithm, or even published software, would far better describe the waterfalls, and associated payment structures."

This is not such a crazy idea, and other serious players agree. Rick Bookstaber was the former head of risk management for Salomon Brothers and Moore Capital, and is the author of *A Demon of Our Own Design: Markets, Hedge Funds, and the Perils of Financial Innovation.*[10] He currently has a full-time job at the SEC as a senior policy adviser for risk, strategy, and financial innovation and knows as much about the weird, opaque world of derivatives as anyone. Bookstaber calls for opening up derivatives to understand their value. "If we want to go down the path of standardized valuation and comparability in these complex portfolios, we need open derivatives models," he says. "One thing we should have learned from the rating agency debacle is that even if we put aside the issues of monopoly power and conflict of interest, we cannot stop with having the proprietor of such models say, 'Trust me, I know what I'm doing.'" For Bookstaber, "Valuations based on a black-box do not get us the transparency we need. I don't care what a trading desk uses for its decision making, but when it comes to valuations that carry beyond the firm, we need to be able to see and critique the models that are being used." And peer collaboration is key: "If a model is to become a standard, if it is going to be used for regulatory or other benchmarking purposes, it should be transparent and subject to peer review."[11]

The industry pundits raise a good point. It is time for bankers to ask themselves a question that most have never considered: "Why should the technology, data, and risk assessment models that are used to value products today be kept proprietary?" Can investors and others ever again believe the stated profits or losses of any financial institution, its purported capital base and financial soundness, when these numbers are based on secret and opaque models that are derived from mathematics so complex that even the company's executive management does not understand them?

Going forward, the mathematics behind the value and risk calculations for new financial instruments should be open and vetted by a crowd of experts, applying the wisdom of many to the problem. They should know, for example, whether the VaR (Value at Risk) analysis is based on information from only a couple of years, which would not cover the consequences of a once-in-a-generation event. The underlying data and the algorithms for complex derivatives such as collateralized debt obligations should be placed on the Internet, where investors could "fly over" and "drill down" into an instrument's underlying assets. With full data, they could readily graph the payment history and correlate information such as employment histories, recent appreciations (or depreciations), location, neighborhood pricings, delinquency patterns, and recent neighborhood offer and sales activities. The trading of credit derivatives should be moved to open public exchanges. Clients deserve full access to the information, including the data and the analytics that the investment banks used to construct the products they are selling.

Is the Crowd Wise Enough to Solve This Problem?

Surely, you might say, outsiders are poorly positioned to look into complicated instruments like CDOs. After all, the average investor has trouble deciphering a financial statement, especially when the key information is often hidden in the notes at the back of the book. But there are plenty of financially sophisticated people around the world who do not necessarily work on Wall Street or in the City but have the requisite skills. For starters the tens of thousands of analysts, traders, and other financial services professionals who lost their jobs during the financial turmoil comprise a rich talent pool eager to get back in the game. Similarly, thousands of academics, doctoral students, and industry experts who study the workings of the financial markets currently have no way to contribute directly to the system. The digital crowd can also offer significant help to regulators who make rules and enforce them. If the financial system becomes more transparent, this can turn into a powerful network that links the digital crowd, the regulators, and the enforcers. Through these digital networks, they can pool intelligence in the same way the law enforcement and national security services do.

But can the same digital networks help decipher some of the most complex financial assets out there? As we write these words, there are still untold billions (some say trillions) of dollars in assets (mostly mortgage-backed securities) floating in limbo. Nicknamed "toxic assets," they are poisoning the financial health of many banks and other financial institutions that currently hold these ugly liabilities on their balance sheets. Most would prefer to get them off, but there is a problem: no one knows how to determine their value. Despite this, new complex instruments are beginning to emerge again which look very similar to the securities that caused the financial turmoil in the first place. It is clear that a nonconflicted, independent valuation process of these assets is urgently needed.

The Open Models Company (OMC) is doing just that with a platform for open, collaborative, peer-reviewed valuation of financial products based on many of the principles articulated in this book. Call it the Linux of financial modeling. Open to any qualified contributor, it is using the new Web to create a global community of experts dedicated to establishing credible valuation and risk assessments for credit securities and contracts such as CDOs, CDSs, or other derivatives.[12] Chuck Bralver, the director of Open Models and a senior associate dean of international business and finance at the Fletcher School at Tufts University, says, "I think the most important point is that even if the valuations that the banks use to value these assets today are now reasonably accurate, the opacity in which this is done is actually a bigger problem than the potential inaccuracy of those valuations." To fix that, OMC has built a system, and an open process for academics, industry experts, quantitative analysts (affectionately known as "quants"), banks, and investors to collaborate to determine such securities' worth. Unlike a three-letter rating system that few still trust, it allows participants to input their assumptions on future economic activity and do the what-if analysis necessary to price these assets consistent with their market assumptions. Craig Heimark, a longtime industry veteran and another member of the OMC team, likens it to the scientific peer review process. "In the scientific world when people publish something, they don't just publish their results, but also the steps in the process, their methods and assumptions so that they can be vetted by others. Currently in the financial industry the creators of these models either self-publish, meaning they declare the price that their models arrive at, or in some cases get third parties like Moody's to certify the result without disclosing the details."

OMC's business model is designed to evaluate both existing assets, such as the problematic mortgage-backed securities, as well as new offerings brought to market by investment banks. Asset holders or sellers would approach OMC for help in valuing the instrument. Open Models would then use an independent network of community modelers and experts who would come up with a value and more importantly comment on the underlying assumptions. The whole process would be documented in a wiki-style format and open to the community. Up to a quarter of the firm's revenue would be devoted to compensating the external experts according to their contributions. The founders believe that the flexibility of the community approach will attract thousands of contributors ranging from those looking to break into the industry to seasoned veterans who are looking for a better work-life balance.

Will Such a Radical Idea Take Hold?

So will the banks and other financial institutions accept this level of openness voluntarily? Not likely at first. The OMC approach is nascent and has an uphill battle against legacy behaviors, systems, processes, and cultures. Many banks view their financial models as a source of competitive advantage and many are bound to fight back to protect the proprietary nature of their models or invent new products that escape the regulators' attention. As one Goldman Sachs executive said to us off the record: "We're a very private company. The less people know about us and pay attention to us the better." In commenting on the U.S. government fraud charges against Goldman, Roger Martin, dean of the Rotman School of Management at the University of Toronto, says, "Sadly for Goldman transparency is not an attractive option. The better job Goldman Sachs does in explaining exactly what its business is, the more outraged regulators and the public will be."[13]

Some financial institutions, such as hedge funds, will likely insist on using proprietary models to make their own bets on behalf of their wealthy clients. For sure, the rating agencies will view the open modeling approach as a direct competitive threat, and rightly so. It is a challenge to how they do business and even the rating business itself. But a combination of pressure from investors and regulators may force banks and other important players to open up, at least part of the way. Over time, the big banks, whose future growth still depends on regaining the public's trust, may come to see the benefits of a more open and transparent approach. Those that are willing to subject their valuations to public vetting will be trusted, and those that are not will soon find their cost of funds increasing and their customers going someplace else.

Banks once spoke of their fiduciary responsibilities, and through that their behavior established trust. Today it will take actions, not words. A process such as that proposed by Open Models to vet complex assets could go a long way toward restoring confidence in the system. Indeed, if financial innovation got the industry into trouble in the first place, perhaps forward-thinking initiatives such as Open Models may show a way out by challenging long-standing practices and assumptions. To be sure, Open Models is facing an uphill battle and success is far from assured. But any amount of transparency and openness that permeates the modus operandi of the industry will have a positive influence. The industry will not turn on a dime overnight, but innovations at the fringes of today's system are showing the way forward.

RETHINKING VENTURE CAPITAL

Venture capitalists are often heralded as the Heroes of Entrepreneurship or the Gods of Capitalism, and for good reason. VCs, as they are called, provide capital to cash-poor start-up ventures or help small companies to expand when they can't tap into other sources of funding. In doing so they constitute a critical link in a long chain of innovation where nascent ideas and technologies morph into market-ready products and services. For starry-eyed entrepreneurs, attracting venture capital is often seen as both a rite of passage and badge of honor validating their work and their viability. For the broader public, venture capital is a largely invisible but enormously powerful engine for job creation, innovation, and growth.

A recent survey produced by the National Venture Capital Association found that venture capital has financed companies in all sectors of the American economy, including innovative giants like Facebook, Twitter, Apple, Starbucks, eBay, FedEx, Google, Genentech, Medtronic, Microsoft, Home Depot, and Intel. Companies backed by VCs generated more than $2.3 trillion in revenue, and 17.6 percent of U.S. GDP, the report found. They account for 9.1 percent of all U.S. private sector jobs, including almost two million workers in computers and related sectors that represent 94 percent of the industry's total jobs.[14] Global Insight, which conducted the survey, concluded: "The nation's innovative and cutting-edge venture capital backed companies sustain jobs and revenue across diverse industry sectors. The nation's venture capital industry plays a paramount role in nourishing the U.S. economy by bringing innovative concepts and business models to life."[15]

The world needs innovation, and for years innovation has been linked to venture capital. But like the rest of the financial services industry, the current process for fostering early stage innovation is not working as well as it could. "The problem," according to a 2009 report by North Venture Partners, "isn't the number of opportunities investors are presented with, but it is rather the lack of an efficient means of filtering the options." Throughput, not supply, argues the report, is placing unnecessary constraints on today's innovation system. "How do investors find, filter and fund the most promising new opportunities in a deep ocean of possibilities? The truth is they can't," the report concludes.[16]

One reason for the swelling pool of promising investment candidates is the proliferation of low-cost business infrastructures that make it cheaper and easier for companies to get up and running. A study by Dr. Robert Hendershott, a professor at the Leavey School of Business at Santa Clara University, found that the availability of open-source tools, cloud computing, and the rise of virtual office infrastructure has driven the cost of launching an Internet venture down from $5,000,000 in 1997, to $500,000 in 2002, to $50,000 in 2008.[17] Lower costs to launch should mean more new companies and more innovation for VCs to fund. But it doesn't. It actually makes it harder for companies to find the money and also the attention they need. Why? More companies receiving investment means more companies to supervise and more demands on the investor's attention. After all, VCs usually add more than just money—they make introductions, assist with strategic sales, and help recruit top talent.

One solution is to make lots of small investments but spend little time with each one, in the Silicon Valley equivalent of speed dating. This is the strategy of the eponymous fund recently launched by Marc Andreessen (from Netscape, eBay, Facebook, Twitter, and Ning) and Ben Horowitz (from Netscape, OpsWare, and AOL). They plan to invest more than $300 million over ten years, starting with seed investments beginning at $50,000 a pop.[18] This speed-dating-for-capital approach results in some matchmaking, but the problem of postinvestment execution still remains: a VC only has so much time on his or her hands. Plus, Andreessen and Horowitz's speed-dating model fails to address an even bigger issue: the globalization of innovation.

New hot spots for innovation are increasingly flourishing outside of the traditional sphere of VC operations rooted in established innovation centers like Boston, New York, and Silicon Valley, causing VCs to expand their circles. Yet many VCs won't even look at a business plan, let alone hear a pitch, from someone who doesn't come highly referred. The theory is that if a VC relies on a trusted network of contacts she or he can effectively lessen the number of crappy ideas that she or he has to sift through to find a golden nugget. But in today's global knowledge economy, the next Facebook, Google, or Tesla Motors is as likely to be born in Tel Aviv as it is to be born in Silicon

Valley, as likely in Bangalore as in Boston. And while the "old boys' network" is good if you are an old boy, it's not so good if you are a young woman from Brazil with a billion-dollar business innovation.

Unfortunately, an unfavorable economic climate has prevented many VCs from exploring new opportunities. Due to the recession, fewer companies are going public and the ones that do are taking longer to become profitable. VCs have been forced to keep their money locked in for longer than usual, which means fewer start-ups can get funding. Even before the recession the industry was suffering from having grown too fast during the boom years, according to a study by the Ewing Marion Kauffman Foundation. The report concludes that the venture capital industry's declining returns are due in large part to the rapid expansion in venture capital assets under management during the past decade. Since the influx, VCs have struggled to figure out how to adapt and allocate the new funds in profitable ways. Dr. Paul Kedrosky, the author of the Kauffman report and an adviser to many VC funds, explains: "Professionals in the venture industry have gotten comfortable with the way their industry is set up in terms of size, structure and compensation. It has been a profitable business for many. However, our study indicates venture participants now need to overcome their resistance to change, so they can most effectively fund entrepreneurs and offer investors competitive returns."[19] But rather than embrace change, many VCs have chosen to hunker down, investing in fewer start-ups and focusing instead on larger private equity deals that require $10 to $25 million in capital. Companies below this threshold are in danger of falling by the wayside in today's investment climate.

Venture Capital 2.0

If the current model of venture capital is flawed, what's the alternative? Sean Wise, a management professor and VC himself, leads one of a growing number of outfits determined to prove that a form of community-powered venture capital can both filter the global wealth of opportunities and channel more intellectual horsepower into making each investment successful. And he's not just talking about it; he's doing it—cofounding a company, VenCorps, that uses mass collaboration at every stage of the process. "For Venture Capital 2.0 to succeed," he says, "there will need to be exponentially more people involved." Just as Goldcorp used mass collaboration to locate their drilling sites, or Wikipedia crowdsourced the publication of expert articles, or Threadless works with customers to design T-shirts, VenCorps is leveraging collaboration. Wise is deploying the power of mass collaboration, not just in the process of choosing which start-ups to fund, but to help grow those start-up ventures after the investment is made. "VenCorps is basically wikinomics applied to venture capital," says Wise.

The money being invested by VenCorps in small companies comes from their own fund, but the choice of where to invest it belongs to the VenCorps community. Founders from around the world log on and register their start-up at VenCorps.com. There, they can upload a video elevator pitch, share some biographic details, and/or post an executive summary. The community at VenCorps (made up of thousands of entrepreneurs, scholars, scientists, angel investors, service providers, and government officials) then reviews and ranks each entry using a five-criteria weighted scorecard. During a challenge the top nine start-ups (as determined by the community) go on to the next round, where they can win an investment (typically $50,000). That may not sound like a lot in typical VC terms, but it's enough to kick-start a small enterprise, as some of VenCorps' early successes have demonstrated.

On a sunny spring day in 2009, for example, Wise and others from VenCorps were meeting with IBM to discuss how the VenCorps platform might be used to solve some really big societal problems. Only a few days later VenCorps and IBM's Smarter Cities program launched The Congestion Challenge, whose tagline, "Help Make Traffic History," became a rallying cry for the VenCorps community. For the next sixty days, 124 teams from around the globe competed to develop smart solutions for reducing urban congestion. Each entry was carefully reviewed and rated by the VenCorps community, whose feedback was weighted and aggregated creating a subjective comprehensive community-based rank for each entry. After two months, the top nine ranked entries—finalists came from Hungary, Canada, Israel, Ireland, the United States, and the Netherlands—went on to the showdown round. Over the ensuing thirty days, each finalist posted slide decks, founder bios, and pitch videos and each participated in an open due diligence call, fielding questions from both experts and the crowd. In the end, the winner was a Seattle-based start-up called iCarpool. iCarpool, a ridesharing platform servicing consumers, corporate clients, and urban infrastructure providers, took home more than 60 percent of the community support.

So what did they win? iCarpool's founders scooped the $50,000 in cash but they also got a trip to ITS World Congress in Stockholm, Sweden. In addition to a supreme networking event, the founders got a chance to demo iCarpool in front of urban planners, government advisers, researchers, and consumers worldwide.

But VenCorps challenges are not just about the money or the exposure.

In fact money is seldom seen as the best reward. iCarpool also won fifty thousand virtual dollars (called VenCorps points) that the founders can use to incentivize the community to participate in their company's growth. For example, community members who participate in focus groups, share insights, and/or make introductions through their personal networks earn points from the start-ups they support. The community members in turn can use the points to bid on a wide variety of items in the VenCorps store, including gift certificates, exclusive lunches with business gurus, cash, stock options, and even physical goods like T-shirts and iPods. According to iCarpool's founder, Lakshmi Krishnamurthy, the community support and feedback benefited iCarpool even more than the cash and the promotion: "The potential to tap into a fast growing diverse community of founders, funders, and technology and business experts—to be able to discuss ideas and exchange feedback—is amazing," she says. "I believe community capital has a bright future. VenCorps is the *American Idol* for entrepreneurs."

By creating this virtual economy, an ad hoc meritocracy if you will, VenCorps harnesses its community of passionate participants, to leverage both the wisdom of crowds and the production capability of the masses. This duality is at the heart of wikinomics. The new collaboration is not just about the community picking the winners; it is also about the community building the winners together.

Was iCarpool the best choice? Did it have the best technology and solution? Did the community get it right? In some ways this is a moot point, at least from the perspective of the VC. In the end the winner, iCarpool.com, may not have been the most technologically advanced solution, but it did win the support of more than half of the VenCorps community, and that may be the key to iCarpool's long-term success. After all, ideas and technology don't change the world; passion, community support, and execution do. These latter traits personify iCarpool. In a very short time (less than ninety days), iCarpool went from being a Seattle-based start-up that few had heard of, to the darling of the intelligent transportation world with an exciting future.

More to the point, VenCorps (and their partner IBM) became interested in a couple of the companies that the crowd did not select. At the time of writing, both these companies were talking about investment and partnership opportunities. "The Congestion Challenge showed the true power of mass collaboration applied to Venture Capital," said Wise. "By leveraging collective intelligence and peer production, we should be able to generate higher returns than other seed funds."

While VenCorps is unlikely to challenge major VC firms anytime soon, the company will provide a chance for many more promising ideas to reach the stage where larger VC investment may be warranted. The collaborative model does appear to address some of the challenges facing today's seed venture capital system. In fact, smart mainstream VCs are excited about the concept. Kevin Kimberlin, the chairman of Spencer Trask (Thomas Edison's venture capitalist), calls VenCorps "21st century barn raising." "Instead of relying on three experts to put in 1,000 hours each, you rely on 1,000 people putting in three hours each." VenCorps also shows how wikinomics-inspired innovation in financial services can not only fix the current broken system but also drive new growth by helping to discover and nurture promising ideas.

THE NEW PEER-TO-PEER BANKERS

She called it "the proverbial first shot fired in an American debtor's revolution against the usury and plunder perpetrated by the banking elite . . . and the federal government." This debtor's "declaration of independence" was first posted by Ann Minch on YouTube in September 2009.[20] "There comes a time when a person must . . . take a stand for what's right," Minch wrote. Then she told her story: "I wasn't over the limit or behind in payments . . . yet recently Bank of America jacked up my interest rate . . . to a whopping 30%! I could get a better rate from a loan shark." Within a few weeks there were over half a million hits, and Ann, from Red Bluff, California, was an instant celebrity with interviews on *FOX & Friends* and on MSNBC. Not surprisingly, Jeff Crawford, an SVP from Bank of America, quickly called and renegotiated the rate.[21]

With stories like Ann's filling the popular media, radio, and blogs, anger at the current state of the economy, debt, and the banks is at an all-time high. For some, a new self-organizing peer-to-peer solution is providing an alternative to the traditional banking industry, and an increasing number of people are taking advantage. They have names like Prosper, Zopa, and Lending Club, and they're gaining ground. According to estimates by analyst firm Gartner, peer-to-peer will grow from $500 million to $5 billion in outstanding loans from 2007 to 2013.[22] Although these are teeny amounts compared to most Wall Street titans, the model is striking at the core of the banking industry.

Zopa, which claims to have been the first social-banking platform, describes social lending as a process "where people lend and borrow money with each other, sidestepping the banks." Loan rates are determined by a bid and ask process, sort of like eBay, though it spreads the risk of any borrower, across fifty or more investors. Prosper in the United States, Smava in Germany, and Qifang in China have very similar models.

As to the performance of these services, relatively well-capitalized Zopa claims that lenders have made returns of "7.9% before bad debts." This of course is meaningless. It's like saying you made a decent return in theory, but not in practice. What Zopa does offer is transparency. "Lenders can see where their money is going and borrowers, where their money has come from." It turns out, though, that Zopa looks a lot like a conventional bank. It takes a fee up front, in this case £118.50, and uses the credit bureaus (in this case Equifax) to check the identity and creditworthiness of its borrowers. If a loan defaults, like a bank, it reports back that default. It even offers federally insured savings deposits.

What most of these P2P networks do that is very different is to let you align your investments with individuals or causes that you believe in. As of this writing, Prosper accepts investments of as little as $25 and estimates its returns to be from 6 percent to 16 percent. That's a lot better than putting your money in a bank and watching it get gobbled away in fees. Prosper encourages you to bid on loans one at a time, and borrowers can post their personal stories, endorsements from friends, even group affiliations to win the hearts, minds, and dollars of potential investors. It's grown fast. With nearly a million members and $193 million in funded loans, it seems that many lenders think of Prosper as an uninsured, potentially high-yield savings account.

Indeed, social lending works to the advantage of both lender and borrower. For example, if one party is now receiving 1 percent on a savings account and another is paying 29 percent on a credit card, a 10 percent mutually agreed rate is a match made in heaven. It gives the lender a tenfold increase in return, while affording the borrower a chance to begin paying down the principal.

Consider, for example, a $10,000 loan with $300 payments per month. Over the course of the loan, the borrower will have saved over $5,000 in interest payments and be debt free a year and a half earlier. The lender will have received over $2,000 more in interest payments over the same period. There is a risk, of course. These loans have no cushion or FDIC insurance, but it seems a risk an increasing number of people are willing to take. Scott Langmack, a former Microsoft marketing executive, tells Bloomberg that he has invested more than $600,000 in loans through this process. "I can reliably get 12 percent, worst case 9 percent," he said.[23] Though the actual returns achieved by the lenders have been inconsistent, and not fully transparent, there is no doubt that these sites are building powerful momentum in a storm of bank customer disenchantment.

Although P2P banking as we've described it originated in Europe and North America, entrepreneurs like Calvin Chin see the relatively underdeveloped banking markets of emerging economies as a potential opportunity too. Chin recently launched Qifang, a peer-to-peer lending platform in China that focuses on the student loan market. "We're taking what has traditionally been an offline informal activity, borrowing between individuals, and bringing it into the digital age," he says. "So a prospective student can now borrow not just from family and friends but from people across China."

Every year, approximately 6 million students enter four-year university programs in China. But according to Chin, this represents only 20 percent of the potential student population, as many families simply can't afford the fees. Given the relatively immature state of the student lending market in China, Chin sees an opportunity to facilitate educational loans for poorer students by soliciting potential lenders in more affluent Chinese communities. With Qifang, potential students fill out a profile outlining their educational expenditures, future plans, and even references. Interested lenders can browse the listings and bid on loans for individual students. On average each loan will have eight to twelve different lenders, says Chin, each contributing as little as US$7. To prevent fraud the funds are directed to the educational institution where the student has been accepted. In the year and a half since Qifang's official launch, the company facilitated over three thousand transactions. The next step for Chin is to expand the lender base with partnerships with traditional banks as well as nonprofits and philanthropic organizations.

Is this the beginning of an outright social movement? Peer-to-peer lending will certainly not displace the retail lending divisions of the big banks anytime soon. That said, social banking clearly offers many advantages both in the developed markets as well as in the rising economies. If some of the early hurdles can be ironed out, the phenomenon could have a promising future. The sheer growth of the sector has certainly chipped away at the skepticism surrounding it and reinforced the viability of a more cost-effective way of lending.

THE AGE OF COLLABORATIVE FINANCE

The financial services industry has come a long way in the past twenty years. Technological innovation has revolutionized what was a paper-based business and today the industry is one of the most digitized and networked on the planet. The financial crisis, which was caused by ill-conceived policies, misaligned incentives, and surely in some cases greed, has shown the critical role that banks and other institutions play in the workings of the global economy. Now is not the time to simply return to business as usual. The industry needs to take stock of what happened during the crisis and work to restore the trust and confidence of investors, regulators, and regular citizens.

A new movement is beginning, and it's inspired by the public anger at a host of things, from the behavior of Wall Street and massive bank bonuses to the widening gap between the interest rate offered to savers and the rate charged to borrowers. It's enabled now by the growth of mass collaboration via the Internet. And it's an alternative to the whack-a-mole game that regulators are forced to play in the ever changing financial services landscape. Innovators such as Open Models, VenCorps, and the myriad P2P lenders are just the first sign of what's to come. Opportunities for wikinomics-style collaborations abound. Financial regulators, investors, entrepreneurs, and banking customers are increasingly seeing the benefits of transparency, openness, and sharing. While financial innovation earned a bad name during the crisis, we feel that at the end of the day innovation will help the industry overcome its current challenges. As Gord Nixon puts it, "developing new products where you lend money to people who don't have income is not innovation, that's just bad risk management and faulty regulation. Developing new products to find better and less expensive ways to service your customers, that to me is innovation."

4. BOOTSTRAPPING INNOVATION AND WEALTH CREATION

Jay Rogers, a Marine veteran of the recent war in Iraq, is going back into battle. But this time the mission is not being waged in a foreign land and Rogers won't be donning combat fatigues. No, now Rogers's target is closer to home: the sclerotic industrial age automotive bureaucracy that has uniformly disappointed consumers, employees, and investors and is just barely clinging to life thanks to a multibillion-dollar taxpayer-funded bailout.

The way Rogers sees it, Detroit's inability to transform itself from a lumbering manufacturer of gas guzzlers into an innovative, world-beating producer of clean transportation solutions is one of the reasons why he and his fellow soldiers found themselves fighting to secure oil supplies in the Iraqi desert. So now he's set out to build a revolutionary car company called Local Motors with a mission to set things straight.

To say Rogers is up for it is an understatement. Already hardened by his years of service, he narrowly escaped death on three occasions and lost numerous colleagues when their caravan came under direct attack, killing two men in the next Humvee and seriously injuring two more.

While Rogers was risking his life in the desert, Detroit was teetering on the brink of collapse. Sadly, that fact didn't prevent chief executives from flying around in private jets and enriching themselves with lavish bonuses and generous perks. Their arrogance was enough to convince Rogers that the real war America should be fighting was not in Iraq but back home, where the country's decaying manufacturing base and the worst recession since the Great Depression have left millions of Americans destitute.

Revolutionizing the auto industry, thought Rogers, could reduce American dependence on foreign oil and bring jobs back to the country's declining manufacturing sector. So he asked his general for permission to apply to Harvard Business School, where he hoped to accumulate the contacts and the know-how required to execute a plan that he readily admits is ambitious and perhaps even a little naive. Two years later, in 2007, Rogers graduated at the top of his class and subsequently wasted no time founding Local Motors, a radical new form of car company. Indeed, if you want a broad-brush idea of what Local Motors is all about, picture GM or Chrysler, and then imagine its polar opposite.

Rogers doesn't employ a design team and he doesn't do any in-house R&D. Instead, he has an online community of almost five thousand designers from 121 countries that participate in contests and collectively design next-generation cars. The car parts? Rogers doesn't plan to make any of them himself, except for the state-of-the-art composite frames. He sources everything else on the secondary market from Ford, BMW, Mercedes—whoever has the right piece at the right price. Unlike its larger rivals, Local Motors doesn't own massive 3 million square foot manufacturing facilities and it doesn't have a conventional supply chain. Instead, Rogers is building a network of thirty-five microfactories around the country, each one employing local people and producing cars designed for that particular geography.

Local Motors doesn't even have dealerships: they sell cars directly from the microfactories where they are made. "We cut out the crap and boil it down to essential parties: the person who makes the car and the person who uses the car," says Rogers. Since the pie isn't divided numerous times along the chain, Local Motors can retain more of the sticker price. According to Rogers, that extra margin more than makes up for its comparatively higher production costs. And, while Detroit's incumbents have carved out only the most circumscribed roles for car buyers, Local Motors' customers are at the heart of everything. They help design, buy, service, and recycle their vehicle at a local microfactory in their area. True enthusiasts can even elect to spend a couple of weekends helping to put it together!

Using this decentralized mass collaboration model, Rogers was able to design his first car in less than three months rather than the two years it would have taken Detroit. And rather than the typical six-year production cycle, it took just another fourteen months and about $2 million to transform a sketch—chosen from tens of thousands submitted to his Web site from across the world—into the Rally Fighter, an extreme off-roader built for high-speed dirt racing. Who would buy such a thing? Rogers is targeting race teams that compete in endurance competitions like the Baja 1000 in Mexico, and other lunatics in the western states with $50,000 to spend on a lightweight, street-legal race car. Local Motors is confident it can sell the 250 Rally Fighters needed to break even, and will produce as many as 2,000 in total. Subsequent models developed by community members include the Boston Bullet, a car designed for a smooth ride in the city's narrow streets, and the slick Miami Roadster for urban racing enthusiasts. Some of these models will include engines that run on electric batteries and alternative fuel sources.

Of course, with its current focus on niche markets, Local Motors will not revolutionize the auto industry singlehandedly. But then Rogers is not the only fresh-thinking entrepreneur trying to capitalize on Detroit's weaknesses. Unburdened by the legacies that encumber incumbent manufacturers like GM, American start-ups like Tesla Motors, Fisker Automotive, V-Vehicle, and Coda Automotive are rolling out electric, hybrid, and other innovative vehicles. The California-based Tesla has already delivered over one thousand of their all-electric roadsters and is planning to ramp up production, expanding into Australia and parts of Asia.[1] Fisker is getting ready to release their first luxury plug-in hybrid, called the Karma, in the fall of 2010. The company has even bought GM's old assembly plant in Delaware where it plans to produce up to one hundred thousand cars annually.[2] Buying a car from Tesla or Fisker will set you back close to $100,000, but on the other end of the spectrum, companies like Coda are preparing to launch an affordable all-electric sedan near the end of 2010. Unlike Local Motors, which emphasizes local manufacturing and assembly, Coda has outsourced its manufacturing to a global ecosystem of partners and suppliers—including Lishen, China's leading producer of rechargeable lithium-ion batteries.[3] Together these companies are filling the void left behind by industrial age mammoths such as GM and Chrysler while creating new business models by involving their customers, creating genuine partnerships with suppliers, and injecting much-needed innovation into the whole process. In the meantime, Rogers and others are betting that car owners are hungry for a different experience and that Detroit and other auto giants will be too slow to adapt to customer expectations. "The whole manufacturer-centric paradigm hasn't changed since the Model T," says Rogers. And that inertia is at the heart of the industry's problems.

So what's the new model? Rogers sums it up. "It's about designing cars on the Web and building them out of advanced materials in factories of the future. There is a better, cooler American auto industry just ahead but it requires a complete paradigm shift. We don't need better record stores; we need to start designing iPods."

COLLABORATIVE INNOVATION MOVES TO THE NEXT LEVEL

Local Motors is just one of many examples of a powerful new paradigm for innovation. We've called it wikinomics: the art and science of collaborative innovation. Twitter, Facebook, and Wikipedia may have captured the popular imagination, but as Local Motors shows, collaborative innovation goes a lot further. It's a new way for people to socialize, entertain, and transact in peer communities of their choosing and it opens up exciting new opportunities for individuals to participate in the economy—from designing and selling T-shirts through Threadless.com to helping NASA build a better astronaut glove.

Indeed, for the first time in history, individuals and small businesses can harness world-class capabilities, access markets, and serve customers in ways that only large corporations could in the past. Small and medium-size enterprises, for example, can make and sell products to a global market without having to manufacture anything themselves. Thanks to new services such as Ponoko, based in New Zealand, you can arrange to have your products manufactured and delivered directly to the customer, virtually anywhere in the world. Upload your design to the Web site, select the materials, and Ponoko does everything else. Entrepreneurs who are just getting started can even post their products to Ponoko's marketplace. Chief strategy officer Derek Elley says, "It's a bit like low-cost global manufacturing and peer-to-peer commerce straight from your living room." Creators can now turn their ideas into tangible offerings with less risk, lower costs, instant scalability, increased control, and less complexity. In turn, consumers get lower prices for individualized products, and perhaps most important, the proposed manufacturing model promises to reduce the environmental impact tied to production—mostly through the elimination of intermediaries and a reduced need for transportation.

Could a company built on Ponoko's micromanufacturing network compete directly with a large-scale industrial manufacturer of widgets? The answer is "not yet," for now. But the sheer possibility raises deep questions about the viability of old industrial age models of organizing economic value creation when networked models can do the same thing, only better. So what if you happen to be leading one of those industrial age businesses, the kind where new products are produced in closed-off laboratories in an incremental, even slow and predictable rhythm? The message is that your monolithic, self-contained, inwardly focused enterprise is dead or dying. Regardless of the industry or sector you operate in, or whether your organization is large or small, internal capabilities and a handful of partnerships are not sufficient to meet the market's expectations for growth and innovation. Say you're leading an $80 billion company like Procter & Gamble, where organic growth of 6 percent, for example, is the equivalent of building a profitable new $5 billion business every year. For years, companies pursued acquisitions, alliances, joint ventures, and selective outsourcing in the quest to build top- and bottom-line growth. But in today's economy these conventional tools are simply too rigid, and not scalable enough, to drive growth and innovation at a level that will make companies truly competitive. As P&G's former head of innovation Larry Huston put it, "Alliances and joint ventures don't open up the spirit of capitalism within the company. They're vestiges of the central planning approach when instead you need free market mechanisms."

So if the conventional tools won't drive growth, what will? To answer that question, first consider the changing nature of today's talent landscape. While North America and Europe still boast the largest pools of highly skilled creative professionals, it is unlikely to be the case for much longer. As we go forward the talent required to produce market-leading products and services will increasingly reside in locations such as Brazil, China, and India—largely because a seismic demographic transition unfolding today places the locus of growth in the global economy squarely in these developing markets. Two billion people from Asia and Eastern Europe have already joined the global workforce over the past decade and more will soon follow. Indeed, while developed countries worry about growing dependency ratios, most of the increase in world population and consumer demand will take place in developing nations. What's more, many of the brightest talents want to work outside the confines of traditional corporations, in small enterprises or as freelance contractors and consultants.

Conventional wisdom says firms should find those people, hire them, and retain them by way of money or perks. When transaction costs were high, this made sense. But today, the Web provides a feature-rich platform that radically drops collaboration costs and thus makes accessing the global marketplace of ideas, innovations, and uniquely qualified minds increasingly cost-effective. Companies that do intensive R&D—like P&G, for example—can leverage a platform like InnoCentive or NineSigma to tap into global scientific and engineering communities using the Web. This doesn't necessarily replace the people they employ currently, as we explain later. But it can vastly increase the raw brainpower at their disposal without greatly increasing their fixed costs. Or consider the recent successes of the consumer electronics giant Apple. The iPhone is no ordinary phone; with the advent of the App Store, the iPhone has become an innovation platform that creates economic opportunity for third-party developers, while giving customers access to a larger variety of apps than Apple could if it was stocking its App Store all on its own. A growing number of companies are now following Apple's lead. The opportunity to turn ordinary, off-the-shelf products into platforms where large communities of customers and partners co-innovate value is simply too irresistible not to.

Although large, recognizable companies are getting the message, collaborative innovation is particularly beneficial for small companies. Consider a start-up like Local Motors, with twelve employees working around the clock. "No matter how smart you are and no matter how hard you work, you've got limited time, limited resources and, frankly, limited creativity," says Rogers. But through mass collaboration, Rogers can tap a talent pool that far exceeds anything he could afford to employ full-time. "By allowing people the opportunity to come in and contribute to a design vision or an engineering vision or an accessory vision or even a branding vision," he says, "we can turn the crowd of automotive enthusiasts—which we all know are already out there—into a tribe of people that really can queue behind a brand and a message and an idea, and that brand doesn't even have to be a corporate identity, it could be the individual designer of the car that takes the lead." In other words, the broader community—including the customers and partners that participate in value creation—becomes an integral part of the enterprise itself and a driver of competitive advantage. Indeed, while products and services are easily imitated, a loyal and engaged community is much, much harder to replicate.

By using the Web as a platform for innovation, Local Motors, Apple, and others have tapped into the winning formula laid out in *Wikinomics*. Successful companies today have open and porous boundaries and compete by reaching outside their walls to harness external knowledge, resources, and capabilities. They're like a hub for innovation and a magnet for uniquely qualified minds. They focus their internal staff on value integration and orchestration and treat the world as their R&D department. All of this adds up to a new kind of collaborative enterprise—one that is constantly shaping and reshaping clusters of knowledge and capability to compete on a global basis.

To be clear, collaborative innovation is not about "everybody doing everything," as critics such as Jaron Lanier have suggested. Nor is it a wholesale replacement for cutting-edge R&D or the art of a good marketing campaign. It's not about putting product duds in the public domain and hoping that someone will turn them into gold. Nor is it about enticing smart and talented people to give away their valuable ideas for free. Sure, a number of companies have exploited so-called crowdsourcing to get marketing and other services on the cheap. But schemes like these are not sustainable and rarely provide the foundation for a dynamic and fertile business ecosystem. In the most successful instances of mass collaboration, companies carve out meaningful roles for contributors and allow community members to share in the ownership and fruits of their creations. They make their core products modular, reconfigurable, and editable. They set a context for co-innovation and collaboration by providing venues for discussion and supplying user-friendly tool kits that make it easy for collaborators to add value. The end result is a superior value proposition for customers and a more powerful engine for driving innovation than the conventional closed enterprise.

In *Wikinomics*, we documented numerous examples of collaborative innovation where firms and other organizations harvest external knowledge, resources, and talent to enhance profitability and competitiveness. We showed how once closed and secretive companies like P&G learned how to reach beyond their corporate boundaries to co-innovate hundreds of new products with external partners while shaving over $1 billion off their annual R&D costs. We saw how Linux, the open-source operating system, evolved from a hacker experiment to a multibillion-dollar ecosystem, and how IBM embraced open source at the core of its business in a way that few organizations of its size and maturity had dared. We also told the stories of companies like Amazon, Google, and Apple that pioneered highly successful open-platform strategies and left their competitors in their dust. In the pages that follow we will revisit some of these companies and learn about some new ones. For those who have not read our previous book, this chapter provides an essential primer. Veteran readers will be intrigued to see how the nascent phenomenon we described in 2006 has exploded into a powerful model of economic value creation that is disrupting industry after industry. But first, let's get back to Local Motors, a radical new breed of car company that collaborates openly on vehicle design, gives customers meaningful roles in the process, and then uses a network of microfactories to build cars that meet the needs and tastes of the local markets.

LOCAL MOTORS: PROSUMPTION DOWN TO THE PLANT FLOOR

Now you may be asking: if the U.S. auto manufacturers badly lag their competitors when it comes to pumping out sleek designs, where did all of the design talent harnessed by Local Motors suddenly come from? It turns out that it was always there—but nobody bothered to tap into it, at least until Jay Rogers came along. Upon his return from Iraq, he visited a number of industrial design schools. He discovered that most transportation design students can't get jobs in the auto industry even when it's healthy. They end up taking jobs at Martha Stewart or Home Depot, but at night they are still drawing cars. "After traveling to these schools," Rogers says, "I realized that there's a pent up supply [of people] looking for a place to put their work."

His ah-ha moment came when he was interviewing the head of the transportation design school at Art Center College of Design in Pasadena, one of the top design schools in the world. The school was preparing to graduate sixty transportation design students, so Rogers asked how many of them had already lined up jobs. "Here I was coming from Harvard Business School, where there's a 98 percent or 99 percent placement rate," says Rogers. "The director said, 'We have seventeen people that are going to have jobs.' I was like—oh, seventeen, excuse me, but that's a miserable placement rate, isn't it? The director said, 'No, actually, it's quite normal. That's what we get almost every year.'" And this was before the economy got dragged into a downturn.

Tapping a Latent Talent Pool

Rogers had only one question left: how was he going to tap into this latent talent pool? After an afternoon on Twitter he had a revelation: he needed an online platform for collaboration. "The tools for this collaborative environment are already there," he concluded. "We didn't need to recreate them. We just needed them to hold up a sign that says 'if you have a great idea for a car, we've got engineering resources, and we've got customers that want to hear your great idea.'"

Today, local-motors.com is a mecca for car designers and enthusiasts from around the world. Although the Massachusetts-based company works very closely with a community of local innovators, its design community is open to anyone with access to a computer and an Internet connection. Collectively, the community has generated 44,000 designs that are shared via the creative commons. Participants vote on the designs they like and then get to work on codesigning the most promising of the community's top picks. Designers aren't doing all this for free, though. Sangho Kim, designer of the Rally Fighter concept, collected $10,000 after his initial sketch won one of Local Motors' regular series of car design competitions. The company kicked in another $10,000 after putting the Rally Fighter into production. Once a car is designed and fully engineered, customers can visit one of Rogers's microfactories and build their own, with a little help from Local Motors of course. And, in the spirit of openness, Local Motors has committed to "liberating cars to be a platform for the specialty automotive equipment market," which means the innovation process will continue long after the cars have rolled off their shop floor.

Designing and building cars with a cast of thousands does have its challenges. Rogers says it can be hard to satisfy the 4,500-strong community of

Internet designers and car buffs who congregate on the company's Web site. When Rogers decided to use tail lamps from the Honda Civic coupe instead of a custom set that would have cost the company a staggering $1 million, the fans didn't take it well, Rogers said. Still, he's just happy to have the engagement. "Our core value proposition is education because we believe that what was missing from the car industry was a sense of knowledge about what's actually going on underneath my foot pedal and beside my leg." Better information and more engagement would make for better car owners, and it might even help make us better citizens. "If I understand more about today's cars," says Rogers, "I can engage better in environmental policy, I could be a better buyer, I could be a better repair person and I could be a better steward of my car."

In the end, it all boils down to a new pact with the customer, and Local Motors' commitment to source and build cars locally is critical to that relationship. "If you're doing co-creation and you want to be real, you need to be hyperlocal," says Rogers. "To co-create with our customers—whether that means helping make the products better, helping improve our processes, or helping improve how our products are serviced—we need to be in touch with them, and I mean physically in touch. It's actually critical to our ability to create a family-like company, if you will."

"Pimp My Ride"

Although the mainstream auto industry has been slow to catch up, customization is big business—and Local Motors is in pole position to tap into it. Car enthusiasts—mostly males in their twenties—spend $4.1 billion a year dressing up and boosting the performance of their vehicles, up from an estimated $295 million in 1997 thanks in large part to MTV television shows such as *Pimp My Ride*.[4] Meanwhile, the Specialty Equipment Manufacturers Association's annual trade show has exploded in popularity, rising from around 3,000 booths and 50,000 attendees in 1990 to over 100,000 attendees and 10,000 booths spread over one million square feet today.[5]

Despite the growing momentum behind the increasingly large communities of auto enthusiasts and aftermarket specialist shops, the big twelve auto companies consider the innovation and amateur creativity that takes place in user communities a fringe phenomenon of little concern or value to their core markets. Even when such innovations look promising, the corporate processes are too rigidly adapted to the manufacturer-centric paradigm to make use of them. So rigid, in fact, that even tier one suppliers struggle to get new ideas firmly implanted in the system. Johnson Controls, which makes seats for car companies around the world, knows just how rigid the system can be. Its engineers had an innovative idea: allow customers to order custom-molded seats that would provide maximum comfort and orthopedic support. It was a great idea, but there was just one problem: there was no place in the hierarchical automotive value chain for Johnson's designers to deal directly with the customers who wanted to take advantage of this luxury option. First, Johnson had to convince the original equipment manufacturer to buy into the program, and then it had to convince the distributor and then the dealers who interact directly with the customers. Getting buy-in at all these levels proved impossible. So yet another great idea died on the vine.

If anything, the Johnson Controls saga underscores why radical new business models are so much more likely to emerge from an upstart like Local Motors than an incumbent like Ford. In fact, you can think of Local Motors as a bit like a Dell for the auto industry, with its highly configurable direct sales model, low inventory, and active customer engagement. Take a peek underneath the hood of the Rally Fighter and you'll find parts fitted in some of the world's best-selling production cars. The power train—a twin-turbo diesel and six-speed automatic—is straight out of the BMW 335d, as is the fuel tank and instrument panel. Attached to thirty-two-inch knobbed tires is a Ford F-150 rear axle, and the long-travel shocks above it are similar to those on Ford's own dirt-racing special, the Raptor. At the front of the 3,200-pound Rally Fighter is a crumple system used in Mercedes vehicles. Call it the Lego block car.

As we look forward, companies like Local Motors are edging us closer to a world where we develop physical goods using methods that increasingly mirror those used to produce intangibles like knowledge. It's already the case that everything from an Apple iPhone to an Airbus A380 to an Intel chip set combines components and services from multiple firms—often hundreds of them. For the firms in charge of pulling the strings in these sprawling webs of value creation, innovation is less about inventing and building physical things and more about orchestrating or coordinating good ideas.

To be sure, Local Motors is not the only auto company to think of the car as a series of component pieces that they can pull apart and recombine as necessary. After all, a growing number of cars are not even made by car companies, at least not the companies most consumers recognize. Companies like BMW focus on marketing, partnering, and customer relationships, and maintain any engineering expertise they deem critical. But suppliers like Magna make most of the components, and increasingly they assemble the final vehicle. The management consulting company Mercer estimates that among the premium brands, electrical systems and electronics already account for more than half the vehicle's value. What's more, by 2015, suppliers, not the automakers, will conduct most of industry's R&D and production. Are companies like BMW worried? No, it suits them just fine because it frees up their creative talents to focus on designing in-car software, polishing their brands, and piecing together the complex electronic gadgetry that runs today's high-performance cars. Everything in the middle will be outsourced or managed through one form of collaboration or another. Welcome to the global plant floor for automobiles.

We like Local Motors because it takes the traditional automotive design and manufacturing and gives it a radical twist. Rather than have engineers make all of the design decisions, it takes the next logical step: let customers mix and match the components and, while they're at it, why not let them pitch in on the final assembly? If Rogers succeeds he'll have proven that the collaborative methods of open-source software developers are just as amenable to cars and airplanes as they are to software and encyclopedias. Mass collaboration in the production of physical things may still be coming of age, but smart companies like Local Motors are already well ahead of the game.

THE PEER PIONEERS: HOW LINUX CONTINUES TO CONQUER THE WORLD

Strolling into the annual Linux Symposium in Tokyo in October 2009 was like walking into a rock concert, starring the gods of Linux. It was the first time the event had been held in Asia, and it marked an important milestone in the project's history: Linux had become truly global. To prove it, over one thousand Linux developers and users from all over Asia showed up for a chance to talk shop with Linux luminaries like Linus Torvalds and Andrew Morton.

Of course, when Torvalds first posted a fledgling version of Linux on an obscure software bulletin board in 1991, no one—apart from the most die-hard open-source evangelists—would dare have predicted that open-source software would be much more than a short-lived hacker experiment. And yet, within just a few years Linux became the largest software engineering project on the planet and spawned a multibillion-dollar ecosystem that upset the balance of power in the software industry. As Linus himself put it when we caught up in Tokyo: "You just can't plan for success like this. It would have been impossible to predict in advance."

He's not kidding. Today, Linux is used in everything from the smallest consumer electronics to the largest supercomputers. It helps run Germany's air traffic control systems. It also runs a number of nuclear power plants, though for national security reasons we can't disclose which ones. If you drive a BMW, chances are it's running Linux. And, at the time of writing, over 500 million users of set-top cable boxes, TiVo, Android phones, and other home appliances use Linux, and over 1.5 billion people use it indirectly whenever they access Google, Yahoo, or myriad other Web sites. As Linux Foundation executive director Jim Zemlin likes to say, "Every person in the modern world uses Linux in some form or another, multiple times every day." Those are impressive credentials for a loose network of programmers who aren't necessarily out to make a profit, just a better operating system.

For us, Linux is part of a larger story. It was and still is the quintessential example of how self-organizing, egalitarian communities of individuals and organizations come together—sometimes for fun and sometimes for profit—to produce a shared outcome.

A New Mode of Production Goes Mainstream

We revisit Linux because it continues to raise a number of important questions that are central to this book. If thousands of people can collaborate to create an operating system or an encyclopedia, what's next? Which industries, and which aspects of society, might be vulnerable to mass collaboration and which organizations might be positioned to benefit? Should smart managers try to bury the phenomenon, the way the music industry has tried to quash file sharing? Or can they learn to exploit the creative talents that lie outside their boundaries, the way technology giants like IBM, Google, and others have harnessed open-source software?

Questions that might have sounded academic a few years ago are suddenly vitally important. Today, millions of connected people around the planet can cooperate to make just about anything that requires human creativity, a computer, and an Internet connection. Unlike before, where the costs of production were high, people can collaborate and share their creations at very little cost. This means that individuals needn't rely on markets or capital-intensive firms to make or trade all of the goods and services they desire. In fact, a growing proportion of the things we value (including newspapers, mutual funds, jetliners, and motorcycles) can now be produced by us or in cooperation with the people we interact with socially—simply because we want to. Yochai Benkler, Harvard Law School professor and author of *The Wealth of Networks*, estimates that the current crop of one billion people living in affluent countries has between 2 billion and 6 billion spare hours among them, every day! If even a small fraction of this creative capacity could be harnessed to produce high-quality information-based goods, the output of these voluntary efforts would dwarf the output of today's knowledge-intensive industries. All it takes is the desire to create and the tools to collaborate, and both of these are increasingly in abundance. Now imagine the productive capacity of billions of people self-selecting for tasks with little regard for organizational, national, cultural, or disciplinary boundaries.[6]

Some economists and business leaders lament the fact that barriers to entry have fallen and they dislike open source, arguing that what goes into the commons takes away from the mouths of private enterprise. But as Linus Torvalds aptly put it, "That's like saying that public road works take away from the private commercial sector." Even if public ownership of key aspects of the transportation network forecloses opportunities for private profit, the gains to the rest of the economy make these losses look minuscule. For Torvalds, Linux is like a utility. It provides the basic infrastructure on which software developers can build applications and businesses. "It allows commercial entities to compete in areas that they really can add value to, and at the same time, they can take all the 'basic stuff' for granted," he says.

"This is especially important in software," he continues, "where proprietary source code at the infrastructure level can actually make it much harder for other players to enter the market. So if anything, open source is what makes capitalism in software possible at all. Without open source, you'd have just a set of monopolies: effectively, economic feudalism." In fact, he finds it rather ironic that those favoring proprietary software would attack Linux as unfair. "At minimum they should accept it as fair competition. We don't have proprietary lock-in, financial capital, government subsidies, distribution systems, or other advantages of private companies," he says. "This is not socialism, it's the opposite—it's free enterprise."

For IBM's Joel Cawley, who, as VP of strategy, has been influential in shaping the company's open-source strategy, contributing to shared infrastructures like Linux does not decrease opportunities to create differentiated value; it increases them. It's just a matter of thinking about value creation differently. "One of the things you can get confused about in doing strategy," says Cawley, "is losing sight of where real value comes from. If you are constantly creating new value then you have opportunities to harvest that value." In other words, shared infrastructures that grow and evolve constantly force firms that contribute to them to grow and evolve constantly too. And so long as they add value, there will always be healthy profits. With that promise comes a warning. "At some point you may stop creating new value," says Cawley, "and when you do, if you continue to harvest, then at some point you're no longer earning the harvest and, in fact, you may not even be doing yourself much good, because you're no longer off creating new value; you're stuck in a rut."

Profiting from open-source communities like Linux may never be quite as direct or straightforward as profiting from more conventional products and services. It's a new skill that requires companies to recognize and seize opportunities to build new products and services on top of vibrant open ecosystems—ecosystems where new value is always being created for a variety of ends and motivations. Companies need unique capabilities to work in these environments. To leverage the benefits faster and more effectively than competitors, for example, companies need capabilities to develop relationships, sense important developments, add new value, and turn nascent knowledge into compelling customer value propositions. Joel Cawley calls it "the ongoing process of regeneration, of creating new sources of value." And that is what a lively functioning enterprise is all about.

Writing a 300-Page Book in One Day

Back in Tokyo, Jim Zemlin is taking the stage. To kick off the Linux Symposium he spends a few moments revealing just how prolific the Linux community has become. It turns out that some 10,923 lines of code have been written by a worldwide team of developers and carefully filtered into the system each and every day for the last four years. That's the rough equivalent of writing a 300-page book every day, or about 1,460 of those books over the full four-year time span. Day after day, those same dispersed developer teams find the time to remove 5,547 lines of code that have become redundant and modify another 2,243. When you add it all up, code-crazed open-source diehards add something in the neighborhood of 2,700,000 lines to Linux every year! It's a stunning accomplishment. "Such rates of change have never before been witnessed in a software project of any size," says Zemlin.

By this time the audience is in rapture and Zemlin is waxing lyrical. But it turns out he's just warming up when he throws out another large number for the audience to chew on:

$10,800,000,000.

That's the estimated amount of greenbacks you'd need to shell out if you wanted to recreate Linux in its entirety using conventional software development methods. And, if that's not big enough, try this one:

$50,000,000,000.

That's the estimated size of the Linux economy, including Linux-related hardware, consumer electronics, and related services. That's up by a factor of five since 2006 and it's more than the GDP of some small countries like Costa Rica, Lebanon, and Bolivia.

Finally, for good measure, consider two considerably smaller numbers: 0 and 1. Zero is the cost of using Linux and one is the number of people it took to kick-start this incredible process.

On the Right Side of History

Although much in the Linux community revolves around the persona and leadership of Linus Torvalds, the truth is that the Linux community is now a highly sophisticated organism. While Linux may not have stock options, corporate campuses, or free haircuts, its community includes a core of five thousand developers and a much broader ecosystem of users and contributors. Large companies like IBM, Motorola, Nokia, Philips, Sony, and Google were among the first to dedicate serious resources to Linux development. But today they coexist with two hundred other firms that contribute regularly to the Linux code base.

At least some of the growing momentum can be attributed to the fact that companies that used to be passive users of Linux have recently become active contributors. After years of developing their own custom Linux solutions internally, they soon realized that they were missing the point. By keeping those proprietary extensions and modifications to themselves they were missing an even bigger opportunity to take advantage of the community's resources. And it turns out that maintaining enterprise Linux installations is a lot cheaper and easier when you've got five thousand of the world's best programmers at your back. Zemlin explains that there used to be very little communication between core Linux developers and the millions of end users. But now that the value of participating in Linux is clearer, the community has hit a tipping point. This, in turn, really ups the ante on collaborative innovation and foreshadows some big structural changes in the economy. Indeed, as companies assign serious resources to open-source communities, questions about when to contribute and how to harness the commons are arriving at the heart of corporate strategy. It's no longer just about boundary decisions (what's in, what's out), but big questions about where to play, when to cooperate on shared infrastructures, and when to differentiate and compete.

There are other signs that Linux is maturing. With the help of companies like IBM, Red Hat, and Novell, the Linux community has organized a shared legal defense system and a patent commons to help protect developers from lawsuits. It's a critical step because no company in their right mind would embed Linux in their products if they felt there was a high probability that a patent infringement suit could quickly derail their product. But by pooling their software patents and making them available for free, major corporate participants in the Linux community are sending a signal to users, developers, and would-be adopters that Linux is open for business. Indeed, if they successfully lift the legal cloud over Linux, the stage could at last be set for open-source software to step it up a gear. Enthusiasts are already counting their chickens, speculating that Linux will not only dominate today's server market, but just about everything from medical devices and consumer electronics to home appliances, automobiles, and traffic lights. "We're on the right side of history," says Zemlin. We couldn't agree more strongly.

TAPPING THE GLOBAL MARKETPLACE FOR IDEAS, INNOVATIONS, AND UNIQUELY QUALIFIED MINDS

Last time we caught up with Larry Huston in 2006 he was still rolling out a radical new innovation model at P&G called "Connect and Develop." We told the story of how a company once renowned for its insularity learned to see the global marketplace as an untapped resource for ideas and a pillar of its innovation strategy. At the core of the strategy was an ambitious goal set by CEO A. G. Lafley. He claimed in 2005 that the company could source 50 percent of its new product and service ideas from outside the company by 2010. It was widely perceived by industry watchers as a gimmick back then. After all, P&G already had nine thousand world-class researchers busy inventing new things—were they all suddenly going on holiday?

But what the world didn't realize was that it didn't matter how much money the company pumped into internal R&D—even nine thousand of the best of the brightest were not going to generate enough innovative new products to keep P&G competitive. The company needed an infusion of talent and fresh ideas. In fact, research by Huston showed that for each of the nine thousand top-notch scientists inside P&G's labs, there are another two hundred outside who are just as good. That's a total of 1.8 million people whose talents it could potentially tap into. But how could P&G access this vast talent pool? Well, for starters it needed to think very differently about innovation. Rather than invent everything in-house, no matter the cost or effort involved, P&G could scour the globe for proven products and technologies that it could improve, scale up, and market, either on its own or with its business web. And rather than try to employ a vast multitude of researchers full-time—after all, what company can afford to employ 1.8 million people?—it could seek to leverage other people's talents, ideas, and assets quickly and move on.

In one case, that meant going to a small bakery in Bologna, Italy, where P&G acquired a technology for printing edible images on cakes and cookies that enabled it to launch a new line of Pringles potato chips with colorful trivia questions and animal pictures emblazoned on each chip. In another case, P&G needed a solution for keeping cotton shirts wrinkle-free. An internal review turned up nothing, so it posted the problem on InnoCentive, a global network of 200,000 scientists who collect fees ranging from $10,000 up to as high as $1 million for providing solutions to innovation-hungry companies like P&G. A solution was soon found, but it came from an unusual place. The answer was sent in from the laboratory of a professor studying polymers on behalf of the semiconductor industry. His idea, when applied to garments, neatly solved P&G's wrinkle problem.

This kind of multidisciplinary cross-pollination is often central to producing breakthroughs. Indeed, InnoCentive solvers have been known to yield surprising results that lead organizations down paths they would never have considered. Take the Oil Spill Recovery Institute (OSRI), which was tasked by Congress with the responsibility of cleaning up the *Exxon Valdez* oil spill off the coast of Alaska. For twenty years it has been trying to figure out how to remove the eighty thousand barrels of oil still sitting at the bottom of Prince William Sound. The problem is the subarctic waters are extremely cold, which means the oil is almost solid and can't be pumped out of the water using conventional technology. The oil might have sat there forever, except one day in 2007 the Oil Spill Recovery Institute put a challenge on the InnoCentive network, where it didn't take long for a construction engineer named John Davis to suggest a solution. Of course, the construction industry has long wrestled with the need to keep cement liquid during mass pours. Cement companies solve this problem with high-frequency vibrations that prevent cement from forming solid concrete too soon. With a few modifications to widely available equipment on a recovery barge, it turns out the same approach works for near-frozen oil. OSRI had its solution, and Davis picked up $20,000 for his troubles. InnoCentive cofounder Alf Bingham says solutions like these are a property of making the problem accessible to many, many minds. "It produces a diversity of thought about the problem that can often make the solution rather unique."

The ability to tap this broader network allows organizations like P&G

and the OSRI to solve problems more efficiently. But just because it saves time and money doesn't mean researchers can't profit from this model of innovation too. More than 40 percent of registered "solvers" come from Brazil, Russia, India, and China; 30 percent from the United States; and the remainder from over 150 other countries. Many have jobs in university labs where seemingly obscure research pursuits could have applications that they hadn't thought of or hadn't yet monetized. In fact, InnoCentive president and CEO Dwayne Spradlin is keen to ignite more entrepreneurial activity among seekers and solvers by expanding the tools available to users to manage rights, communicate with other registered users, and self-organize into ad hoc freelance organizations. "Think 'government research retirees' or 'Chinese nanotechnologists' or 'ABC Corp's contract research partners,'" says Spradlin. "Then envision engaging those groups in specific challenges of interest to them."

Indeed, where innovation problems are highly integrated, it may be preferable to offer problems to skilled external teams rather than or in addition to posting them in an open market. Imagine, for example, USAID seeking ready-made solutions for bio-latrines and housing in the aftermath of another Haiti-scale natural disaster, or the city of Phoenix seeking help to sustain its fresh water resources in the face of dwindling supplies. The fact that InnoCentive structures its problems in a modular way already means that there are opportunities for individuals and organizations to build a business around this model of innovation.

Back at P&G, Lafley and Huston's radical project continues to gather steam today. Through Connect and Develop in the first two years alone, P&G acquired over one thousand external ideas and technologies and launched more than one hundred new products for which some aspect of execution came from outside the company. In the process it has increased R&D productivity by nearly 60 percent. Its innovation success rate has more than doubled, while the cost of innovation has fallen.[7] R&D investment as a percentage of sales was down from 4.8 percent in 2000 to 2.5 percent in 2010. And in the ten years since the company's stock collapsed in 2000, P&G has doubled its share price and built a portfolio of twenty-three billion-dollar brands, which makes it the industry leader. Although Lafley admits that going outside the company for new ideas was a fairly radical step, he considers it to be so critical to P&G's future that he recently made "proudly found elsewhere" a mantra for employees. Employees are now being rewarded not just for inventing new products, but for commercializing them, regardless of whether the idea originated inside P&G or somewhere else.

Infostructure and the Virtual Collaboratory

Of course multinational firms have long tapped diverse environments for new ideas, but even today they are mostly using outmoded industrial age models to do it. In India, where the population of skilled talent is exploding, just about every major company is trying to stake a claim in this emerging talent pool. Companies like GE have recently invested heavily in bricks-and-mortar facilities in emerging hubs such as Bangalore and Mumbai—investments that Huston now argues have been misspent given the alternative that Internet-enabled models present. Sure, face-time is important when you're trying to manage people and intellectual assets across cultures, disciplines, and organizational boundaries. But for Huston, GE and others made a big mistake. Rather than put down bricks and mortar in China or India, smart companies increasingly just tap into a global ecosystem online. When it comes to tapping into global talent pools, argues Huston, "It's more infostructure than infrastructure."

GE's Innovation Center in Bangalore is a sprawling complex that occupies fifty acres of land, employs four thousand people, and cost about $175 million to build.[8] "But in a world of flat R&D," says Huston, "I don't need to be on a bench in the upstream lab, when I can dig into these global ecosystems from the two flat-screen monitors sitting on my desktop." P&G decided against building an innovation center in India and instead sent sixty people on a mission to fan out across the country and leverage the intellectual insights and capacities that already exist there. India has an extensive system of national labs and boasts the world's largest database of botanical ingredients—an attractive asset for a company whose skin care products are among the most profitable in its portfolio. "Why not piggyback on that and save the effort of having to build all of that capability myself," says Huston.

So if bricks-and-mortar R&D centers represent the old school, what's the new school? Huston calls it "the virtual collaboratory." Think of it as a vast ecosystem of commingled innovators working on complementary lines of research and development that may intersect in moments of serendipity. This ecosystem doesn't exist in a single physical space but in an online network that makes accessing these individuals and companies and building relationships about as simple as setting up a Facebook group.

Larry Huston wants to put this capability on every innovator's desktop. His company inno360 has built a platform that includes everything from a network visualization interface through to the tools for designing innovation briefs and managing intellectual property. "Investments in infostructure give me speed, far less investment, a diversity of ideas that I'm not going to get anywhere else," he says.[9] According to Huston, a typical company's innovation ecosystem has four major components: its full-time employees, its suppliers' employees, its extended social network (including all of the personal connections to colleagues and collaborators outside the firm), and all of the people who are working on related activities that aren't currently connected, but probably should be—call this the latent network. Put it all together, and for a company like P&G, it adds up to about 2 million people. Its top fifteen suppliers alone have a combined R&D staff of fifty thousand and represent a significant source of innovation. But the actual scope of its ecosystem is much deeper and broader. After all, P&G now spends close to $2 billion on R&D across 150 science areas and 300 brands that span multiple product categories. That means it is constantly rubbing elbows with a vast talent pool that could potentially contribute to its virtual R&D department.

Many companies don't even realize the potential they have at their fingertips. Their people networks are too fragmented to give them a complete, company-wide picture, let alone one that encapsulates an entire ecosystem. The ability to visualize these connections on an enterprise-wide basis—along with references to relevant scientific literature, patent databases, and other data sources—would be welcomed by most companies. Say, for example, you are leading a food company that for many years has peddled an unhealthy concoction of carbohydrates, trans fats, sugars, and high-fructose sweeteners. And let's say your future in being a food company is all about using healthy ingredients and creating healthy foods. That is a whole new social network and a whole new industry ecosystem to plug into, one that likely involves building networks of nutritionists, organic farmers, and food chemists who know how to make quinoa into a tasty breakfast cereal! "At a certain point," says Huston, "the recognition sets in that 'Wow! I can't do this myself, it's going to take me forever.'" But with a strategy tied to open innovations, ecosystems, and social networks it's highly probable that the transition could be much faster and more successful. "Without a mass collaboration capability you're not ever going to develop the technical know-how to do this—not in the time that you need to get this done," says Huston. This reality drives a whole new set of boundary considerations and a whole new way to think about strategy based on the fact that you need to quickly leverage new intellectual assets and technologies to remake many of these industrial age businesses.

The World Is Your R&D Department

Learning how to tap the potential of a global talent pool means turning the traditional innovation process on its head. Companies should no longer invent first and ask questions later. They should ask: "What do our customers really need?" and then scour the global marketplace for ideas for the necessary inventions and technologies. Companies will need to be ambidextrous: building on core capabilities internally, while acquiring the greatest, most complementary ideas externally. The deep-rooted "plan and push" modality will need to give way to a new approach in which companies engage and co-create with the best available talent.

For companies with a long tradition of in-house innovation, these changes won't be easy or automatic. "It requires a lot of trust to believe that you can accomplish your goals by relying on freelance scientists to come up with solutions," explains InnoCentive founder Alf Bingham. "Most people at big companies are not ready to entertain that idea." After all, the modus operandi of R&D departments is to invent, not to acquire outside ideas. CEO leadership and a commitment to appropriate staffing, incentives, and organization in R&D will be critical to success. For starters, companies need new capabilities to create, transfer, assemble, integrate, and exploit knowledge assets. Sensing external opportunities is just the beginning. The really hard work starts with conceiving the ultimate customer offerings and then executing smart decisions about acquiring IP and partnering to design, assemble, and deliver the final value. As Larry Huston reminds us, "Once an external idea gets into the development pipeline it still needs R&D, manufacturing, market research, marketing, and other functions pulling for it." Similar rules apply farther down the chain, in domains such as marketing and customer support.

In the end it's no surprise that smart people remain as valuable as ever, if not more so in today's economy. But for smart companies, the notion that you have to motivate, develop, and retain *all* of your best people internally is old. Collaborative innovation provides companies and other organizations with a new way to tap the creative power of the global economy whenever they need complementary skills or knowledge to create a new product or solve a problem. Of course, great organizations will still need great internal talent. But increasingly, assume that many of the best people are to be found outside your corporate walls. With a global platform for collaboration, however, a massive reservoir of talent would be a few clicks away.

III

REINDUSTRIALIZING THE PLANET

5. REVERSING THE TIDE OF DISRUPTIVE CLIMATE CHANGE: A NEW GLOBAL POWER EMERGES

Greg Asner and Carlos Souza, two scientists at the forefront of forest science, are hoping to uncover the location and rates of deforestation around the world and link them to climate change. But instead of traversing through vast tracts of jungle in Indonesia or Brazil, they have been using a tool available to anyone with a PC and an Internet connection—Google Earth. The scientists are working with Google's team to analyze satellite images that can shed light on the status of the world's forests, without the need for expensive field studies.[1] In fact, the idea over time is to gather together all of the earth's raw satellite imagery data—petabytes of historical, present, and future data—and make it easily available through the Google Earth platform to anyone who cares to make use of it. The evidence accumulated to date is already helping scientists, governments, and conservationists to assess the scale of the deforestation problem on a global basis. For example, we now know that emissions from tropical deforestation are comparable to the emissions of all of the European Union, and greater than those of all the cars, trucks, planes, ships, and trains on the planet. And thanks to the work of economists such as Nicholas Stern, we also know that protecting the world's standing forests is one of the most cost-effective ways to cut carbon emissions and mitigate climate change.[2]

While free tools like Google Earth empower the world's scientists and policy makers, they also make information that was once inaccessible and hard to understand available to the broader public. Indeed, by displaying that information in bold visual formats, these tools help communicate complex phenomena in a way that most laymen can easily grasp. Whether mapping the world's oil spills, simulating the effects of sea-level rises, tracking mammals on the verge of extinction, or showing national per capita CO_2 emissions, Google Earth, along with the data-crunching capabilities of Google's server farms, provides an ideal platform on which to enhance our understanding of humanity's impact on the biosphere.

Of course, fostering understanding on its own is not enough to heal our imperiled planet: we need innovative solutions and we need to take the necessary actions to put those solutions in motion. On January 31, 2008, millions of American students in nearly two thousand academic institutions stepped up to the plate as part of a day-long civic engagement discussion called Focus the Nation that focused on climate change, its consequences and potential solutions. Joining these students were some powerful players, including sixty-four members of Congress, fifteen state governors, and countless local politicians. The event created a groundswell of activism and helped spur government policy makers into action. Thousands went online to vote on a list of possible solutions—with the most votes going to a proposed increase in federal spending on clean energy R&D. The proposal is now gaining traction as part of the U.S. Congress's upcoming climate bill. Few other issues have garnered the attention of a million-plus participants, and few initiatives, if any, have gone as far as Focus the Nation has in convincing colleges, universities, and secondary schools to commit to lending an entire day of instruction to just one topic. Focus the Nation has now become an ongoing campaign to engage young people to find and promote solutions to climate change using the Internet.[3]

What do scientists like Asner and Souza and these students have in common? On one hand, they share a belief, one increasingly backed by scientific evidence, that problems like pollution, water shortages, habitat degradation, and climate change threaten the ecological systems that sustain our economy and our way of life. On the other hand, they are part of a broader mass movement—a growing network of people with the motivation and the capacity to enact the deep changes required to bring our modern way of life into balance with nature's support systems. In fact, on the balance of evidence presented in this chapter, one could argue that we are in the early days of something unprecedented. While there are still fierce debates about both the scope of the problem and the appropriate solutions, there is a nearly universal consensus that polluting the air, water, and soil and destroying natural habitats are not good for humanity. And what's more, for the first time we also have one global, multimedia, affordable, many-to-many communications system that we can use to better understand the causes and consequences of climate change and develop appropriate responses with input from a global array of citizens, governments, and businesses. Around the world there are already hundreds, and probably thousands, of collaborations occurring where everyone from scientists to schoolchildren is mobilizing to do something about carbon emissions. There is no guarantee that these bottom-up initiatives will scale to meet the mammoth tasks ahead. But this chapter shows that we already have the tools, creativity, and social connectivity required to do the job. In the end, it will come down to our willingness to embrace change. A growing number of pioneers are showing the way forward, but it will ultimately be up to us to turn this wave of concern into a massive global effort to fight climate change.

A KILLER APPLICATION FOR MASS COLLABORATION: SAVING THE PLANET

The debate over climate change may not be over, but just as most average citizens believe that pollution has negative consequences, virtually all scientists now agree that the risks associated with a substantial warming of the planet are far too great for the world to do nothing about the soaring levels of CO_2 emissions we pump into the atmosphere. Rising average surface temperatures combined with rapidly expanding deserts, melting Arctic sea ice, and ocean acidification already provide what many of the world's top scientists believe to be unequivocal evidence that human activities are fundamentally altering the Earth's climate.[4] Although we cannot fully predict the repercussions, the risk is that if we fail to rein in greenhouse gas emissions there will be devastating social disruption caused by extreme weather events, water and food shortages, mass migrations, unpredictable disease patterns, and loss of biodiversity.[5] Moreover, the effects of climate change will be felt unevenly, with the burden falling most heavily on those least able to cope with the consequences, resulting in untold human misery.

For many people, those consequences seem distant and it's certainly true that the worst of it will be inherited by future generations. Given the short-termism that dominates our political systems, our economy, our capital markets, and our individual daily decisions, it is easy to be dismissive of the notion that humankind will suddenly become motivated by a sense of intergenerational justice to make the deep and difficult adjustments that are required to avert global ecological disruption.

So how will the world make the historic act of stewardship required to bring greenhouse gas emissions within acceptable levels? Conventional policy wisdom suggests that putting a price on carbon (through a "cap and trade" system or a straightforward carbon tax) will help usher in a new mindset among consumers, investors, farmers, innovators, and entrepreneurs that in time will make a big difference. Make people and businesses pay the full environmental costs of what they produce and consume and suddenly every investment and purchasing decision made in retail stores, financial markets, and small and large companies around the world will be made in pursuit of the least-cost low-carbon option. Weaving carbon emissions into every business decision will drive innovation and deployment of clean technologies to a whole new level, and make energy efficiency much more affordable. Industries will need to invent and adopt new technologies that boost efficiency to limit their emissions. And consumers will curtail their own carbon footprints as the prices they pay for things like air travel and exotic fruits begin to reflect their true costs to the planet.

It's true that centrally managed taxes, credits, and incentives provide important levers for steering society toward low-carbon solutions. But the erroneous assumption underlying conventional wisdom is that politicians and other powerful interests can "manage" climate change with new regulations issued from a patchwork of national capitals around the world. Indeed, if the global debacle in Copenhagen in December 2009 shows anything, it's that world leaders can barely forge a basic political agreement to move forward on addressing climate change, let alone negotiate a binding international treaty.[6] Moreover, the risk is that anything that ultimately receives political approval in the key national capitals will be watered down to the point of being practically useless. So while we can't give up on political solutions, we certainly can't depend on them alone.

The missing and largely overlooked ingredient in most of the punditry and policy recommendations to date is the need for a new, more agile way to marshal and fully exploit the collective ingenuity of citizens and businesses—something wikinomics can offer in abundance. We need an approach that relies less on central control and more on a self-organized critical mass of people and organizations working in all sectors to initiate small experiments and social innovations that, under the right conditions, could mushroom into pervasive changes in societal behavior. This approach would not supplant but supplement conventional policy approaches, making it easier for politicians to gain support for the tough measures that will be required to stave off the more drastic effects of climate change.

So what would a wikinomics approach to solving climate change look like? For starters, we need Internet-based platforms for innovation that can unite networks of the willing and committed, allowing new ideas, relationships, and partnerships to surface. Think distributed business laboratories where social entrepreneurs can launch experiments, build communities, and attract funding for their ventures, for example. Or, social networks where peers challenge one another to take actions that reduce emissions and measure their collective progress over time. Imagine a "green technology commons" where industries share intellectual property and other assets that could hasten the transition to a low-carbon economy once widely adopted.

Or, how about Web-based tools for turning raw data into reliable and usable information that would allow everyone from investors to regulators to ordinary citizens to monitor the progress of communities, nations, and corporations toward carbon neutrality?

Imagining is all well and good, you might say, but is any of this achievable, today? As this series of chapters shows, it's not only possible; it's already happening. This chapter examines the case for radical transparency around climate change, including its causes and consequences, as a means to stimulate widespread consumer and business action. You'll learn about what some social entrepreneurs are doing to increase access to information about the climate change impact of everything from industrial activity to everyday household consumption decisions like buying toilet paper or eating out. We'll also take a look at how Nike, Best Buy, and the Creative Commons are pioneering a new form of radical sharing that could dramatically accelerate business efforts to reindustrialize the planet. Chapter 6 lays out the most daunting of challenges: weaning the world off its dangerous addiction to fossil fuels and building a new green energy economy that can sustain human civilization for centuries to come. You'll learn why applying open-source principles to the energy grid is the best way to ramp up the supply of green power. We'll also meet a new generation of energy prosumers that provide a signpost to a future where all homes and buildings generate enough of their own power to live off the grid entirely. Chapter 7 homes in on our broken, industrial age models of transportation. We explain how principles like openness, collaboration, and sustainability can help guide us toward a transportation system that is safe, efficient, intelligent, and networked. We'll also take a peek at the car of the future and we'll get the scoop on what über-entrepreneurs like Shai Agassi (former SAP exec and founder of Better Place) and Robin Chase (founder of Zipcar) are doing to usher in radical new models for personal transport.

There are no "magic bullets" among the solutions we present, and many actions will need to be taken congruently by individuals, businesses, and governments. Given that the costs of climate change mitigation will only grow larger with time, it would be shortsighted to delay action any further. Tim Palmer, a climate scientist at the University of Oxford whose current work focuses on quantifying and managing the uncertainties surrounding climate change, suggests we take an analogy that will be familiar to most homeowners. "We don't have to believe that our house will burn down in the coming year to take out insurance," says Palmer. "Similarly we don't have to be 100% certain that dangerous climate change will occur to take action to cut emissions."[7]

Of course climate change deniers, and those who see the world's attempt to control climate change as a threat to their business interests, will continue to unleash their armies of lobbyists to water down policy, spread bogus science, and block innovations that might threaten their business models. But the best way to counter backroom lobbying and misinformation is not to hunker down as some climate scientists have in the wake of the climategate scandal (see chapter 9), but to foster greater transparency and open debate around the risks of not acting now. For instance, Palmer suggests that everyone concerned about the climate change issue, particularly those who are skeptical, ask themselves exactly how large the probability of serious climate change should be before we should start cutting emissions: 0.1 percent? 1 percent? 10 percent? 50 percent? "Considered this way," says Palmer, "it's clear that the black and white dichotomy between the 'climate believers' versus 'climate skeptics' is indeed a false one."[8] And if you happen to be one of those people who believe that action is merited today, there is no point waiting for stubborn inertia exhibited by our political institutions to recede. Thanks to the Web, we have the most powerful platform ever for people to learn about climate change, inform others, and self-organize.

Going on a Carbon Diet

The world needs to go on a dramatic carbon diet. But it's misleading to assume that only massive system-wide change and heroic feats of engineering will move the needle. In the fight against climate change, simple actions, taken en masse, can add up to a big difference too. A case in point is Jason Karas, a social entrepreneur who quit his gig at a mobile operator to develop a site called Carbonrally that pits teams from around the world against one another in a contest to see who can reduce their carbon footprint the most. Launched in 2007, Carbonrally lays out environmental challenges and keeps score by translating green actions into pounds of carbon dioxide averted. On July 12, 2009, for instance, 4,041 people committed to following a few smart driving rules for one month—gentler accelerating and stopping and driving at optimal speeds—to increase fuel efficiency in their cars. Just twelve days later, "rallyers" from across North America and Europe had eliminated some 247 tons of CO_2 that would have been released into the atmosphere, the equivalent of turning off the electricity in 339 homes for one month. In the process, the average participant saved themselves nearly $20 a month at the gas pump.

Although the competitive element introduces some fun into the equation, the site is based on real science. The challenge was translating the science into digestible pieces of information that can actually influence people's behavior. Karas, who majored in environmental economics at Duke, explains how everything we do that is powered by fossil fuels has a carbon dioxide cost, and it adds up—a bit like credit card debt. Some actions, like commuting in a gasoline-powered car, have obvious carbon costs. Others are less clear but still significant. Take your diet: livestock are responsible for an estimated 18 percent of global carbon emissions, so when you chow down on a hamburger, you're effectively emitting CO_2 as well.[9] Even something as small as an iPod Nano will add to your carbon footprint, thanks to both the energy used to produce and ship it and the energy later needed to charge it (68 lbs. of CO_2 over its lifetime, according to the British design consultancy IDC).[10] One's personal carbon footprint is a rough measure of all the greenhouse gases we're responsible for emitting.

But unlike Karas, not everyone has a degree in environmental economics. And until recently, it's been pretty near impossible for the average consumer to get a reliable estimate of their personal carbon footprint. "Back when we started, most of the information available online regarding climate change or global warming was pretty technical," says Duane Dahl, the founder of EarthLab.com, one of the Web's most used carbon calculators. "If the average Joe wanted to find out the most basic information, inevitably he would need to download a sixty- to ninety-page report that he wouldn't have time to read and probably wouldn't understand anyway." Today, you can go online, answer some questions about your home, energy use, commute, travel, work, and general lifestyle, and EarthLab.com will spit out a comprehensive analysis of your personal contribution to global warming. Then you can compare scores against other EarthLab.com community members in your city, state, country, and even the world.

The problem, according to Karas, is that carbon calculators don't address the underlying issues with motivation. It's one thing to measure your impact. It's quite another to do something about it. Even though many activities that reduce emissions could save people money, the potential gains are spread so thinly across millions of homes and businesses that getting widespread participation can be a real challenge. Even for those who want to contribute something, there's the problem of feeling isolated and relatively futile in one's own personal efforts to make a difference.

Solving the Motivation Problem

"People are interested in the climate change issue," Karas said, "but it's really just too big for them to get their hands around. And even if they understand they're not really sure what they're supposed to be doing. And even if they know what they're supposed to be doing, they're not feeling particularly motivated because they're just one person in a world of billions of people."

Jason Karas thought the motivation problem could be solved if contributing to the solution felt less like a chore and more like playing a competitive sport. "Some people thrive on the competition," he said. Other Carbonrally users like the idea of working together as a social unit. "Sometimes they're just excited to be working with their neighborhood on something that they have control over." That natural urge for affinity highlighted the importance of having a vibrant online space for people to discover and commit to small, positive actions over time in a setting that allows them to see not only their own progress but the collective impact of thousands of other people taking the same actions, at the same time. The community itself has become an important source of carbon-busting challenges. Users propose new ideas for saving energy and reducing carbon dioxide emissions and the community chooses the best ideas to pursue as a team. The site's "challenge workshop" contains almost three thousand suggestions.

Most of the proposed challenges are deliberately low-tech and easy to do. Line-drying one load of laundry per week reduces CO_2 emissions by an average of 4.7 pounds, according to Carbonrally's scientific advisers, who crunch the numbers to figure out the impact of each challenge on CO_2 emissions reductions. Unplugging your computers every night for one month cuts CO_2 by 51 pounds. Installing a programmable thermostat will set you back $50, but will shave 200 pounds off your carbon footprint and save the average homeowner about $32 a month if properly used.

Building a Network of the Willing and the Committed

Carbonrally's overall impact in terms of reducing CO_2 emissions is fairly modest—to date the site's 32,000 users have managed to reduce CO_2 emissions by almost 5,000 tons, the equivalent of taking a thousand cars off the road for one year. Given that global CO_2 emissions due to human activity total more than 28 billion tons a year, that still doesn't add up to a whole lot. However, the long-term potential of initiatives like Carbonrally and EarthLab.com is clear to see. The natural inclination of users to congregate into smaller, interest-based groups suggests there are opportunities to broaden the membership of those communities by catering to their unique sensibilities. For example, a large contingent of moms on his site is motivated by a desire to find very practical information, says Karas. "They realize that they are responsible for the next generation. They're looking for good information, and they are also looking to share good information. They don't care about competition in the way that corporate teams do." A partnership with online communities for mothers like Netmums (which boasts over 750,000 members) could help expose Carbonrally's tools to a wider audience. The teens on Carbonrally are unique as well. According to Karas, they're burning with passion and their sense of urgency can even lead them to be aggressive. "Their attitude is typically 'get out of the way everybody else because you screwed things up and now we've got to deal with this.'" Equipped with tools like Carbonrally, today's teens can turn their passion into results.

Dozens of high schools across the United States have already formed teams that are actively challenging for honors. In fact, a fifth-grade class in Northern California and a seventh-grade class in New Jersey are among Carbonrally's reigning champions, having saved 77 tons and 52 tons respectively. In the spring of 2009, Vermont governor Jim Douglas threw down $15,000 in prize money and challenged schools across the state to reduce their emissions using the Carbonrally platform to track their progress. Fourteen schools competed, making a 114-ton dent in the state's footprint.[11] But what if every school in America took Earth Week as an opportunity to compete in a nationwide challenge, with the winning team getting a chance to have lunch at the White House with the Obamas? Enrolling more institutional support could give Carbonrally the traction it needs to break into the mainstream. Imagine government agencies, hospitals, major retailers, and professional associations all challenging each other in a race to reduce carbon. There is also a potentially endless list of competitive rivalries to be exploited. Microsoft employees competing head-to-head with Google employees or Harvard versus MIT. Or, why not pit British football team Manchester United's 75 million worldwide supporters against the 42 million fans of archrival Liverpool FC?

Until such a time, changing people's attitudes requires direction and a supportive social structure. Emerging tools such as Carbonrally try to provide both with the scalability of a virtual platform. In order to drive change it is also important to simplify complex scientific concepts and calculations and give people the ability to make a choice from a variety of energy-saving actions—not impose solutions. The key is to let people find their own ways to reduce their impact. "We encourage people to approach this problem like a diet: if you want to splurge in one area you have to make up for it in another," says Dahl.

A World Without Oil and the Power of Imagination

Carbon calculators and competitive challenges aren't for everyone. For Ken Eklund, a freelance writer and game designer, interactive gaming experiences provide an engaging alternative where ordinary people can immerse themselves in an experiential process of finding everyday solutions for climate change that drive real-world changes in behavior. Eklund is the creator of a fascinating alternate reality game (ARG) called World Without Oil, an interactive, Internet-based narrative where large numbers of game players collaborate to solve plot-based challenges and puzzles. Unlike a tightly scripted game or media production, the outcome of an ARG is determined almost entirely by the players and their interactions. The premise of World Without Oil was simple and provocative: What if an oil crisis started on April 30, 2007—what would happen? How would the lives of ordinary people change? Players were invited to imagine how their lives and communities would be different and how they would cope if the world's oil suddenly dried up. The "plot" unfolded dynamically. First, the players read the "official news" and what other players were saying. Then, using a combination of blog posts, videos, images, and even voicemails, they told their own stories of the challenges they were facing. As the crisis continued, players updated their stories with further thoughts, reactions, and solutions. The game ended after thirty-two days, having engaged thousands of players around the world and woven the fabric of 1,500 stories into what Ken describes as a "living breathing meganarrative that presented some eerily plausible scenarios, complete with practical courses of action to help prevent such an event from actually happening."

Herein lies the key point. World Without Oil presented players with an engaging, interactive structure in which ordinary individuals could collectively imagine how the world would respond to what we all know is an inevitability: sometime in the near future (and perhaps sooner than we might expect) we will all live in a world in which we can no longer rely on fossil fuels to power our daily existence. Although most individuals would have difficulty discussing the intricacies of "global" solutions to climate change like the cap-and-trade system, they are indeed authorities on something very important—their daily lives. Bringing people together to create and collaborate around solutions is a powerful weapon with which to tackle the problem. What's more, the interactive game format encourages curiosity, learning, imagination, and innovation. "With just a little bit of narrative structure," says Eklund, "the game becomes this incredibly enriching experience, with people learning from other people, joining a community, acquiring new skills, getting access to novel data and information and then using the game experience to make real changes in the way they conduct their lives."

OPENING THE KIMONO ON CLIMATE CHANGE

Carbonrally, EarthLab.com, and World Without Oil bring greater awareness and a sense of community to making ordinary household decisions that can improve our odds against climate change. What about other parts of the economy and society? If we really opened the kimono on climate change, would a world that is more informed about the causes, consequences, and potential solutions to climate change be in a better position to act quickly and decisively? We certainly think so.

In the climate change debate, insufficient information about which economic activities—and, by extension, which companies and nations—are contributing most to climate change undermines society's ability to target remedial actions and assign responsibility for correcting damaging behaviors.[12] The right amount of transparency in such cases can change perceptions, reveal new factors that alter the stakes, or compel other participants to accept the need for and legitimacy of new regulations. Getting our hands on comparable CO_2 emission data for all industrial facilities and other human activities such as logging, fishing, or mining, for example, would be a gold mine for scientists, policy makers, environmentalists, investors, and ordinary citizens. Even better would be the ability to measure, in precise detail, the impact of those activities on our climate in the same way companies apply financial metrics to their investment decisions to understand the bottom-line impact.

Freeing the Data

Some of this information has been collected by scientists and government agencies for years, but most of it is trapped, buried deep within university and government databases. Over the past few years, however, a whole new ecosystem of transparency brokers has emerged to make climate change information more accessible to the public and key institutions, including the investment community, regulators, and government purchasing organizations. Some of these efforts focus on collecting and revealing emissions data, some concentrate on reporting standards, while others assess products for consumer labeling and document the work companies are doing to go "carbon neutral." Together, these initiatives are helping to improve individual and institutional decision making around climate change.

For instance, the Carbon Monitoring for Action (CARMA) project maps the CO_2 emissions of over fifty thousand power plants and four thousand power companies across the world. The data for current and planned installations is easily accessible through a Google Map on the project's Web site as well as through an application programming interface (API). "Our role is to translate," says CARMA's lead researcher, David Wheeler. "Take reams of data which are available out there and translate them into an easily accessible format. There are few other institutions that have the incentive to do this—most scientists don't as it doesn't affect their publication records, and policy people are either too busy or not sufficiently technical to do the work." CARMA's work is particularly important as the energy sector is the single largest contributor to greenhouse gas emissions, at around 65 percent of the world total, with power generation accounting for a large share of that.[13]

The power of the platform became apparent one day when Wheeler received a call from a friend at the World Bank inquiring about a plant being built in Mmamabula, Botswana. It turned out that the installation would be a major polluter, which piqued Wheeler's interest—what else is the World Bank funding? Scrolling over to India he found plans for another coal plant, the Tata Ultra Mega, which ultimately would become one of the biggest emitters of CO_2 in the world. Wheeler's finding led to a large campaign by the Environmental Defense Fund, a not-for-profit, to institute stricter standards at the World Bank, and the following year new legislation was put in place to limit the types of projects that would be eligible for funding.

Making Money Green

Another initiative, called the Carbon Disclosure Project (CDP), targets the people with lots of money to invest and who have enormous influence on the companies in which they invest. Institutional investors—the big mutual and pension funds—are a critical audience in the effort to accelerate business action on climate change because they pretty much own the economy. Paul Dickinson, the site's founder, has calculated that access to capital will become a powerful lever for encouraging companies to reduce carbon once a critical mass of investors and lenders starts attaching risk premiums to companies with climate liabilities and those without sound carbon management plans. The CDP aims to speed the transition by helping the investment community better understand how companies are positioned in relation to the risks and commercial opportunities associated with the transition to a low-carbon economy.

What if you are an investor looking at your local utility company, for example? The Southern Company runs seventy-eight power plants across the southern states and has the dubious distinction of being the largest single source of carbon dioxide in the United States, according to CARMA's database. Collectively, these plants produce a staggering 206,000,000 tons of CO_2 every year. In the very near future, either as a result of a carbon tax or a cap-and-trade scheme, it will cost power companies like Southern at least $10/ton to emit CO_2 into the atmosphere (a conservative estimate given that analysts are speculating the price of carbon emissions in the United States will be closer to $30/ton once the cap-and-trade program kicks in). Two hundred million tons a year times $10 a ton is $2 billion. Subtract that from Southern's bottom line and see how it looks. Not good!

Many investors are already wary of investments in new coal-fired power plants in the United States, and many projects have (thankfully) come to a standstill as financial analysts look at the implications of new climate change regulations. CDP's analysis is based on information it receives from some 2,500 private and public organizations, including many of the largest corporations in the world. The key was getting companies to agree to voluntarily disclose information about the strategies they are deploying in relation to climate change. Dickinson's strategy was to request it on behalf of a powerful stakeholder. "We identified a legitimate authority to request the data," said Dickinson, "475 institutional investors representing $57 trillion in assets!" The CDP now has one of the largest databases of corporate climate change information in the world. Less altruistic operators might have chosen to keep the data proprietary and make money by selling access to institutional subscribers. But Dickinson thinks the public value of exposing the data to a broader audience exceeds the commercial potential. "Our goal is to apply the intelligence of the world to the climate change problem. Anyone that wants to look at the data can go to the Web site and download it."

The Climate-Friendly Consumer

Some of the most important transparency initiatives may be those that help consumers make more climate-friendly choices while shopping. Overpackaged, overshipped, and often unnecessarily harmful to the environment, consumer products are one of the biggest environmental culprits. But in an era when every business wants to tout its green credentials, consumers are flying blind. "Quite frankly we're in the dark ages," says Dara O'Rourke, a UC Berkeley professor and founder of GoodGuide, a site where users can enter the name of a product and obtain a rating of its social and environmental impact. "You go into a store and basically know nothing besides the pricing, the calorie content, and whatever else that brand is willing to tell you about their products. We want to cut through all of the marketing and advertising and tell people what they actually want to know about these products."

GoodGuide builds what O'Rourke calls a product ontology, which boils down to "a detailed assessment of a product and the supply chain behind it." To come up with the assessments, GoodGuide's scientists trawl through 200 data sources and apply 1,100 criteria to eventually land on an aggregate score from 1 to 10, which incorporates individual scores for the health, environmental, and social performance of the product. For instance, Tom's of Maine deodorant gets an 8.6 in part because it has no carcinogens, while Arrid XX antiperspirant rates a 3.8 because it contains known carcinogens. Another click leads to information behind the scores, like whether an ingredient causes reproductive problems or produces toxic waste, or whether the company has women and racial minorities in executive positions or faces labor lawsuits. To date, GoodGuide has scored some 75,000 products.

"Our goal, in two or three years, is to be part of an ecosystem in which every consumer in the world gets the information they need to make better decisions while they're standing in a store or shopping online," says O'Rourke. The ideal scenario would be to have GoodGuide's scores displayed right next to the price tag on retail store shelves. Some retailers like Tesco are already ahead. The UK-based chain started putting carbon-count labels on varieties of orange juice, potatoes, energy-efficient light bulbs, and washing detergent in 2008, allowing shoppers to compare carbon costs in the same way they can compare salt and calorie content. For now, GoodGuide users will need to be content to look up products on their mobile phones. The site already has apps for the iPhone and allows users to scan a bar code to get scores, rather than typing in a product name. The next step is personalization. According to O'Rourke, that means allowing shoppers to set their personal preferences so that the product scores they call up change to reflect how the consumer weighs different issues. So for instance if a consumer is more interested in climate change than animal rights, the product scores will reflect that difference by assigning a stronger weighting to climate change–related criteria.

Opening the kimono on climate change starts with a more informed consumer, but it should end with a more sustainable economy. "What we think of now as green is a marketing mirage," usually based on a single environmentally friendly practice, said Daniel Goleman, author of *Ecological Intelligence*, who switched deodorants and shampoos because of GoodGuide.

The site could potentially "have a revolutionary effect on industry and commerce," he said, by educating shoppers about the ramifications of buying a particular product.[14] But the real challenge lies in educating companies about how to make products that are less environmentally damaging. As entrepreneur and environmentalist Paul Hawken put it, "We assume that everything's becoming more efficient, and in an immediate sense that's true; our lives are better in many ways. But that improvement has been gained through a massively inefficient use of natural resources. We are now heading down a centuries-long path toward increasing the productivity of our natural capital—the resource systems upon which we depend to live."[15]

NATURE'S BOTTOM LINE: TWENTY YEARS TO REINDUSTRIALIZE EVERYTHING

Heading down the pathway that Hawken describes entails making some very substantial investments, and making them fast. Some of the changes include dramatic improvements in the energy efficiency of products, including the processes to make them; shifting to renewable and recyclable materials; increasing transportation efficiency; and running corporate facilities using clean energy sources. Exactly how fast do we need to do all this? A recent report by the WWF (the World Wide Fund for Nature) concludes that the world has a window to reindustrialize everything that will close between now and 2014. More conservative estimates put the point of no return closer to 2025 or 2030, after which runaway climate change becomes certain. Irrespective of which time horizon you prefer, the drive to low-carbon reindustrialization must be faster than any previous economic transformation. To achieve an 80 percent reduction of greenhouse gas emissions by mid-century, the world will have to invest $400 billion annually in green industries by 2025.[16] Every year the economic transformation is delayed will only increase its costs.

So how are we doing? Out of twenty industries tracked in the WWF's research, only three are moving fast enough.[17] Former World Bank chief economist Lord Nicholas Stern, author of the Stern Review, makes it clear that the economic risks of inaction are nearly too dire to contemplate. If no significant action is taken today, the overall costs and risks of climate change will be equivalent to losing at least 5 percent of global gross domestic product (GDP) annually by 2050. That's a bit like trying to weather a fresh global financial crisis each year in perpetuity, with little to no chance of recovery. And it gets worse, according to Stern, when you factor in damage to the environment and human health. Tally up the secondary effects and the estimated economic losses skyrocket to 20 percent of global gross domestic product (GDP) each year, or roughly $12 trillion at current values.[18]

The good news is that avoiding these consequences will cost a great deal less. Stern estimates something closer to 1 to 3 percent of global GDP will do. Nevertheless, the efforts entailed still amount to nothing less than a total reindustrialization of the planet. As Hawken puts it, "there isn't a single thing that doesn't require a complete remake!"[19] But there's something else to consider. The kinds of shifts envisioned will be impossible without instituting radical new business designs based around the key principles of wikinomics. Among other things, the world needs a green technology commons—an online marketplace where companies can connect around sustainability issues, share IP, and build a common set of sustainability assets that the world can build on for free.

Some businesspeople will claim that it's not their job to be environmentalists; their job is to make money. But this is narrow thinking, according to Ray Anderson, a widely celebrated industrialist who set out to make his business sustainable long before being green was fashionable. His carpet company, Interface, is fifteen years into a twenty-six-year journey to get to a zero environmental footprint, which to Anderson means "taking nothing from the earth that's not naturally, rampantly renewable and doing no harm to the biosphere." Indeed, the ultimate rationale for reindustrializing the planet couldn't be more compelling, in Anderson's view. "Companies will either do it or be superseded by those who do. The industrial system that operates today is committing suicide," he says, "because nature is the underlying factor. There's no business that can operate without air and water and food and energy and materials and climate regulation and ultraviolet radiation shields and pollination and seed dispersal and distribution. All of those are supplied by nature. If we kill nature, we will certainly kill the economy."[20] We guess you could call it nature's bottom line.

The Green Technology Commons

So how do we stop killing the planet and avoid plunging our economy and civilization back into the dark ages? Or, to put it more practically, how can we remake every business into one that, like Interface, strives for zero waste and makes zero impact on the climate? It's pretty clear that individual firms, governments, and nonprofits working in isolation will not get the job done. This isn't a "one company or one industry at a time" type of problem. Among other things, we need to invent, scale, and disseminate new sustainable technologies at a rate the world has never seen. And that means a deep and fundamental rethinking of the way industries innovate and manage their intellectual property. Just as the Alexandrian Greeks endeavored to collect all of the books, all of the histories, all of the great literature, all of the plays, all of the mathematical and scientific treatises of the age and store them in one building, we need to take the sum of mankind's knowledge about sustainable technologies and industries and share it for the sake of the planet and the future generations that will inhabit it.

In *Wikinomics* we called this a "pre-competitive knowledge commons," but in this instance we think "the green technology commons" is more apt. And, yes, we agree it's a bit of a mouthful, but we're talking about something big—a new, collaborative approach to research and development where like-minded companies (and sometimes competitors) create common pools of industry knowledge and processes upon which new sustainable innovations and industries are built.[21] Although this may sound a bit wacky, we can assure you that it hardly means the end of capitalism as we know it. On the contrary, the history of capitalism is replete with examples of how the material success we enjoy today is directly attributable to the evolution of openness in science and private enterprise and the rapid technological progress this unleashed. Today's challenge is to ensure that our material progress remains sustainable for generations to come. And to do that, we need a new generation of Alexandrians to take the lead.

Right on cue, up steps Kelly Lauber, the global director of Nike's Sustainable Business and Innovation Lab. Lauber spends most of her days thinking about how Nike can reach a point at which it will never use another new raw material ever again. Old shoes would be recycled and turned into new ones. T-shirts, handbags, and sports apparel would all be wholly recyclable too. Customers could come back for as many new designs as money can buy without Nike extracting another scarce resource or worrying about hundreds of millions of well-worn cross-trainers taking up space in the world's landfills. "This isn't just corporate social responsibility," says Lauber. "Our lab is focused on much bigger industry shifts—we're making investments in initiatives that will take us to the new green economy."

So far, so good, right? Except, here's Lauber's problem. Nike is developing a whole portfolio of sustainable technologies, many of which are not core to Nike's business model. Lauber knows that countless other companies are investing in new green technologies too, but right now there is no easy way of sharing their discoveries. "We want to move our sustainability efforts forward as quickly as possible and I've got no interest in reinventing the wheel," says Lauber. But try finding others' innovations? "Next to impossible," she says. "And even if you could find them . . ." pausing for a second, "Once you get legal involved, it's all over."

And that highlights a broader problem. Green technologies are being developed by traditional companies just about everywhere, but these novel ideas and technologies are not being leveraged to their full potential. Too often this results in activities that are isolated and ineffective, in conflict or compromised, or diluted by duplication of effort. For example, Nike spent an entire year creating an index to measure the sustainability of their products only to discover that companies ranging from Marks & Spencer and Patagonia to Adidas and Wal-Mart had invested similar amounts of time and effort to develop their own proprietary indexes. "Collaboration," says Lauber, "would not only have gotten us there quicker, it would have resulted in a better index." What really gets Lauber are the tough problems—issues like climate change and water scarcity that are moving Nike out of its comfort zone. Manufacturers like Nike tend to treat inputs like water as free. But what happens in a world where fresh water is scarce and costly? "Some of these issues are so big," says Lauber, "they're bigger than any one company can tackle."

Now entering stage left is John Wilbanks, director of the Science Commons, a spin-off of the Creative Commons founded by our colleague Larry Lessig. Together with Lauber and colleagues at Best Buy, Wilbanks is developing a new model of patent licensing that would make it as easy for companies to share sustainable technologies as it is for digital artists to make their media content open for reuse and modification. It's called the GreenXchange, and the idea is to create an online marketplace where companies share not only cutting-edge practices but also the technology and underlying intellectual property needed to make things happen. As the initial tenants, Nike and Best Buy have agreed to contribute their green tech assets to the pool and in turn they hope to benefit from innovations contributed by new participants as they join.

Lauber says the benefits of exposing their IP to the world are already accruing. Take Nike's investments in water-based adhesives and biodegradable rubbers. Nike will never sell these technologies directly, and they don't really confer much competitive advantage on their own. In fact, technologies like these will only get better and cheaper for Nike the more other companies use them, their competitors included. "We need economies of scale," says Lauber. "Nike alone cannot create enough demand for recycled rubber to achieve price points that are comparable to nonrenewable sources today."

Although Nike is giving many of its innovations away freely, there is something it wants in return. "We'd like to understand who is using them, how are they using them, and whether they have made improvements," says Lauber. "And, of course, we'd also like to get credit for what our innovations have done for the world." Wilbanks says the GreenXchange will help solve the attribution problem by recognizing the intellectual leadership behind the technologies that are shared and by ensuring that companies that license the technologies give credit where credit is due.[22] Lauber thinks it's the ability to quickly cross boundaries and cross industries that will ultimately move the needle on the big sustainability challenges facing the company and the world. "My vision is to have the GreenXchange become a destination for sustainability," she says. "Not just a database but a way to see what's happening in sustainability around the world: what IP is available and who's using it, how are they using it, and can it be modified. If I can take all this innovation to the next level and really let collaboration rip in this space, that would be success to me." It seems the New Alexandrians are now coming of age.

IT'S UP TO US

It is quickly becoming clear that climate change will be the biggest issue that human civilization has ever had to cope with. Unfortunately, it is also evident that today's institutions are ill equipped to deal with the scale and complexity of the problem, and there is a desperate need to infuse new thinking and new principles into our responses to this unprecedented challenge. For government and for business there will be new rules of engagement, and increased scrutiny will be the norm. Fortunately, we have an immensely powerful set of communication and collaboration tools at our disposal and that bodes well for our ability to tap into the collective ingenuity of society. After all, the answers to climate change will not be found in one company, at UN headquarters, or in the lone mind of Al Gore, as brilliant and committed as he may be. The new system will be based on openness, collaboration, transparency, and other wikinomics principles. In fact we can already see the contours of the new system taking shape in emerging communities like Carbonrally, GoodGuide, and the GreenXchange. These diverse initiatives prove that leadership can and will come from anywhere—we just need platforms to unleash our talents. Tackling climate change will not only require unprecedented transformations in our systems of commerce and industry, it will also require fundamental changes to our way of life. Luckily, we already have most of the necessary tools as well as the strongest motivating factor: the lack of alternatives.

6. WIKINOMICS MEETS THE GREEN ENERGY ECONOMY

Humanity's ability to transform raw materials into energy powered the rise of modern civilization and shaped the fortunes of nations throughout history. James Watt's coal-fired steam engine was the spark that set off the industrial revolution in Britain and triggered a period of enormous technological, social, and economic transformation. Roughly a century later, the invention of electric power and lighting furthered the rise of industrial capitalism and helped sweep a youthful United States into international prominence. After all, electrified factories fueled by oil set the stage for mass production and the rise of large-scale business enterprises. It was an era that brought large double-digit increases in productivity and transformed the United States into the modern economic powerhouse that it became. In fact, the *New Scientist* estimates that the energy in one barrel of oil is equivalent to that of five laborers working nonstop for a year![1]

Today, the world stands at the brink of a new energy revolution—one that will fundamentally transform the ubiquitous but largely invisible infrastructure that powers every home appliance, every medical device, every light source, and virtually every industrial process, from agriculture to construction. The fossil fuel–based economy is coming to an end and a new green energy economy is emerging in its place. Like past energy revolutions, there will be great payoffs for the countries and companies that master the new technologies early. The opportunity for new product and service innovation is huge, as is the potential for smart firms to create hundreds of thousands of new high-skill jobs in fields ranging from solar engineering to software. But to really tip the scale in favor of green energy we need an infusion of wikinomics principles. The need for cross-sector collaboration in developing and scaling new technologies is paramount. But we can and should go further. Truly opening up our energy infrastructure could catalyze new sources of supply, provide a platform for new energy services, and help foster a culture of energy prosumption whereby household and business users become active producers and managers of energy, not just passive consumers and ratepayers.

The business opportunity in green energy will be key to driving innovation and progress toward this vision, but there are a number of other important reasons why this new energy revolution couldn't come a moment too soon. We know now, for example, that the huge leaps in population and economic growth enjoyed over the past century have been based in large part on the illusion of cheap energy. In fact, one way to think about fossil fuels like oil and coal is to liken them to a geological savings bank that has been accumulating deposits for hundreds of millions of years. The problem is that since industrialization we have been withdrawing those deposits ten thousand times faster than nature can replace them. Most people would never plunder their personal savings accounts with such reckless abandon, but that is exactly what we are doing with oil and we are beginning to see the consequences. Not only are we well on our way to virtually exhausting the equivalent of an ancient geological endowment fund, our voracious appetite for dirty and unsustainable fossil fuels is irrevocably altering the climate and our environment with potentially devastating implications for decades and even centuries to come. To make matters worse, most of the world's remaining oil reserves are either in ecologically sensitive regions or in parts of the world that use their oil wealth inappropriately—to fund despotic governments, antidemocratic societies, and even terrorism. With energy demand set to nearly double in the next thirty years, we can be pretty sure that the global scramble for new oil reserves will only intensify in the years ahead.[2] The problem is that by continuing on a business-as-usual path with respect to fossil-fuel usage, we could end up extinguishing life on Earth as we know it and sabotaging the hopes and aspirations of future generations.[3] Turn the other way, however, and there is still time and ample opportunity to head down a green energy path that will lead to greater prosperity, innovation, health, and sustainability. But can the world really make the leap to a green energy paradigm?

The answer no doubt depends on whom you ask. But spend a bit of time in Al Gore's company, for example, and one can hardly resist his infectious sense of optimism. He's been issuing a lot of bold calls to the world, having recently reinvented himself as a leading figure in the fight against climate change. And when it comes to energy, the former U.S. vice president is not shy about laying down the gauntlet. Just as John F. Kennedy set his sights on the moon, Al Gore is challenging the nation to produce every kilowatt of electricity through wind, sun, and other climate-friendly energy sources within ten years, an audacious goal he hopes the Obama administration will embrace. Skeptics are rife and even people in the green power industry are calling it "ambitious." Given the prognosis, is there any cause for optimism that Gore's challenge can be met?

Pat Mazza, a research director for Climate Solutions, points to a dizzying array of new energy technologies that are reaching or nearing the marketplace. "Electrical power is entering its greatest revolution in a century," he says, comparing the coming era of renewable energy to Thomas Edison's creation of the world's first electrical grids and George Westinghouse's origination of long-distance power transmission. "Newer choices to generate electricity include fuel cells, wind turbines, solar cells and microturbines. Energy storage is approaching practicality, for example, through reversible fuel cells and flywheels. Under development are smart home appliances that can sense and adjust to grid conditions and commercial heating-ventilation-air-conditioning systems that allow remote diagnosis and control."[4] It's true, solar power is expensive today, but its long-term prospects are very good. Each year, the sun sends enough solar radiation our way to cover our current level of energy use ten thousand times over. By harnessing that solar power throughout the economy, we could eventually dispense with fossil fuels. Indeed, if most office buildings and homes become net producers of clean energy, we could even tear down what future generations may only see in the history books: an industrial age relic called "the grid."

Such breakthroughs would be truly exciting, but let's face it: we are talking about a long shot here unless we radically rethink our approach. Even optimistic scenarios from the U.S. Energy Information Administration put renewable energy's share of world consumption at a mere 14 percent by 2035. The same forecast made in 2010 suggests that fossil fuels will still supply at least 80 percent of energy consumption, down from eighty-five percent today.[5] And there are some serious wild cards in the deck. With its ample coal supplies, China, for example, can meet its soaring electricity demand only by opening more coal-fired power plants—approximately three or four per month according to some analysts. Even with the best available technology, each megawatt-hour of power created by a coal-fired power plant creates a minimum of 1,600 pounds of carbon dioxide. In total, China is now pumping more than 6 billion tons of carbon dioxide into the atmosphere each year, to say nothing of the sulfur dioxide and particulates that pose a major public health threat in many cities. If its energy usage remains unchanged, scientists estimate that China will need more than 100 billion tons of standard coal in 2050, exceeding the Earth's capacity to sustain and far more than the 16.1 billion tons of standard coal the entire planet consumed in 2008.[6]

On the other hand, there are clear signals of China's green intent: it wants to lead the world in developing new green industries and is beginning to tackle its energy dilemma with the same gusto and muscle that transformed its agrarian economy into a global manufacturing powerhouse overnight. A $34.6 billion investment in the clean energy economy in 2009 places China at the top of the clean energy investment league and well ahead of the United States, in second place with an annual investment of $18.6 billion.[7] Though $34 billion sounds large, it is still only 10 to 20 percent of the annual investment in renewable sources that some analyst reports claim will be necessary to contain China's greenhouse gas emissions while still meeting rapidly growing domestic energy demands.[8] Nevertheless, China's stake in the ground is evidence that it will be a major force in determining how the green energy economy evolves in the coming decades.

Where does all this leave the United States, the world's largest consumer of energy? Will it fall behind in the race to develop new industries, address energy security, and reduce emissions? Or will it seize the moment to race ahead? Green energy enthusiasts like to point to past efforts at reindustrialization as inspiration for how to mobilize government and industry behind the new energy revolution. After all, it's true such grand feats of engineering have been accomplished in the past. Oft-cited examples include the retooling of American automobile factories to build 300,000 aircraft during World War II and the 47,000 miles of interstate highways built over thirty-five years and seven administrations. But these instances of transformation were arguably only possible in a world that no longer exists, except perhaps in places like China. Both the Interstate Highway System and the military-industrial complex grew out of industrial age thinking and institutions. Factors that were present then—like an overriding sense of common purpose, a singular objective, and authoritative command and control leadership from business and government—are lacking today. Even if they were sufficiently present, they wouldn't work in a world where the knowledge, authority, and capability to move rapidly to clean energy sources is widely dispersed in society.

In short, society needs a new model of reindustrialization for an age of networked intelligence—one built on a platform of openness that mobilizes not just large utility companies, but a whole ecosystem of small-scale generators and household producers, software developers and business leaders. With the right mix of bottom-up collaborative action and inspired leadership, we can wean the world off its dangerous addiction to fossil fuels and build a new green energy economy that can sustain human civilization for centuries to come. Of course, the future of energy is a vast topic, and we can't cover every problem or solution in one chapter. So we have picked up on issues and opportunities that lie at the intersection of today's energy challenges and the new world of macrowikinomics where a number of countries, companies, and social innovators are already showing the way forward.

THE OPEN-SOURCE GRID

In August 2003, the largest power outage in decades swept across much of the northeastern United States and Ontario, plunging 55 million people into darkness for as long as forty-eight hours in some areas.[9] The event triggered fears of a terrorist attack, but the actual cause was more banal: three failed transmission lines caused by overgrown trees in northern Ohio were sufficient to unleash cascading failures that wiped out 508 generating units at 265 power plants spread across two countries. President George W. Bush called it a "wake-up call" as officials lamented that North America's centralized grid system, now stressed to its limits, had become vulnerable, increasingly brittle, and inefficient. Indeed, if a few overgrown trees in Ohio can take nearly 20 percent of the continent's customer base offline in a matter of seconds, what would a major natural disaster or a full-on terrorist attack do?

The incident reveals a sad record of stagnation that has left the electrical grids in most countries around the world antiquated and ill equipped to meet the demands of an economy based on renewable energy. Blackouts, rising prices, congestion, reduced capital spending, and unsatisfying returns for investors all point to disintegration at a time when the need for innovation in the way we produce and consume energy is paramount. According to industry analysts, the faults lie in a series of systemic flaws that all link back to planning traditions that emphasize centralized models of grid design, regulation, operations, and profit making. From an engineering perspective, everything about our current electrical systems—from the transformers, meters, and breakers right through to our household appliances—is designed on the assumption that power flows one way from large-scale generators to the consumer. Unfortunately, this centralized approach is vastly inefficient. In coal- and gas-fired power plants, almost two thirds of the energy produced by converting fuel into kilowatts escapes as heat. Another 8 percent, on average, dissipates as the electricity travels over transmission lines to get to your home. A further problem arises from the fact that the grid is designed to handle peak consumption—the points throughout the day when local industry is at full throttle and household consumption spikes. The system has no built-in storage capacity (e.g., a nationwide network of electric car batteries). That means a lot of money is misspent building and operating large-scale centralized generating plants that are only called into action to meet those rare instances of peak demand.

On top of this, the grid is remarkably opaque. The average utility company has no visibility into real-time demand for electricity and often no way to know if there is a power outage in the network, until a customer calls to alert them. So utilities produce as much as they think they need and hope that they neither overload the system nor leave consumers going without. When this system fails, it gets expensive. Blackouts cost America an estimated $80 billion a year, according to a study by the Lawrence Berkeley National Laboratory.[10] Consumers fare no better, with little or no way to assess their usage until presented with an aggregate bill at the end of the month. Homeowners rarely get information about pricing considerations. Nor do most people know what proportion of their power was generated by nuclear, coal, gas, or some form of renewable energy, or what emissions were produced in the process.

The irony is that while Alexander Graham Bell would not recognize today's telephone network, Thomas Edison would feel right at home running today's electrical grid. "Since Edison passed away," said Leonard Gross, vice president of telecommunications engineering at Ontario's electrical utility Hydro One, "we've created a compact fluorescent light bulb. Nothing else has happened."[11] Indeed, most research into new energy technologies is largely occurring at the edges of the electric network, not at its core, the utility companies. And there's a reason. It's because the incumbents most often don't have the answers and neither are they particularly motivated to find them, irrespective of what incentives governments offer.

So how do we break the gridlock?

Opening Up the Grid

What if there was a way to integrate new sources of renewable power, including the power that homeowners, businesses, and buildings generate themselves? What if you could also provide better tools and better information to allow consumers to manage their energy usage and even pump energy they generate back into the grid? This same system would allow utilities to monitor and control their networks more effectively and make new business models and dynamic pricing schedules possible for the first time. And, on top of all that, you could also sharply reduce greenhouse gas emissions and help save the planet.

A mere fantasy? It's not as far-fetched as it sounds. We just need an energy grid that is intelligent, decentralized, and transparent, and where people and devices everywhere create capability and value. It's a grid for the age of networked intelligence. Call it the open-source grid. After all, there is already increasingly broad agreement that our electrical systems should do more than carry electricity. They should carry information. And once the grid carries information, there are few reasons, if any, why it shouldn't benefit from all of the rich possibilities for innovation, collaboration, and wealth creation that the Internet has fostered in other sectors of the economy.

In many ways, the argument for a smart grid based on open standards parallels the argument for an open Internet. The old power grid is analogous to broadcast media with its bias toward centralized, one-way, one-to-many, one-size-fits-all communication. A smart grid, if it could be built, would leverage the Internet's connective tissue to weave millions, and eventually billions, of household appliances, substations, and power generators around the planet into an intelligent and programmable network. And, just as open standards and "edge intelligence" helped unleash unparalleled creativity on the Internet, a similar ethos of openness will ensure that the new energy grid becomes a platform for a vast array of new energy services, not just a computerized pipeline for delivering cleaner electricity.

Treating the grid like an open platform would, for example, allow software developers to build applications to help you conserve energy the same way developers build apps for the iPhone. A straightforward application could include a service that analyzes a household's electricity usage data, identifies inefficient appliances or practices in the home, and offers tips on how to reduce energy or provides special discounts on efficient appliances or electronic equipment. So you need no longer worry if your son or daughter forgets to turn the lights out when they leave the room, no matter how many times you remind them. A smart grid equipped with sensors in your home will follow your instructions to turn the lights off automatically when it's 2 a.m. and no one has moved in the house for the last hour! An intelligent grid can also change consumer behavior with smart appliances that would save money automatically. Armed with more information about tariffs, for example, the dishwasher would wait for the price to fall below a certain level before switching on, or the air-conditioner would turn itself down when the price goes up. "If you're asking me to make the analysis of what times of day and night I use my hot water heater, and to turn it down accordingly, it will never happen," says Zipcar founder Robin Chase. "If you ask the gas company to do an analysis of people's water heaters and then ask me, 'Robin, do you want us to turn it down and save yourself $40 over the year?' I'll say, 'Of course.'"

To date, 8.3 million homes in America have been equipped with smart meters covering 6 percent of the population. The number is set to grow to 33 million by 2011, while the worldwide total will reach about 155 million.[12] Cisco Systems estimates that by the time it all gets built out, the energy grid will be one thousand times larger than today's Internet.[13] Meanwhile, a vast and growing number of companies are already lining up to offer consumers tools to help them make sense of the smart meter data.

Typically leadership does not come from the companies that dominated the old industrial era of energy, but from a new generation of companies that understand the age of networked intelligence. Predictably, Google is in the vanguard. Google's PowerMeter is one of these much-anticipated tools. "There is enormous potential to take near real time usage data to create contests, build applications and enable social networking," said Google's Niki Fenwick. Fenwick says users will be able to compare their usage by neighborhood, zip code, or even with friends on Facebook. Like other tech players in the emerging energy economy, Google is actively lobbying for open standards so that consumers are able to buy smart appliances, thermostats, or energy monitors from different companies and have them talk to each other.

Personal Carbon Markets

Pilots under way in Europe show how far the open-source grid concept could go. Homes across Europe, including those in Manchester, Birmingham, Bristol, Ruse (Bulgaria), and Cluj (Romania), have been equipped with advanced smart meters and sensor networks that track energy usage, efficiency, and overall household emissions to generate a real-time carbon footprint. Users pull up a Web-based interface to analyze the sources of their emissions, compare their home with the neighborhood, forecast household savings, or control their energy use remotely from a PC or a mobile phone.[14] Like Google's PowerMeter, the system developed by the Manchester City Council and its partners is an open platform, which means it can be seamlessly integrated with other applications for mobiles, TV, and social networks.

However, the real action starts in 2010, when each household in the pilot project will be assigned a "personal carbon allowance" and participate in a household emissions trading market, the first of its kind. The carbon allowance sets a cap on household emissions and the marketplace allows households to buy and sell quotas, according to their carbon budgets. Policy makers in the United Kingdom have been contemplating whether such household emissions trading schemes could set the stage for the introduction of a comprehensive nationwide cap-and-trade system that would apply to individuals, not just businesses. As in the pilot, each adult citizen would be assigned a carbon allowance that would determine how much carbon dioxide they can emit driving, flying, and keeping their homes. Emitters who exceeded their quota by relying on big cars, living in large houses, and taking lots of plane journeys would buy additional allowances from people who have allowances to spare because they emit less.

British politicians such as David Miliband have argued that low-income people would be net winners. "People on low incomes are likely to benefit as they will be able to sell their excess allowances," he said in a speech in 2006.[15] Exploratory research by the UK Department of Energy and Climate Change found that 71 percent of low-income families would, in effect, get paid for having lower than average emissions. But critics complain that such a complex system of wealth transfer between millions of individuals would be expensive and could be undone by blunder-prone computer systems. Privacy advocates worry that the need to track each citizen's individual consumption decisions would provide the government and unlawful entities with new ways to snoop on individuals. Perhaps surprisingly, there are skeptics among environmentalists too. While few green campaigners dispute the need for individuals to take greater responsibility for cutting emissions, they also are extremely wary of proposals that would put added burdens on citizens, particularly at a time when governments and industries still could do a lot more to switch to clean, renewable forms of energy. That may be true, but from our point of view these innovative proposals are noteworthy.

The mere fact that neighborhood trading schemes and personal carbon allowances are even being debated is a sign that the efforts under way to make our infrastructure more intelligent and interactive will pay large dividends. As we argued in the previous chapter, it's easier to remain aloof about climate change when the connections between our actions and the climate seem vague and hard to measure. But it becomes harder to simply ignore one's personal responsibility when the smart meter on your wall not only shows you your real-time carbon footprint, but also compares your score to the neighborhood average and offers you tips on how to improve. Coupled with a real price for carbon, this new transparency and interactivity provides the fuel for truly creative responses to some of the world's great challenges. And while the focus is on energy and climate change today, there are equivalent opportunities in many other sectors.

Smart Everything

A growing number of companies are developing intelligent infrastructures to measure everything from water to natural gas flows. For example, SAS business analytics allow utilities such as Brazilian utility Cemig to precisely forecast electrical demand in both the long and very short term. Cemig can even predict how demand will change when a World Cup soccer game starts or ends and televisions are turned on or off. Others, like IBM, are piloting schemes to monitor entire systems such as supply chains and transportation networks. The company has developed sensors and RFID tags that can track foodstuffs such as meat or other horticultural products from the producer all the way to the supermarket shelf.[16] Armed with this data, retailers can ensure the quality of supply while customers can make smarter purchasing decisions.

The drive to make all things "smarter" by connecting electrified objects to the Internet will, within a few years, result in a flood of new data that can be aggregated and analyzed, providing a powerful engine for energy dashboards and trading platforms that help households and businesses optimize their consumption. Think of the pilots in Europe and imagine similar personal carbon markets operating on a regional or global scale. As initiatives like CARMA and Carbonrally demonstrate (see chapter 5), nascent platforms can facilitate new levels of transparency and collaboration around reducing our impact on the planet. Early results also show that, armed with data and a few simple suggestions, individuals will begin to change their lifestyles to become more sustainable. Studies have found, for example, that when people are made aware of how much power they are using, they reduce their use by about 7 percent. With added incentives, people curtail their electricity use during peaks in demand by 15 percent or more.[17] A report released by The Climate Group estimates that the application of digital technologies to enable smart grids and smart buildings has the potential to avert 3.71 gigatons of CO_2 equivalent global emissions by 2020, delivering some $464 billion in global energy cost savings to businesses, taxpayers, and consumers.[18]

RISE OF THE ENERGY PROSUMER

Perhaps the most striking opportunity for innovation in the new energy economy pertains to the changing role of the modern energy consumer. In a green energy economy, most households will no longer be passive consumers and ratepayers, but informed and increasingly self-reliant producers of their own green energy. The idea is that sometime in the not too distant future, just about every household and institution that currently consumes electricity will also produce it and even sell it back to the grid, perhaps even turning a profit. For example, rooftop solar panels and backyard wind turbines will, at times, produce more energy than we can store. That energy can flow back into the grid. As we discuss in the transportation chapter, electric vehicles and plug-in hybrids will also return power to the grid as they sit idle in parking lots around the globe. Some households and institutions will be net producers. Others will be net consumers. The great thing about an open-source grid is that it generates what we need and lets us use what we generate ourselves.

Travel to places like 75 Ravina Crescent, in the east end of Toronto, and it's possible to glimpse the future of North America's energy economy in action. While it may be an eighty-year-old house on a leafy residential street, its electrical guts are thoroughly futuristic thanks to the efforts of retired software developer Gordon Fraser and his wife, Susan. The Fraser household is on the verge of becoming one of the first homes in Toronto to generate all its electrical power using its own resources. And, by taking their energy destiny firmly into their own hands, the Frasers have joined a small but growing tribe of people we call the modern energy prosumers.

The Frasers' journey toward energy self-sufficiency got started shortly after the blackout in 2003, which provided all the incentive they needed to consider their own vulnerability. It turns out the outages were not just a wake-up call for industry, but they also set off alarm bells for household consumers who became concerned about the reliability of their power supplies. Susan Fraser recalls worrying about what might have happened had a similar blackout occurred during the depths of winter, when temperatures in Toronto can fall to as low as -30 degrees Celsius. "Would we freeze to death," she said. "I wanted some kind of alternative power source in case something else happened to the grid." A self-described autodidact, her husband, Gordon, began tinkering with solar power. Then, in 2006, he and Susan turned their home into an experimental test bed, calling it the Ravina Project. The couple, now in their sixties, installed a 1.5 kilowatt solar panel on their roof that stores power in a large lead battery and sends any surplus energy back to the grid. An on-demand, 95 percent efficient, computer-controlled boiler now heats hot water for their home's faucets and radiators. Although all of the components are "off-the-shelf," this is not your typical household solar power installation. The Frasers' solar panel is mounted on a tiltable axis that follows the sun across the sky in order to maximize energy production. The stand uses a 1995 version of an integrated satellite receiver and descrambler (IRD) to control the movement. But rather than rotate a satellite dish, Gordon has programmed the IRD so that it moves to keep receiving sunlight at a 90-degree angle throughout the day, "the sweet spot" for solar generation.

Despite these investments, the Frasers are not entirely off the grid yet.

In 2007, they generated 35 percent of the energy they used over the course of the year. In winter months like December, when the days are short and cold, solar-generated energy accounted for only 7 percent of total household consumption. But in the month of May, solar power supplied over 94 percent of what they needed. By doubling the battery storage, the Frasers claim they could live off the grid for four months out of the year. And with the addition of a wind turbine generator, they could get off the grid completely. Not bad for Canada, where household heating is the largest single source of carbon emissions and a large share of the average family's monthly budget.

One of the spin-offs of the Ravina Project is a treasure trove of public data that will help inform other consumers looking to become energy prosumers too. Gordon keeps meticulous records that document every aspect of their energy production and consumption: how much power they generate, the amount of gas they use, the amount of grid power that comes in from the street, and a detailed inventory of how much electricity they use on a daily basis. The Frasers release all of this data on their Web site and publish daily generation updates on their Twitter feed (@ravinaproject). Want to know how much reinsulating an old home is likely to produce in savings? The Frasers have it all mapped out, in incredible detail, in their paper on "household thermodynamics." Gordon calls it their "digital legacy" and hopes their data will be a resource for others seeking to press forward with their own green power projects. Jon Worren, a green energy analyst for MaRS, a Toronto-based hub for scientific research and innovation, thinks the Ravina Project could change people's perceptions of what's possible. "The observations that 1) you can produce solar energy 12 months a year on a Toronto roof-top and 2) that you can generate significant savings, even with a 1920's home, are very important in terms of helping other urban residents understand that making these investments can make a big difference."

When the Frasers got started researching in 2004 they found reliable information on green power hard to find. "Prudent people will want to investigate the issues and find as much relevant information as possible before making large investments," said Gordon. "By publishing the answers to the questions we asked ourselves when we got started, we hope to improve dramatically the availability of relevant data." It's not only consumers who stand to benefit from the Frasers' fastidious efforts to document their experience, but also researchers and analysts looking for the most efficient methods to give the green energy economy a shot in the arm. "Science is so serendipitous," Gordon said. "You just never know when you're going to stumble across a new idea or some treasure in the data. If there's anything of value in that data, it will be other people who will find it."

From Experimentation to Scale

Could all homes and buildings generate enough of their own power to live off the grid, or will households like the Frasers' remain an anomaly in an energy economy that overwhelmingly favors centralized models of power generation? There is no simple answer, as it depends on factors like one's energy consumption, the capacity of the solar array, and one's access to sunlight throughout the day. But, in Toronto at least, some interesting conclusions are surfacing. The Frasers' data suggests that stimulating millions of green energy experiments on urban rooftops is a more economically viable method of creating a green energy economy than building large centralized wind and solar farms. "We need a new paradigm," said Gordon. "We can't go on borrowing the same underlying assumptions and design traditions that underpin the industrial age model of power generation." Naturally, the utility companies—who have little to gain from distributed generation—are not particularly enthralled with the Frasers' arguments.[19] Taken to its logical conclusion, distributed generation would unleash massive disruption and potentially make redundant a large part of what power utilities do today. Rather than selling power—a majority of which would be produced locally by homeowners and businesses—the core of the industry would shift to selling green energy installations and energy services like smart grid apps that help households and businesses save money. These shifts, in turn, would give an advantage to a new generation of energy companies that already specialize in microenergy solutions and don't bear the huge physical legacies of the incumbent utility companies.

Of course solar generation and other renewable sources are still far more expensive than conventional methods, which means very little will happen in the short term if we don't get serious about creating a context in which it becomes economically attractive for both consumers and producers to invest in retooling our energy infrastructure. After all, it's unrealistic to expect any significant proportion of the business or consumer population to pursue green energy as a losing proposition. Take the consumer's perspective. Total costs for the Ravina Project, for example, have reached about $45,000 to date, including the solar panels, batteries, electrical retrofitting, and numerous energy conservation measures like beefing up insulation and installing new windows, doors, and light bulbs.[20] At that price point, only a small portion of the urban population could afford to undertake such an elaborate green energy project. It's also highly unlikely that the Frasers will see a financial return on their investments anytime soon, although it should be noted that making money was never really the Frasers' goal. "We're running an urban energy research project," said Gordon, "and, as unique as it is, our data has great value, far beyond the limited range of our pocketbooks." Moreover, three things have changed since the Frasers invested in solar panels in 2006. First, the cost of solar panel installations has fallen by 50 percent and will likely continue to decline as demand grows and solar technologies are produced on a larger scale. Second, governments have begun to offer serious incentives that could make decentralized energy production and self-sufficient homes and buildings a viable, even lucrative, proposition. Third, social entrepreneurs like One Block Off the Grid have come up with collective purchasing arrangements to help further lower the costs of solar technology for consumers. By using the Web to aggregate hundreds of potential solar customers in a given city, the San Francisco–based start-up claims it is able to lower the cost of installations by 15 percent. The company completed six hundred installations in ten cities in 2009 and expects to complete over five thousand in 2010.

Even with innovative initiatives like One Block Off the Grid, there is still a major role for government in scaling green energy. Ontario, for example, passed legislation in May 2009 with incentives to help accelerate the development of new renewable energy generation across the province. The goal is to double renewable energy production, triple energy conservation, and help homeowners install 100,000 solar roof systems in homes across Ontario. If successful, these actions should reduce the emissions of Canada's most populous province by 15 percent below 1990 levels by 2020.[21] While jurisdictions like Ontario are in the early stages of a systemic change, there are more mature examples in European countries such as Denmark, Germany, and Spain, where green energy now accounts for roughly 30 percent of overall consumption. Indeed, when it comes to large-scale transformation, there are few better examples than Denmark.

DENMARK'S PATH TO RENEWABLE ENERGY

Ever wondered what smarter energy policies could have bought had they been implemented thirty years ago in the wake of the 1970s energy crisis? Here's a remarkable fact. Since 1980, Denmark has managed to grow its economy by almost 80 percent while maintaining the same level of energy consumption! Over the same thirty-year period, Denmark changed from being an economy that was entirely dependent on energy imports to a net exporter of both electricity and energy technology.[22] Now imagine what

Americans could have bought themselves with all the money saved had the U.S. followed Denmark and maintained its energy consumption at 1980 levels!

Granted Denmark is a country of five million citizens and it's geographically much smaller than the United States. But its achievements are nevertheless remarkable, especially in light of its starting point. The 1970s oil shock hit Denmark harder than many of its European neighbors. At the time, every last drop of oil was imported. When prices rose dramatically the entire economy was crushed. Anne Højer Simonsen of the Danish Ministry of Climate and Energy recalls the sacrifices. "When I was a child, each Sunday, you could not ride on the highways. There were car-free Sundays. Only emergency vehicles rode. It was terrible. It was very cold all winter. We couldn't have more than sixteen or seventeen degrees [Celsius] in our living rooms." It took an entire decade for Denmark to recover. But the experience created a burning platform that unleashed an unstoppable drive toward self-sufficiency.

Creating a Context for Self-Organization

In the old model of government, politicians and bureaucrats not only steered, they rowed. Government was supposedly omnipotent and tried to do everything from soup to nuts. This model may have worked for the old industrial age model of power generation, but it doesn't work for a new era with decentralized production and smart grids, as well as active and informed energy consumers. In the new model, the government creates a context for self-organization. It holds on to the policy role, but it empowers other sectors by laying the enabling infrastructure and making sure the incentives are right. This is the model Denmark chose when the crushing effects of the oil crisis ignited a new fire under a slumbering nation's belly.

Under strong leadership from its energy agency, Denmark invested heavily in domestic energy sources, including natural gas, wind, and biomass. It worked with household consumers and businesses to implement aggressive energy efficiency campaigns. And through a partnership with the private sector, it built a "state-of-the-art" energy grid that taps into a distributed network of small-scale green energy producers. Today, Denmark has a world-class renewable energy sector, thanks to years of early investments in green energy, including funding for R&D, start-up grants, and incentives for businesses to collaborate with each other. Denmark's "feed-in tariff" would drive many U.S. conservatives bananas, but it works. The policy, which requires utilities to purchase renewable power at above-market prices, has helped a lot of new innovations get off the ground and achieve scale that could never have come to fruition in an environment where fossil fuels enjoy an overwhelming advantage.[23] Feed-in tariffs have also had considerable economic development benefits: 80 percent of Denmark's wind turbines are owned by local residents.[24] The key with feed-in tariffs is setting the right price for renewable energy, a price high enough to attract investors without being so high it generates windfall profits. Once the sector reaches critical mass, the feed-in tariff rates are scaled back.

Critics of feed-in tariffs claim that the subsidies are wasteful and that cash-strapped governments should simply let market forces dictate the speed at which we transition to renewable sources. Other critics are not necessarily against government subsidies, but suggest they would be better spent building nuclear power plants or subsidizing large-scale research into clean coal technology or carbon sequestration. But the critics should consider this: markets simply allocate scarcity in the present and near term; they are virtually blind to scarcities that may arise in ten or twenty years. The fact is that oil is only going to get more expensive and the threat of climate change will only escalate over time. Solar power is a proven technology whose costs have much further to fall. It doesn't need costly, drawn-out research programs to prove its viability and, unlike nuclear power plants, the infrastructure doesn't take a decade to build and there are no radioactive by-products that need to be stored for thousands of years. The cost per watt hour of solar has already decreased from $25 to $30 in the 1980s to less than $5 today.[25] It's also a proven job creator. Like the Danish investment in wind power, German investment in solar has produced a thriving industry, pushing Germans to the forefront of solar technology, creating thousands of trained engineers and installers, and shifting public perception toward renewable energy and the role that energy plays in their lives. It has pushed down the cost of solar photovoltaic (PV), making it more affordable for the world, and these trends will accelerate as more governments get on board. Of course, as Denmark's experience shows, solar is not the only solution and any jurisdiction should consider its natural assets before choosing it over other green alternatives like wind, wave power, and geothermal. Finally, policy makers should not underestimate the importance of shifting the culture of energy consumption. One advantage of microsolutions is that they encourage, indeed require, participation. When consumers have access to information and the tools to manage and even produce their own energy, they can become part of the solution rather than a simple pawn to be manipulated by incentives, taxes, and rate changes.

Accelerating Green Energy Adoption

The ultimate proof of the pudding, as they say, is in the eating. Denmark is recognized as a world leader. And its energy sector, once almost nonexistent, now makes up 10 percent of the country's exports.[26] Of course, Denmark's energy transformation was thirty years in the making. It was an orderly transition that enjoyed high levels of support from all sectors. Other countries won't have the luxury of taking a full thirty years to shift paradigms, and arguably more radical action will be warranted in light of rising energy costs and the specter of runaway climate change. At a minimum, a lot of infrastructure will have to be retooled and there will have to be substantial reductions in the costs of green technology for consumers, while carbon-intense energy sources like oil and coal will have to become more expensive. Countries such as Denmark and Germany have started solving these problems with incentives, feed-in tariffs, and substantial public and private investments in new infrastructure. Many other jurisdictions across North America are poised to do the same.

At the end of the day, it is certainly true that greater political leadership is required across the board, particularly in the face of strong energy lobbies. But political leadership alone is nowhere near sufficient to create the conditions required to accelerate green energy adoption around the world. One lesson from Denmark and other leaders is that much more attention should be paid to mobilizing all of the world's knowledge and capability around this problem. By creating a context for self-organization and by leveraging the ideas and skills of other sectors, we might just have a chance at succeeding.

TWO PATHS AND TWO FUTURES: WHICH WILL THE WORLD CHOOSE?

In the face of climate change and diminishing stocks of fossil fuels, the world can go one of two routes. One path—the path we're on now—leads to escalating prices, energy shortages, environmental catastrophe, and conflict between the great powers. The other path—call it the wikinomics path—leads to growth, global cooperation, and an abundant supply of clean power delivered through a smart energy grid that enables consumers to become active and informed managers of their energy consumption. The choice between the conventional path and the wiki world path is obvious. Indeed, there is near universal agreement that shifting to 100 percent renewable energy sources is not just desirable, but highly necessary. The technology needed for smart grids and green power production is largely here, but we need to massively increase our investments and share our intellectual property. Average consumers are perhaps not as informed as they should be, but we can help solve that problem with intuitive energy management tools for households and businesses. And we've even learned a few things about creating a political context for self-organization, thanks to leaders like Germany and Denmark.

The fact that all these factors are present and growing makes it all the more aggravating to watch the world's leaders nonchalantly flirting with disaster. The risk is that a perverse psychology is taking hold: as fears of an imminent energy crisis mount, the world powers are buying a futile insurance policy by engaging in an all-out scramble for resources. As the saying goes, scarcity is the enemy of restraint. With shortages of conventional fossil fuels looming in the near future, it's safe to say that some of the world's leading nations have hardly been exhibiting their best behavior. The United States has fought two wars in Iraq to secure control over oil in the Middle East. The latest Middle East adventure, according to economists Linda Bilmes and Joseph Stiglitz, will cost U.S. taxpayers an estimated $1.5 trillion in direct costs and perhaps an equivalent amount in indirect costs if one factors in the unrealized economic potential forfeited by not spending the money at home.[27] By comparison, U.S. government R&D spending on energy amounted to a paltry $5 billion in 2008—a tiny fraction of what is being spent on the Iraq war.[28]

In the meantime, oil firms from Britain, China, France, India, and the United States are flocking to Africa, seeking to exploit the continent's underdeveloped fields in countries such as Nigeria, Angola, Chad, Sudan, Equatorial Guinea, and the Democratic Republic of the Congo. Revenues from their extraction should provide funds for badly needed development, but instead have fueled state corruption, environmental degradation, poverty, and violence. Rather than being a blessing, Africa's natural resources have largely been a curse.[29] Meanwhile on the European continent a newly resurgent Russia already uses its vast resources as an instrument for carrying out domestic and foreign policy, having twice threatened to shut off supplies to Europe to make a political point. The energy superpower controls one third of the world's natural gas, is the second-largest oil exporter after Saudi Arabia, and will be the principal source of natural gas needed by Europe in 2030.

Indeed, the great irony of the world's current climate predicament is that instead of encouraging creative alternatives to fossil-fuel consumption, global warming may in fact lead to a renewed drive for untapped sources of petroleum. As the planet continues to heat up, the Arctic will continue to melt, and a vast, pristine reservoir of oil will be unveiled to the energy-hungry great powers. Will the five Arctic players—Russia, Canada, the United States, Denmark, and Norway—end up on the front line of what could be one of human history's most gratuitous displays of resource-inspired conflict? Or is another model possible: one where the Arctic ecosystem is managed as an international commons, with the proceeds from sustainable exploitation going toward R&D funds for alternative energy sources and transferring technology to developing nations? For the time being, the major players are adopting a defensive pose, seeking to advance their national energy interests at any cost—even at the risk of unleashing a new type of Cold War, except this time fought over energy.

These are the kinds of questions to which we need new answers, not yesterday's answers. A peaceful and prosperous green energy future is just around the corner, and it will be built on key principles of wikinomics like openness, collaboration, sustainability, integrity, and interdependence. Fortunately, we can take at least some comfort in the fact that the final chapter in today's energy revolution has not been written, at least not yet. And perhaps most important, the choice among paths and final destinations is not just up to a few elite decision makers in the executive suites of major oil companies and governments around the world; it's up to all of us, and wikinomics shows the way forward.

7. THE TRANSPORTATION REVOLUTION: MOVING AROUND IN THE TWENTY-FIRST CENTURY

Imagine that you've landed on a planet a lot like Earth a century ago and you've been given an assignment to design the dominant mode of transportation to move people around. Your design criteria are as follows: The system must maximize the consumption of fuels and the surface area of the planet—using up as much farmland and other space as possible. Your system should produce the most toxins and use more physical materials (steel, glass, rubber, leather, synthetics) than available alternatives. It must be the system that will result in the largest possible number of deaths and injuries (hint: have free movement vehicles and make every pilot an amateur). It should also be the least predictable system, giving passengers little idea how long a trip home might take, and it should slow down to a crawl, not speed up, the more people use it. Plus you get bonus points for pitting inhabitants against one another and, in extreme cases, causing travelers to fly into an uncontrollable rage. It's hard to imagine a better solution to these design criteria than an automobile powered by an internal combustion engine.

Heretical analogy, you say? To be sure, the car is the centerpiece of the American Dream. Flashing down the Pacific Coast in a red convertible has been featured in the songs, films, and fantasies of North Americans—and much of the world—for good reason. Cars give us mobility, utility, status, and a sense of independence. Anybody who's owned a car doubtless revels in the freedom the car gives them. And there are many, many more people on the planet who want one.

For all its goodness (and we're speaking here as car owners and car enthusiasts), the car has created substantial problems, problems we'll have to work hard to solve for generations. Simply manufacturing a car requires one to blow up mountains for steel, iron, and aluminum, and to drain oil patches for rubber and plastics. Add to this the cost of energy to process these raw materials, and then the energy costs to assemble the car itself. On average, one car generates 28 tons of waste and pollutes 1,421 cubic meters of air, just in its manufacturing.[1] The costs continue after the car rolls off the assembly line. The 600 million cars on the road today account for 10 percent of global CO_2 emissions, not to mention the cocktail of noxious fumes that make cars one of the single largest sources of atmospheric pollution and a major contributor to respiratory problems and other public health challenges.[2] Cars are also expensive for drivers, consuming 18 percent of the average American income, a figure that will surely increase as one factors in the rising costs of fuel, parking, tolls, car maintenance and insurance, not to mention the lost productivity resulting from increased traffic congestion.[3]

Of course, cars only work—that is, they only give us freedom of mobility—when we have roads to drive on. And this creates still more trauma to the planet. In the developed world, roads consume about 40 percent of land in urban areas and even more in North America.[4] In 2008, the world's expansive road network covered some 70 million kilometers—enough roadway to build 180 expressways to the moon.[5] Urban planners have long known that high rates of car ownership encourage urban sprawl—sprawl that further entrenches our car dependency while increasing roadway congestion. Fewer people recognize that roads also disrupt ecosystems, interfere with species migration, and compromise natural drainage.

Given that cars are one of the most popular forms of transportation in the developed world, cars inevitably become a nexus for death. In 2004, the World Health Organization reported that an estimated 1.2 million people die and that 50 million people are injured in road accidents each year, making car accidents the tenth leading cause of death worldwide. The WHO projects these figures to increase by about 65 percent over the next twenty years unless we make changes.[6]

Sounds bad enough, right? And yet, the truth is our car troubles are just beginning. The emergence of a consuming middle class in China and India is driving a sharp rise in the demand for cars. Those cars in emerging countries drive on even more congested roads and use cheaper, older engine technologies, creating an immense demand for fuel and tremendous amounts of car emissions. Fifteen years ago, there were virtually no private cars in China. By the end of 2007, the number had reached 15.2 million.[7] And by 2050, the number could be as high as 700 million.[8] Meanwhile other high-growth countries like India and Brazil will push the worldwide car fleet up to total a staggering 3 billion.[9] Unless a radical new personal transportation solution comes to market soon, today's 2.8 billion tons of tailpipe emissions will easily double within twenty-five years.

Given the comparatively nascent state of their car markets, countries like China have room to experiment with alternatives. Planners can still make smart infrastructure choices to accommodate more intelligent, efficient, and sustainable transport options.[10] Unfortunately, for countries with a long history of personal car ownership, our choices for pursuing new transportation options are more constrained, if not determined, by the choices we have made in the past. As Zipcar's founder, Robin Chase, put it, "Infrastructure is destiny," referring to how U.S. transportation policy over the years has favored the automobile over investments in rail and public transport systems. "Think about how we built out the national highway interstate network in the fifties," she says. "We built highways, we ripped out all the trolleys, and we didn't build any trains. We created our destiny as a car-dependent nation because that's the infrastructure we built up." We're massively invested in our outmoded infrastructure and we're emotionally and socially wedded to our cars too. The irony is that of all of the transportation options available to us, the car is the least sustainable, consuming excessive energy, affecting the health of populations, and delivering a declining level of service despite increasing investments in building and maintaining expensive roadways. Many of these negative impacts fall disproportionately on those social groups who are also least likely to own and drive cars. So while car ownership may look attractive to individuals and households at least in the near term, the long-term personal, economic, and social results of a car-dominant lifestyle make ownership highly irrational.

The risk is that when we do decide to change, it will come too slowly to make a difference. The United Nations Environment Programme estimates that even under optimal conditions, new technologies and more sustainable alternatives will take *forty years* to reach market, given the current climate of cost reductions, technology gaps, and the decade-long life cycle of a new car.[11] The world can't wait forty years, which is why we must dramatically accelerate our efforts to build the transport systems of the future. A good start would be to recognize the scale and urgency of the problem and mobilize public and private resources accordingly. But arguably more important, we must weave principles like openness, collaboration, and sustainability into the fabric of our new institutions for transport. And it turns out that the digital revolution is bringing us a world of fresh possibilities—both in the way we mobilize the world to solve this problem and in the solutions themselves.

A new generation of tech-savvy transportation entrepreneurs is at last bringing fresh thinking and new approaches to a transportation sector mired in its own unfortunate legacies. Contest-driven models of innovation, for example, are ratcheting up the investment required to help get us from radical breakthroughs in technology to mass-market deployments of superefficient vehicles in two years instead of ten. New forms of green transportation infrastructure are being built and tested in various countries in hopes that such infrastructure could eventually power a worldwide electric car fleet and end our dependence on oil. Innovative car sharing services like Zipcar are beginning to reshape car ownership, while a ubiquitous transportation data network, should it be built, could help optimize our roads and unleash a large interoperable cloud of new services that do everything from real-time traffic shaping to in-car shopping and entertainment.

These initiatives by no means exhaust all of the possibilities, nor do they address all of the challenges associated with different means of transport, such as air travel. Any sensible transportation strategy should include a greater emphasis on improving public transportation options, including light rail systems to connect urban centers and facilitate travel within them. That said, we have focused our analysis on the future of the personal automobile based on the assumption that a combination of sunk costs, intransient car owners, and recession-ravaged government coffers will make major public transportation projects politically unpopular and difficult to finance, particularly in North America. In the future we believe that shifting people's mind-sets and habits away from personal automobiles will naturally lead them to demand, and in turn pay for, public transportation options. Above all we have tried to explain why the new paradigm for transportation should be built on principles like openness and why new models of collaboration will be key to making the shift to a transportation system that is safe, networked, intelligent, efficient, and sustainable. The journey starts with fundamentally reinventing the car itself—and we're not just talking about a hunk of metal, with a steering column, gasoline engine, and four tires.

THE CAR AS AN OPEN PLATFORM

Conventional wisdom says that "being open" is rather like inviting your competitor into your home only to have them steal your lunch. That may have been true of the old industrial age automobile industry. But it's not true in a world where the big twelve automakers are no longer at the front of the pack when it comes to designing new sustainable transportation solutions. It turns out there's a diverse and growing ecosystem of problem solvers working on everything from advanced batteries to the future telematic features of electric cars. These problem solvers aren't all working in Detroit; they're scattered across thousands of small companies, research institutes, nonprofits, and consultancies around the planet. And when innovation is fast, fluid, and distributed, conventional thinking about openness will be challenged.

The key lesson emerging from other industries—including software, pharmaceuticals, entertainment, and online retailing—is that openness, done right, is a powerful force for growth and competitiveness. As long as you're smart about how and when, you can blow open the windows and unlock the doors to build vast business ecosystems on top of what we call platforms for innovation. Such platforms can include physical products ranging from a video game console to a cell phone—virtually anything that runs software. The idea is that tens of thousands of interoperating agents converged on a shared platform can marshal more bandwidth, more raw intelligence, and more requisite variety than the largest organization.

The big opportunity initially has been to turn ordinary, off-the-shelf products and services into vibrant platforms for innovation on which large communities of customers and partners can innovate and create value. At the time we were writing *Wikinomics*, leading platforms included Web services such as Google maps and Amazon.com's e-commerce system for warehousing, purchasing, and distributing goods. Amazon's platform success is legendary: some 200,000 external partners get a cut of the retail action by building compatible applications and services that drive more traffic, more clicks, and ultimately more purchases through Amazon's online system. The next generation of platforms is being built on mobile phones (Apple's App Store) and social networks like Bebo and Facebook. But if a platform for innovation can include virtually anything that runs software, then why not think of the car of the future as a platform for innovation too? After all, the car is not just a vehicle for moving around; it's a place for work, learning, and entertainment with a series of software programs connected to a wireless network. Now imagine a car with a set of open APIs allowing thousands of programmers and niche businesses to create a cloud of applications for your car—from remote personal assistants to navigation and geospatial search applications to on-demand movies and music, and why not throw in mobile Skype for good measure.

Optimizing Our Infrastructure with a Ubiquitous Data Grid

A rich cloud of in-car services can do more than inform and entertain us; it could help optimize our entire transportation infrastructure. Accompanying your iTunes service, for example, would be an infinite number of applications that enable you to fundamentally change the way you use your car. Some apps could facilitate ridesharing. Others could ease congestion and keep you safe by distributing road traffic more evenly or selecting optimal routes for reducing air pollution on days when concentrations reach dangerous levels. Or, rather than sole ownership, there could be applications to facilitate shared car ownership, with dynamic pricing models that take into account environmental factors like location, time of day, traffic congestion, and seasonal demand patterns. These networked cars will be enriched with sensors that share data about weather, traffic, and road conditions with other drivers in the vicinity. Drive over a pothole and that information will be relayed to the local municipality. Get into an accident and your car will instantly alert all drivers behind you. One day, cars will just self-organize in a fashion that makes the whole notion of "driving" redundant. Your "car" will just be another node in an intelligent ad hoc network of vehicles that carts people and goods around the planet with zero noise, zero fuss, and zero emissions.

But is any of this practical now? For a perspective we spoke to Zipcar's founder, Robin Chase, one of a handful of visionaries who understands the far-reaching potential that the car-as-an-open-platform presents. "The only way we will change our driving habits is when we're paying the real cost of driving: including the full cost of carbon, the cost of congestion, the cost of building and maintaining the roads," says Chase. "And the easiest way to be able to pay the real cost of transportation would be if we had ubiquitous wireless data connections. It's the same thing that we talk about with the smart grid: dynamically priced power consumption, with the real price of what it's costing." In other words, the intelligent network we need for electricity can also turn cars into nodes. She sees automobiles as just another network device, one that, like the smart grid, should be open and net-based.

"Cars are network nodes," she says. "They have GPS and Bluetooth and tollbooth transponders, and we're all on our cell phones and lots of cars have OnStar support services. That's five networks. Automakers and academics will bring us more. They're working on smart cars that will communicate with us, with one another, and with the road. Put them together—network the networks—and for the same exact infrastructure spend, you get a ubiquitous, robust, resilient, open communication platform—ripe for innovation—without spending a dollar more."

Prying Open the Door

What will it take to realize Chase's vision? "Technology has a potential of making data exchange between everything simple and easy, but it's siloed," she says. "Transportation technology today is closed and proprietary, so only GM gets to decide what goes into an OnStar when there are a thousand things that could be interesting to do with a more open platform." In other words, we come back to the need for a massive rethinking of the way the industry approaches technology and intellectual property. Chase has even been doing some lobbying in Washington to raise awareness about technology investments in many sectors, including the smart grid, education, health care, and transportation. "Instead of producing closed proprietary devices and closed proprietary networks, I'd like our political leaders to realize that it benefits everybody and lowers costs if devices and networks are multipurpose," she says.

Are the automakers listening to people like Chase? They should be. The plummeting costs of collaboration and the advantages of harnessing a larger talent pool are causing many to rethink their assumptions about innovation. Companies like Apple, Amazon, and Google were among the first innovators that really learned the subtle art of making open platforms for innovation work. The result is that the Web looks less like a walled garden than a global stage where millions upon millions of users add value and establish synergistic businesses. It's time now to do something similar for the car.

Shai Agassi, who has long thought about the advantages of openness reaching back to his software days at SAP, poses the essential question. "Do you take your core assets and processes and keep them to yourself," he asks, "or do you expose them to every software company on the planet and entice them to come in and help develop those assets?" For Agassi it comes down to a basic principle of our networked world: there are always more smart people outside your enterprise boundaries than there are inside. "Whether we're talking about software for enterprises or software for cars, a large pool of innovative companies can now provide customers with additional solutions with integration by design, not integration as an afterthought," he says.

FAST-TRACKING THE ELECTRIC CAR

Former SAP exec turned green transport entrepreneur, Shai Agassi is thinking about more than software these days. Recently he's been busy inventing a new class of sustainable infrastructure for an age of intelligent transport. If he's successful, his new company, Better Place, will kick-start the electric car industry and help wean America and other countries off foreign oil in as little as a decade. That's a big task. So what's the plan?

"Forget hybrids," says Agassi. "Only a truly emissions-free car will allow for the expected growth in car ownership without disastrous consequences for the planet." Out of all the alternatives—including hydrogen and biofuels—he's betting on electric cars. Not only do they generate fewer greenhouse gas emissions, they have lower fuel costs than gasoline burning cars, even when oil is cheap. That's because electric motors, unlike engines, do not generate friction or heat, providing 90 percent–plus efficiency in converting electricity to motion. Generating energy on a large scale (in nuclear plants or wind farms, for example) is also less wasteful than doing it on a small scale (by burning gasoline in an internal combustion engine).[12] Down the road, there's a big economic incentive for consumers too. Electric car batteries may be expensive today, but within a decade a single year's supply of gasoline will cost more than the energy required to power an electric car over its entire lifetime, even when taking the cost of batteries into consideration.[13]

Sounds good, right? But unfortunately there's a catch. The market for electric cars today is like selling MP3 players in 1998. They are likely to revolutionize the automobile as we know it, but it is still unclear who will develop the equivalent of the iPod. Electric cars, as well as all their critical components, are produced in small runs, not on commercial scale. Most major auto manufacturers have been holding out for a battery that can last for ten years and provide enough energy to safely drive a car for 500-plus kilometers, on the likely assumption that no mass infrastructure for recharging cars would be put in place in the near term. Since such a battery does not exist, and most likely will not for another fifteen to twenty years, automakers emphasized hybrids and pushed their EV plans into niche solutions focused on fleets of cars that run predetermined routes and come back to home base after 100 to 150 km, such as postal delivery trucks and taxis.[14]

When the Electric Car Meets the Open-Source Energy Grid

All this leads Agassi to conclude that the future of the electric car rests with better infrastructure, not necessarily better batteries. And he's determined to be among the first to set up a pervasive infrastructure of charging spots and battery swapping stations that could eliminate the electric car's Achilles' heel: lack of battery life and longevity.

Transport statistics suggest that a battery that gets forty miles to a charge would probably suffice to cover the daily driving habits of about 70 percent of the U.S. population, assuming drivers recharge their vehicles at home.[15] But most drivers would probably feel more comfortable with a vehicle that could travel at least two hundred miles without stalling, even if they only made longer trips infrequently. The solution, according to Agassi, is to see the car battery as part of the infrastructure system, not part of the car, much as the SIM card inside an iPhone is part of AT&T's network infrastructure. So Agassi's company, Better Place, is proposing the creation of a ubiquitous infrastructure (called an electric recharge grid) that automatically charges the car when it's parked, and on the exceptionally long commute, drivers can use an exchange station where an empty battery is replaced with a full one in automated lanes resembling a car wash. "Build renewable energy sources for the recharge grid," says Agassi, "and you get a sustainable transportation energy solution which will go practically forever with no reliance on oil and no emissions." Put together the charge points, the batteries, exchange stations, and the software that controls timing and routing and you get a new form of infrastructure, a whole new set of business models, and a new category of companies that will install, operate, and service customers across the network.

Better Place does not make the electric cars. Nor does it produce the electricity needed to charge them. "We make the solution that connects the two in a convenient way that is more affordable for consumers," says Agassi. That solution sees electric car owners subscribe to the Better Place network in the same way that cell phone owners subscribe to a wireless network. In exchange for monthly fees (which are carefully calibrated to be below the cost of operating a gasoline vehicle), they get the right to swap their batteries or charge up at home or at work. To some, the notion of buying a subscription for battery charging seems odd at first. But, in reality, it's not all that different from what we do with gasoline. "Filling up your gas tank is just like a prepaid card for your cell phone," Agassi says. "You fill up twenty gallons' worth of gasoline and depending on your car, that twenty gallons will take you anywhere between four hundred and six hundred miles. When you run out of miles, you come in and you fill up again."

Agassi's company is gearing up to enter markets in Israel, Denmark, and Australia with other pilots in Japan, Canada, and the United States. In Tokyo, Better Place is building a network for the world's first fleet of electric taxis with switchable batteries. Japanese taxis represent a mere 2 percent of all passenger vehicles on the road in Japan, yet they emit approximately 20 percent of all carbon dioxide (CO_2). The outcome of the Tokyo pilot is being closely watched by other urban centers. But Better Place's biggest deal to date is in Israel, where the concept has been embraced with zeal by the public and has political support from the country's president and prime minister. Israel Railways signed an agreement to develop an electric vehicle charging infrastructure in train station parking lots, and there are arrangements to build recharge stations at many workplaces too. Agassi estimates that some 50,000 to 70,000 charge spots will be required to service the entire country. In the meantime, the company has signed up nearly 150 car fleet owners—

mostly multinationals like Cisco, IBM, Intel, and Orange—that have agreed to convert a portion of their gasoline-based fleet into electric vehicles with Better Place batteries and mobility subscriptions.

Built on Openness

Unlike other industrial age infrastructures, Agassi insists that the recharge grid should be open. With open standards, anyone could operate an exchange station and drivers could avoid lock-in. Imagine the frustration if recharge stations were like mobile phones—have you ever tried using your BlackBerry charger to juice up an iPhone? To fix this, Agassi argues that each recharge network should be compatible with all of the others. Openness may invite in new competitors, but he is willing to trade off a more competitive playing field if it means the infrastructure will get built out more quickly. "You got to have roaming between those networks," he says. "We cannot build a proprietary network to try and compete against the biggest monopoly on earth, which is oil."

Just as a rising tide lifts all boats, an open recharge network creates a larger market and enables new business models in the process. Once a grid reaches critical mass in a given region, Agassi thinks car owners could subscribe to a complete commute solution: car, energy, and maintenance contained in a single predictable monthly price. This could be welcome news for consumers struggling to cope with the damage fluctuating gas prices inflict on their household budgets. Depending on the length of the subscription, service providers could even subsidize the cost of the car, much the same way wireless network operators are invariably expected to offer a basic handset for free with any new subscription. Will the costs of batteries and clean electricity decline to the point where electric vehicles will be given for free to long-term subscribers? It's a radical proposition, but it's happened once before in the wireless phone industry.[16]

GETTING FROM CRAZY IDEAS TO RADICAL BREAKTHROUGHS

The emerging transportation system needs to be built on the principles of the wiki world. However, it also turns out that these principles and the techniques of mass collaboration can be critical in the journey to get us there. To be sure, the challenge of producing a worldwide fleet of electric cars (along with the infrastructure to charge them) is more than one company, and even one country, can do on its own. Indeed, if the X Prize Foundation is successful, there will soon be a dozen low-emission vehicles ready for the mass market and a host of new entrants poised to nibble away at the heels of the automotive industry incumbents. "It's time to shake things up," says Peter Diamandis, founder of the X Prize, a unique foundation that organizes large-scale innovation contests as a way to turn nascent ideas into radical breakthroughs that will benefit humanity. "We're still using the internal combustion engine after a century and we're still getting twenty miles to the gallon—just like we were forty years ago!"

Citing a lack of mainstream consumer choices for clean, superefficient vehicles, Diamandis decided that what the world needs now is a dramatic demonstration that cars powered by alternative fuel sources can be just as reliable, safe, affordable, and beautiful as any other car on the market. That dramatic demonstration came in the form of a high-profile competition that is pitting teams from around the world in a race to develop a superefficient car. Diamandis sees the contest as an ideal catalyst for much-needed investment into new automotive technologies and alternative fuel sources.

The competition has drawn interest from emerging start-ups, universities, independent inventors, and even India's largest automaker. In 2010, twenty-eight teams passed on to the second stage and will soon be lining up their vehicles for a series of races designed to simulate real-world driving conditions. The winner will walk away with a $10 million prize purse. The contest rules have three broad components: efficiency (cars must get at least 100 miles per gallon); emissions (cars must produce less than 200 grams of greenhouse gases per mile); and economic viability (mass production of the cars has to be feasible, and the company has to have a plan to make 10,000 a year).[17] It's this last point—that a winning vehicle has to be safe, comfortable, and ready to be mass-manufactured at a reasonable cost—that will separate the fantasy-mobiles from those that could actually be put into production and sold for a profit. "If we do this right," says Diamandis, "we're going to draw a line in the sand and say all the cars we drove before this date are relegated to the history museums."

Contests as an Innovation Model

It's not the first time high-profile contests have been used to motivate creative solutions to grand challenges. In 1927, Charles Lindbergh won the $25,000 Orteig Prize for flying nonstop from New York to Paris—a seminal event that helped launch the modern aviation industry. After Lindbergh's nonstop solo flight across the Atlantic, the number of people who bought airplane tickets in the United States went from about 6,000 to 180,000 in eighteen months: a thirty-fold increase.[18]

The first X Prize, modeled after the Orteig Prize, captured the public's attention in 2004, when the contest awarded $10 million to designer Burt Rutan and Microsoft's cofounder Paul Allen for building the first privately funded spaceship that traveled 100 kilometers above the surface of the Earth. Today, the X Prize Foundation is just one of many organizations that have latched on to incentivized challenges as a way to unleash fundamental breakthroughs in society. Richard Branson, the founder of Virgin, will part with $25 million of his own money in exchange for a commercially feasible way to remove greenhouse gases from the Earth's atmosphere. Netflix has issued a global challenge to anyone who can improve the company's automated movie recommendations algorithm, while Google's Lunar X Prize will go to the first private venture to send image-transmitting rovers to the moon. Why do these competitions work? "We are genetically bred to compete," Diamandis explains. "It's when we do our best business many times, we do our best sports, and I believe competition extracts the best out of individuals." Diamandis argues that competitions also bring out the best in small teams. "When a team has a clearly focused objective goal, an external competition really hones them to collaborate well together."

But competitions have other advantages as well, particularly when you're up against an entrenched bureaucracy. The problem for mature companies is that the very commercial success of their products increases their dependency on them. Making radical changes in the product's capabilities, underlying architecture, or associated business models could cannibalize sales or lead to costly realignments of strategy and business infrastructure. It's as though popular and widely adopted products become ossified, hardened by the inherent incentive to build on their own successes. The result is that entrenched industry players are generally not motivated to develop or deploy disruptive technologies, as Harvard Business School professor Clayton Christensen has pointed out.

So success breeds complacency. R&D departments are discouraged from investigating alternative technologies and so channel their resources into refining components, adding new features, or tweaking their existing product architectures. This strategy of marching down a well-defined product road map may pay dividends for some time. But complacency creates enormous vulnerability when disruptive innovations emerge that may threaten the product road map itself. It's this vulnerability that the X Prize Foundation is trying to exploit by offering a financial incentive and a platform on which new entrants can successfully launch disruptive automotive solutions.

Moving the Needle

Another advantage of the contest model is that you only pay the winner. That may sound ruthless. But for a results-driven organization, the upside according to Diamandis is that "you don't pay someone who tries really hard but fails to deliver the goods, and you only pay on delivery." Those teams that work hard, but wind up second, third, or fourth, still stand to benefit, not just from the exposure but from the access to capital flowing from interested investors.

Most important, Diamandis estimates that well-designed challenges can catalyze new investment totaling between ten and fifty times the amount of the prize purse. The foundation's suborbital space flight challenge leveraged a $10 million purse into $100 million of team expenditures. But winning an X Prize is only the beginning. "It's also about launching new industries that attract capital, that get the public excited, that create new markets." And since the Ansari X Prize was won by Burt Rutan and Paul Allen, more than $1 billion has been invested in the suborbital market. Though there has yet to be a commercial flight into space, hundreds of private customers have bought tickets for suborbital flights.

Will the Automotive X Prize prove decisive in moving the needle in the same way? It depends on how you measure success. Ten million dollars sounds like a lot of money, on the one hand, but it's peanuts in the auto industry. General Motors, for instance, will spend up to $750 million developing a production-ready version of the Chevrolet Volt extended range electric car.[19] On the other hand, changing public perceptions of what is technically and commercially possible still represents a crucial first step toward launching a nascent green car industry into the mainstream. Although contestants such as Tesla Motors, which markets an electric sports car for $100,000, can comfortably carve out a niche, it remains to be seen if any of the current contenders will have the staying power to take any significant market share off the major players. However, if the contest sparks the public imagination and helps generate demand, it will give established automakers greater incentive to get their green vehicles to market more quickly.

SOCIAL COMMUTING

While electric cars are seen as environmentally friendly, the most sustainable choice—the choice that would do the most to improve urban environments and positively improve human health—would be to make a lifestyle change in favor of walking, biking, or public transit. The root problem is that no matter how you describe it, a 4,000-pound car is just not the most efficient means of transportation for several hundred pounds of human beings. Sustainable transport experts have a point when they argue that governments should invest just as much in developing public transit and pedestrian-friendly communities as they do in funding research to support the traditional auto industry. After all, one of the most cost-effective and instantaneous ways to decrease our impact on the planet is dead simple: stop driving, or at least stop driving alone. Just consider this: if we double the passengers per vehicle, we halve the traffic! Imagine the difference that could make in cities like Los Angeles, Tokyo, and Bangalore. The idea behind social commuting is an old one. But the technology to do it right is just coming of age. With a new intelligent transportation platform we can share resources that were once jealously guarded and make much smarter decisions about how we transport ourselves around.

Carpooling 2.0

Although carpooling has often been encouraged by governments to reduce congestion, two problems inhibit its growth. The first issue is trust. How do you know the person you are carpooling with is not dangerous? The other problem is coordination: matching two or more people who are leaving from the same place and going to the same destination at the same time has never been easy. With the emergence of carpooling platforms that make connections and help establish trust between drivers and passengers, these problems are now much easier to address. One of the biggest platforms is called mitfahrgelegenheit.de (rideshare), which operates across Germany and Austria. The popular site has 1 million registered users and facilitates between 20,000 and 30,000 rides every day. With high gas and rail ticket prices in Germany, the service is especially popular with young people. "Our goal is to reduce the traffic on our streets and with that the amount of pollution that is released into the atmosphere," says Michael Reinicke, one of the site's founders. To make connections drivers fill out a profile and input the details of their upcoming trip. A premium service offers extra security with more detailed profile information, driving history, and user ratings. "The network works a bit like Facebook," says Reinicke. "If you had a good ride with someone you can add them to your network and the next time you need a ride their profile is searched first." The network extends to two degrees to increase the chances of a match.

A host of other start-ups are also exploring the space of "social commuting" by developing communities around ridesharing. Some, such as GoLoco and Zimride, rely on the social graph to create groups of friends who carpool. Others such as PickupPal or Carticipate use geopositioning—either mobile or computer based—to match people departing from the same location. One such start-up called Wikit proposes a form of transportation marketplace, where drivers could advertise their daily routes using their in-car GPS device and would-be passengers could publish their current location, desired destinations, and the amount they are willing to pay to get there, from the convenience of their GPS-enabled phone. Wikit's matching software would do the rest, while taking a small cut of every transaction. Zipcar takes the proposition even further. Rather than just share rides, why not share cars too.

A Personal Journey to a New Model of Personal Transportation

When Aaron Hay finished college he moved to Toronto to start a new job. As a young professional with reasonable income and the occasional need for a vehicle, Aaron considered purchasing a small, fuel-efficient city runabout. In calculating the costs, Aaron was faced with monthly costs of a $290 car payment, a $170 insurance bill, about $100 in fuel costs, and an $85 apartment parking fee. Without incurring any maintenance costs, Aaron was looking at about $645 per month to own and run a small car in downtown Toronto.

Then Aaron considered Zipcar—something he had seen all over Toronto but knew little about. (Toronto residents have embraced the Zipcar with considerable enthusiasm. One cannot walk more than a few blocks in Toronto's inner city without sighting several Zipcar parking spots.) He learned that he could join the service for a $35 membership fee, on an "Occasional Use" plan. He calculated that if he took 6 car trips per month, at an average of 2 hours each, for about $13 per hour, he would spend about $156 monthly. Aaron decided the economics of the Zipcar model looked pretty sharp at first glance, so he decided to give the service a try. This decision was not only cost-effective (indeed, months went by when Aaron did not use a Zipcar at all), but completely transformed Aaron's approach to personal transportation. Since Aaron could, quite literally, see the dollars adding up by the minute each time he used a Zipcar, he began to combine trips. For example, a single trip to the grocery store began to turn into grocery-getting, visiting the dry cleaner, picking up a friend from the airport, and returning an ill-fitting shirt. Moreover, Aaron began to plan his routes carefully, to minimize his total distance and hourly costs, even with multiple stops. A 2009 study conducted by the company confirms this: Zipsters, on average, reduce their "vehicle miles traveled," or VMT, by 40 percent.[20] (If you were wondering, the company calculates that 92 *billion* VMT are saved annually, as of 2009.)

Aaron also began to understand that the hourly rate system penalizes him for longer trips—likely what Zipcar intends. Indeed, an overnight visit to a friend's nearby cottage proved expensive. In these cases, he saw how using traditional car rental services made more sense—and he found that paying a flat daily rate of $30 to $40 through Enterprise or National worked well for trips longer than eight hours. Public transportation increases in attractiveness for Zipsters, too—47 percent typically increase their use of public transportation.[21] In Aaron's case, he took several yearly trips to neighboring Montréal using VIA Rail, Canada's national train service, instead of a rental car.

In short, Zipcar has provided Aaron with transparency in the cost of personal transportation, which now allows him to optimize the economics—and the energy consumption—of his vehicle use. If Aaron had indeed purchased that fuel-efficient Honda or Mazda, he would have had to track at least four different costs, on a monthly basis, to determine that he pays around $600 monthly just to own a car he had access to in the Zipcar lot across the street. Aaron remains a passionate Zipster as of 2009—as do his Zipster pals, 63 percent of whom delayed or halted the purchase of a car, thanks to the service.[22]

Putting the Car in the Commons

The concept of "sharing" is central to Zipcar's business model. Robin Chase, Zipcar's founder, has made it her personal crusade to "find and make use of excess capacity." The idea is that we get more out of what we own or have already invested in, or, conversely, to pay only for what we use. The average car costs around $25,000 (making it the second most expensive asset individuals own, after their house). And yet, on average our cars sit idle about twenty-two out of every twenty-four hours! To make much more efficient use of those cars, Zipcar offers them up to consumers as a pay-only-for-what-you-use model that strips away some hassles of ownership while taking away some of its conveniences. Chase likes to say that Zipcar is the car your mother always said you couldn't have—all the fun and none of the responsibilities. For a significant demographic, the trade-off makes a whole a lot of sense. At the time of writing, Zipcar was serving sixty-seven cities in the United States,

Canada, and the United Kingdom, using a fleet of 6,500 vehicles.[23] As the largest car-sharing company in the world, Zipcar has cultivated a roster of 300,000 faithful and enthusiastic members, called "Zipsters," who have shared access to this trendy, useful, and generally fuel-efficient fleet.

Zipcar estimates that for every vehicle added to its fleet, fifteen to twenty single-user automobiles are removed from the road.[24] The claim may be somewhat exaggerated. Not every Zipster necessarily sells his or her primary vehicle or will purchase one in the absence of the service. Even if one takes a conservative approach to the firm's estimate, however, the implications are exciting and profound. The cars that the company claims to eliminate no longer emit carbon, nor do they require servicing and a whole range of caustic chemicals. But what about the infrastructure these cars run on? Most municipalities would thank their lucky stars for a private service that expressly eliminates personal-use automobiles from their crumbling, overburdened roadways. One can surmise that drivers—of cars, buses, and transport trucks—would be fairly happy with this development, too.

Notably, Zipcar not only allows its members to optimize vehicles and traffic—it crowdsources maintenance too. Members are communally responsible for returning vehicles on time, reporting dirty cars, recording their own and others' damage and accidents, tattling on low-fuel returnees, and even rescheduling with Zipcar's central reservation line—via phone or text—if travel itineraries change (as they often do, when dictated by the vagaries of our public road systems). Zipcar goes even further with routine tasks; for example, if Zipsters decide to take a car through the wash, they can submit the receipt and receive a free hour of driving for their work.

Powerful new wikinomics models always force incumbents to pay attention and eventually change their approach as well. The auto industry is increasingly realizing it has little choice but to play ball. Toyota and Ford have already begun exploring ways to work with Zipcar, from using its members to test electric cars to designing vehicles specifically for the sharing market. "The future of transportation will be a blend of things like Zipcar, public transportation, and private car ownership," says Bill Ford, Ford's executive chairman. "Not only do I not fear that, but I think it's a great opportunity for us to participate in the changing nature of car ownership."[25]

FOCUSING THE MIND ON THE FUTURE OF TRANSPORT

When it comes to fixing our industrial age transportation system, we may have only twenty to twenty-five years to complete such a complex technological effort. We've tried to show in this chapter, however, that many of the key ingredients for a more sustainable transportation paradigm are already in place or under development, from a solar-power recharge grid to new models of car ownership and a ubiquitous transportation data network that will help us optimize our infrastructure and unleash a rich cloud of new services. To really move the needle, however, each component of this revolutionary personal transportation paradigm must be built on the five principles of wikinomics.

Openness will reveal the true costs of car dependency while ensuring that data and emerging software services can flow freely across the intelligent network for transportation. Collaboration will bring together the people, ideas, and resources needed to accelerate design and adoption of sustainable technologies and services. Sharing intellectual property and other assets—including the car itself—will help us optimize what currently amounts to a massively inefficient use of resources. Integrity will guide us to make the investments needed today to ensure that our transportation systems are sustainable for future generations. And finally, the concept of interdependence should help us understand that no car owner, company, community, or country is an island and you can't operate a sustainable transportation system without internalizing the negative costs to the environment and to other people. Above all we must not be afraid to challenge conventional wisdom. "I think when we look back at ourselves sitting alone in our 120 square feet of car, driving down these highways with incredible storage costs and incredible operating costs," says Robin Chase, "I think we will look back at how we travel today and be just astounded: astounded at the cost, astounded at the waste. It's such a wacky idea," she adds, "that we'd want to be alone in our cars spending huge sums of money and all that parking space, when it is less fun and more expensive and kind of crazy."

IV

LEARNING, DISCOVERY, AND WELL-BEING

8. RETHINKING THE UNIVERSITY: COLLABORATIVE LEARNING

Encyclopedias, newspapers, and record labels have a lot in common. They are all in the business of producing content. They all own and manage capable producers whom they recruit and compensate in various ways. Their products are composed of atoms—books, papers, CDs, and performers on stage—and are costly to create and distribute. Their products are proprietary and they take legal action against those who infringe on their intellectual property. Because they create unique value, their customers pay them and they have revenue. Their business is possible because of scarcity—quality news, information, knowledge, learning, art.

Come to think of it, they are a lot like universities as well. But there is a big difference. Today the businesses of encyclopedias, newspapers, and record labels are in various stages of collapse. They've all lost their monopolies on the creation and delivery of content. They were killed by the digital age that brought abundance, mass participation, the democratization of production, the rise of new digital delivery channels, the infeasibility of old notions of intellectual property, and completely new business models all enabled by the Internet. The allegedly unassailable attributes of their age-old businesses were erased faster than you can tap "delete" on your iPhone.

Does a similar fate await the university? Despite certain surface similarities, evidence would indicate that universities are indeed different. For starters, university enrollment throughout the world is at an all-time high, indicating that more and more people continue to see value in traditional modes of learning. What's more, the wage premium that employers will pay for a university graduate continues to rise despite the growth in supply of qualified graduates—a sign that the market value of qualified workers is rising faster than our institutions' ability to create them.[1] The competition to get into the most prestigious universities has never been fiercer, and campuses across North America are thriving thanks to the wave of young, tech-savvy learners flowing through their halls and classrooms.

Yet scratch beneath the surface and the picture is not so rosy. A dismal 58 percent of entering freshmen in the United States actually graduate from the same college within six years.[2] More and more students are questioning the "bang for the buck" as college tuition has risen in cost more than any other good or service since 1990, leaving students with $714 billion in outstanding student-loan debt in the United States alone.[3] Students around the world are increasingly choosing alternative models of higher education. In 2007, nearly 20 percent of college students in the United States—some 3.9 million—took an online course, according to the Sloan Consortium, and their numbers are increasing.[4] The University of Phoenix now enrolls more than 200,000 annually,[5] including 16,000 in its online MBA program,[6] compared with 900 at Harvard. Given the huge explosion in MBA courses offered online, many of which are offered from Asia, it's a fair guess to say that most MBA degrees today are either taken online or enhanced by online learning. Yet the proportion of institutions declaring that online education is critical to their long-term strategy has actually declined.[7]

There are more subtle indicators of disruption as well. Students and faculty alike are refusing to pay for academic periodicals and are file-swapping like it's 1999.[8] For many of the smartest students, it's fashionable to try to get an A without going to any lectures—meaning that the cream of the crop is beginning to boycott the basic model of pedagogy. Mark Taylor, chairman of Columbia University's religion department, whipped up a storm of academic controversy with a provocative article in *The New York Times* called "End the University as We Know It."[9] "Graduate education," he began, "is the Detroit of higher learning. Most graduate programs in American universities produce a product for which there is no market (candidates for teaching positions that do not exist) and develop skills for which there is diminishing demand (research in subfields within subfields and publication in journals read by no one other than a few like-minded colleagues), all at a rapidly rising cost (sometimes well over $100,000 in student loans)." One week later, the outcry from fellow academics filled the entire letters page in the Sunday *New York Times*. One of his own colleagues at Columbia said it was "alarming and embarrassing" to hear "crass anti-intellectualism" emerge from his own institution. Another academic accused Taylor of "poisoning the waters of higher education."[10]

Ironically, in an age of unprecedentedly high enrollment, there is evidence that universities are losing their grip on higher learning as the Internet inexorably becomes the dominant infrastructure for knowledge—both as a container and as a global platform for knowledge exchange between people—and a new generation of students requires a much different model of higher education. The transformation of the university is not just a good idea—it is an imperative. Luckily, we are living in a time of great opportunity and there is a stream of proposals for change coming from within academia itself.[11] Some say the Web enables distance learning and the elimination of campuses. Others argue that we need more technology in higher education, or colleges should be opened up and made free to all. There are renewed calls to abolish tenure, and even to replace traditional departments with a new set of problem-focused disciplines. The trouble is that most of these ideas don't address the fundamental problems with the university or show a way forward.

Rather, change is required in two vast and interwoven domains that permeate the deep structures and operating model of the university. First, we need to toss out the old industrial model of pedagogy—how learning is accomplished—and replace it with a new model called collaborative learning. Second, we need an entirely new modus operandi for how the content of higher education—the subject matter, course materials, texts, written and spoken word and other media—is created. Rather than the old textbook publishing model, which is both slow and expensive for users, universities, professors, and other participants would contribute to an open platform of world-class educational resources that students everywhere can access throughout their lifetime. We call it a Global Network for Higher Learning. If universities open up and embrace collaborative learning and collaborative knowledge production, they have a chance of surviving and even thriving in the networked, global economy.

TRANSFORMING THE MODEL OF PEDAGOGY: COLLABORATIVE LEARNING

In the current model of pedagogy, education is largely about absorbing content and being able to recall it on exams. The teacher is essentially a broadcaster: the transmitter of information to an inert audience in a one-way, linear fashion.[12] In today's world, and for today's students, this model of "broadcast learning" is anachronistic, if not obsolete. Yesterday you graduated and you were set for life—only needing to "keep up" a bit with ongoing developments in your chosen field. Today when you graduate you're set for, say, fifteen minutes. If you took a technical course in the first year of your studies, half of what you learned may be obsolete by your fourth year. Of course you still need a knowledge base, and you can't Google your way through every activity and conversation. But what counts more is your capacity to learn lifelong, to think, research, find information, analyze, synthesize, contextualize, and critically evaluate; to apply research to solving problems; to collaborate and communicate. This is particularly important for students and employers who compete in a global economy. Labor markets are now global and, given networked business models, knowledge workers face competition in real time. Workers and managers must learn, adapt, and perform like never before.

The answer for educational establishments is not simply to expand distance learning offerings—though this would help. Nor is it about students being able to access lectures by some of the world's leading professors from free online sites like Academic Earth—though this practice has proven popular with and useful for both professors and students. Rather, with today's technology, it is now possible to embrace new collaborative and social models of learning that change the actual pedagogy in more fundamental ways.[13]

In a 2008 article, John Seely Brown and Richard P. Adler argued persuasively that our *understanding* of content is socially constructed through conversations about that content and through grounded interactions, especially with others, around problems or actions. The implication is that we need to focus less on *what* we are learning and more on *how* we are learning. "Instead of starting from the Cartesian premise of '*I think, therefore I am,*'" they argue, "the social view of learning says, '*We participate, therefore we are.*'"[14] In other words, the real learning begins when students leave the lecture hall and start discussing and internalizing what was just said.

Research shows that mutual exploration, group problem solving, and collective meaning-making produces better learning outcomes and understanding overall. In one important study, Richard J. Light, of the Harvard Graduate School of Education, discovered that one of the strongest determinants of students' success in higher education was their ability to form or participate in small study groups. Students who studied in groups were more engaged in their studies, were better prepared for class, and learned significantly more than students who worked on their own.[15] It appears that when students get engaged they take a greater interest in and responsibility for their own learning. Some, like Seely Brown and Adler, have argued effectively that the Web provides powerful new tools and environments for collaborative learning to occur—everything from wikis to virtual worlds like Second Life.

For example, imagine, as a student, that you could not only read about what it is like to be a scientist, an architect, an artist, an entrepreneur, or an engineer, but also collaborate with fellow students in a safe virtual environment to recreate that experience for yourself. In other words, you could directly participate in and experience the ways a particular discipline thinks about and solves problems. Thanks to the malleability and immersive nature of virtual worlds like Second Life, that possibility is already here today. Peggy Sheehy, a library media specialist in Suffern Middle School in New York State, is one of a growing number of educators who believe that well-supervised virtual environments make a valuable addition to any teacher's toolkit. With the help of a supportive school administration, Sheehy made Suffern Middle School among the first schools to have created a private virtual learning space on the Teen Grid in Second Life (visitors to this part of Second Life's virtual environment must be between the ages of thirteen and seventeen). Ever since, teachers have been working closely with Sheehy to integrate Second Life into their curriculum. One Suffern teacher used a virtual flea market to help teach students about budgets and managing their money. The students were given one hundred Linden dollars (the in-game virtual currency) and a list of items they needed to purchase while staying within their budget. In another example, an eighth-grade English teacher who was teaching John Steinbeck's *Of Mice and Men* brought students into Second Life to conduct a mock trial. "They did research on the American judicial system; they took on roles and then responded according to assigned characters from the book. . . . They were so much more invested in the book, and they came away with such a richer experience," says Sheehy.

Of course, activities like a mock trial can be conducted in the classroom, but one of the surprising advantages of staging experiences in virtual worlds is that students tend to drop many of the inhibitions that might have otherwise prevented them from participating fully in a real-world classroom. "Group identities often fall away; fashion becomes unimportant, since each student is represented by an avatar they created themselves; and teachers can engage their students in a deeper level of discourse than is typical of high school students who are often preoccupied with appearances," says Sheehy.

Sheehy teaches at the middle school level, but university professors are experimenting too. Andrew Lang, a professor of mathematics at Oral Roberts University, had been teaching distance learning classes for years using a set of typically static Web pages and course materials when he first stumbled upon Second Life in 2006. He finds the "publish and browse" model of distance learning frustrating, describing it as "totally deficient compared to a face-to-face class." But discovering Second Life literally opened up new worlds for Lang and his students, deepening the engagement and unleashing a level of creativity and collaboration unsurpassed by anything he had previously experienced as a professor. Lang, a mathematical physicist by training, also has a knack for Internet programming. Using Second Life's advanced scripting capability, he has built some pretty unique and fascinating functionality into his new virtual classroom—functionality that allows students to visualize molecular phenomena and run experiments that would be impossible in a real-world lab or lecture theater. In one instance, Lang built what he describes as a "molecule rezzer." For laymen like us, that essentially means that students can create a 3D representation of a molecule, model a chemical reaction, or even build a molecular machine that will interact with other virtual objects and people. "Students sometimes don't even realize it," says Lang, "but when they are manufacturing a molecule in Second Life, the application is doing some pretty powerful stuff in the background, hitting Web servers around the world, obtaining the connectivity of the atoms, and then building the 3D visualization. And all that can be accomplished in as little as thirty seconds." Sounds complex. But for Lang's students it is proving to be a lot of fun and certainly beats staring at a textbook. Students naturally collaborate as they build their molecular machines, and according to Lang they often leave them in Second Life for other passersby to play with.

Immersive experiences and group learning are two key elements of the new pedagogy, but perhaps just as important is the fact that interactive computer-based courseware allows professors to spend less time lecturing and more time collaborating with students. As Seymour Papert, one of the world's foremost experts on how technology can provide new ways to learn, put it: "The scandal of education is that every time you teach something, you deprive a [student] of the pleasure and benefit of discovery."[16] If this group collaboration is centered around solving real research problems, then chances are this type of interaction will be valued by professors, too.

Even in K–12 there is strong evidence for replacing "instruction" with a more collaborative approach to "constructing" knowledge.[17] We toured a classroom of seven-year-olds in a public school in Lisbon, a city where every child in the classroom is getting a laptop connected to a high-speed network. It was the most exciting, noisy, collaborative classroom we have ever seen. The teacher directed the kids to an astronomy blog with a beautiful color image of a rotating solar system on the screen. "Now," said the teacher, "who knows what the equinox is?" Nobody knew. "All right, why don't you find out?" The chattering began, as the children clustered together to figure out what an equinox was. Then one group leapt up and waved their hands. They found it! The children in this Portuguese classroom discovering astronomy and the solar system barely noticed the technology, the much-vaunted laptop. It was like air to them. But it changed the relationship they had with their teacher. Instead of fidgeting in their chairs while the teacher lectured and scrawled some notes on the blackboard, they were the explorers, the discoverers, and the teacher was their guide.

When educators shift from mass production to mass customization, students' learning outcomes improve. Indeed, some leading educators are calling for this kind of massive change. Richard Sweeney, university librarian at the New Jersey Institute of Technology, says the education model has to change to suit this generation of students. Smart but impatient, they like to collaborate and they reject one-way lectures, he notes. While some educators view this as pandering to a generation, Sweeney is firm: "They want to learn, but they want to learn only from what they have to learn, and they want to learn it in a style that is best for them."[18] There are shining examples of collaborative learning. Dr. Maria Terrell, who teaches calculus at Cornell University, used a collaborative method that's part of a program called GoodQuestions, which is funded by the National Science Foundation.[19] One strategy being used in this program is called just-in-time teaching; it is a teaching and learning strategy that combines the benefits of Web-based assignments and an active-learner classroom where courses are customized to the needs of the class. Warm-up questions, written by the students, are typically due a few hours before class, giving the teacher an opportunity to adjust the lesson "just in time," so that classroom time can be focused on the parts of the assignments that students struggled with. This technique produces real results. An evaluation study of 350 Cornell students found that those who were asked "deep questions" (those that elicit higher-order thinking) with frequent peer discussion scored noticeably higher on their math exams than students who were not asked deep questions or who had little to no chance for peer discussion.

Indeed, the research evidence dates back years. In a 1997 article published in *Educom Review*, the authors wrote: "Compared with students enrolled in conventionally taught courses, students who use well-crafted computer-mediated instruction . . . generally achieve higher scores on summary examinations, learn their lessons in less time, like their classes more, and develop more positive attitudes towards the subject matter they're learning."[20] These results hold for a broad range of students studying across a broad range of disciplines, from mathematics to the social sciences to the humanities.

Today every university student has at his or her fingertips the most powerful tool ever for discovery, constructing knowledge, and learning. Like Gutenberg's invention, it democratizes learning. Rather than threatening the old order, universities could embrace this and take discovery learning to the next step. To better serve today's learners, employers, and society at large, the university needs to break down the walls that exist among institutions of higher education, and between them and the rest of the world.

OPENING UP THE UNIVERSITY: COLLABORATIVE KNOWLEDGE

The phrase "ivory tower" usually carries some harshly pejorative connotations. Since the nineteenth century it has been used to designate a world or atmosphere where intellectuals engage in pursuits that are disconnected from the practical concerns of everyday life. For cynics, it connotes a willful disconnect from the everyday world; esoteric, overspecialized, or even useless research; and academic elitism, if not outright condescension. But setting aside some of these more negative associations, the ivory tower metaphor still captures one of the key flaws in today's system of higher learning. In a world of unparalleled connectivity, especially among today's youth, universities still operate as largely autonomous islands of scholarship and learning and have thus far failed to seize the opportunity to use the Internet to break down the walls that divide institutions, professors, and students.

In our view, the twenty-first-century university should be a network and an ecosystem—not a tower. Indeed, there is an enormous opportunity to create an unparalleled educational experience for students globally by assembling the world's best learning materials online and enabling students to select a customized learning path with support from a network of instructors and educational facilitators, some of whom may be resident at a local university and some of whom may be halfway around the globe. To make this work, universities will require deep structural changes, and educators will need to get going on the partnerships to make this work for students. But given the inertia in the system, is there any prospect of this happening soon?

According to MIT president emeritus Charles M. Vest, the vision is not quite as far-fetched as it sounds. Vest himself offers a tantalizing vision of the future when he suggests that with the growing open access movement we are already seeing the early emergence of a meta-university—a transcendent, accessible, empowering, dynamic, communally constructed framework of open materials and platforms on which much of higher education worldwide can be constructed or enhanced. In this new model, the Web will provide the communication infrastructure, and a global open-access library of course materials will provide much of the knowledge and information infrastructure. Vest argues that a noble and global endeavor of this scale would speed the propagation of high-quality education and scholarship and give teachers and students everywhere the ability to access and share teaching materials, scholarly publications, and scientific works in progress, including Webcasts of real-time science experiments.[21]

We like the direction of Vest's thinking. For universities to succeed, we believe they need to cooperate to launch a Global Network for Higher Learning. It would have three stages or levels. The first is content exchange. Professors park their teaching materials online for others to freely use. The second is content co-innovation, where teachers collaborate and share ideas across institutional and disciplinary boundaries to co-create new teaching materials using wikis and other tools. By level three, the university changes from being a place to being a node in the global network of faculty, students, and institutions learning collaboratively, while at the same time maintaining its identity, campus, and brand. The Global Network for Higher Learning is not a pipe dream. Leading scholars are implementing elements of all three levels today. They know that universities and their faculties cannot continue to operate as islands, constantly reinventing the wheel.

Level 1. Course Content Exchange

The lowest level of collaborative knowledge production is simple content exchange: universities post their educational material online, putting into the public domain what would have traditionally been considered a proprietary asset and part of a university's competitive advantage in the global market for students. MIT pioneered the concept, and today more than two hundred institutions of higher learning have followed suit, including other Ivy League universities, such as Yale.

OpenCourseWare came about when MIT asked its faculty in 1999 how best to leverage the Internet to advance knowledge and educate students. The faculty's reply was to publish, as much as possible, the faculty's teaching materials, such as lecture notes or exams, on the Internet. The process began in 2002, and by 2007, MIT completed the initial publication of virtually the entire curriculum, over 1,800 courses in 33 academic disciplines. The online content may be freely used, copied, distributed, translated, and modified by anyone.[22] MIT's OCW initiative is probably the single most important and cost-effective contribution to the world's knowledge base in the past ten years. Each course costs MIT $10,000 to $15,000 to put online. It has to compile materials from faculty, ensure proper licensing for open sharing, and format materials for global distribution. Courses with video content cost about twice as much. The result is a huge bang for the education buck and a large stride forward for educational innovation.

In addition to helping students study, the OCW material is being reused by teachers around the world. Consider what a change this offers to a typical professor's life. Prior to OCW, faculty members were largely isolated in the content creation process. Say you were a psychology professor assigned a second-year course on behavioral psychology. You chose a traditional textbook, created your own course outline. and then painstakingly went through each module, filling in the blanks. Slowly you built up the lecture presentation materials, reading lists, and so on. You developed material such as essay topics, tests, and exams. This was hard work, because as a professor you know about concepts such as reliability and validity of multiple choice tests, and you knew it was going to take multiple testings to get the instrument right. Sometimes you could rely on the work of other professors, but in general this was discouraged as being unoriginal. You might run your course idea by a few colleagues, but the only people who really benefited from all your diligent work were the students who took your course. Once you had the whole package, if you were a good professor you upgraded it every year or even semester, adding in new research, new examples, discarding ideas that didn't work. You were professor as an island.

The better and more conscientious you were, the worse it got. If you wanted to foster enriching discussions and have students work in teams to solve problems, you needed to invent a framework, methodology, and, of course, the content. And if you decided to create computer-based learning modules for certain sections of the course, to enable students to learn about, say, different types of Skinnerian procedures, the effort and cost of finding the right technology and hiring programmers to develop the courseware could be pretty prohibitive. OpenCourseWare solves the problem of isolation and provides a wealth of materials that others can use and even build on, regardless of their university affiliation.

Level 2. Course Content Co-innovation

On the OCW Web site, MIT's president, Susan Hockfield, exhorts users to "learn from [OCW] and build on it. Find new ways not only to pursue your personal academic interests, but to use the knowledge that you gain—and that you create—to make our world a better place. In the spirit of open sharing, we also encourage you to *share your scholarship with others* [emphasis added]."[23]

Sharing materials is an important first step for sure. But rather than treating OpenCourseWare as just a simple online library where users pick and choose what material they want, it could be a platform for users to collaborate, share experiences, and help improve and add to the content over time. Bear in mind that as the Global Network for Higher Learning gains momentum, the volume of material posted will be overwhelming. After all, we're not just talking about texts and digital books, but about materials such as lecture notes, assignments, exams, videos, podcasts, and so on. Professors and students will need better tools for gauging the quality and suitability of various assets. In the example of the psychology professor, he or she could use the OpenCourseWare platform to join the community of psychology professors and exchange teaching strategies and share insights about particularly valuable and engaging course materials.

Much of the logistics of true collaboration should be built into the platform itself. Indeed, the Global Network for Higher Learning, as we envision it, would include a social network—a Facebook for Faculty. It would need to go beyond the limited capabilities of Facebook, however, to enable much deeper forms of collaboration. During the academic year, for example, professors would record test results to help gauge the students' retention and understanding of the material taught. Professors could compare the effectiveness of the different learning materials and students could provide their ratings too. By pooling this data, psychology professors and students around the world could determine the best material for their own particular use. We're involved in just such a project with the Portuguese education system that includes a growing online community of teachers across the country. The system will use collaborative methods for creating, managing, sharing, and deploying curricula and tracking the results via a sophisticated learning management system. There are many benefits, including much greater collaboration among teachers and a more consistent means to measure the students' progress.

The next level in collaborative knowledge creation goes beyond discussing and sharing ideas to the actual *co-creation* of content. In the same way Wikipedia's distributed editors collaborate to create, update, and expand the online encyclopedia's entries, professors can co-innovate new teaching material (based on work already in the OCW and other repositories) and then share this newly synthesized content with the world. Or consider the approach taken by the Wikiversity, a project of the Wikimedia Foundation. Rather than offer a set menu of courses and materials, Wikiversity participants set out what they want to learn, and the Wikiversity community collaborates to develop learning activities and projects to accommodate those goals. Imagine what a platform like Wikiversity could do if it had the muscle of the world's universities behind it. These are the sorts of projects that should be invigorating the worldwide academic community.

Once again, take the example of the behavioral psychology course.

Psychology professors would work together to design the "perfect course" that pools the collective knowledge of the world's leading thinkers. Of course there can be no complete agreement as there would be various perspectives, schools, and teaching techniques. But as in Wikipedia, the professors could work globally to create core modules that were not controversial and then subnetworks of like-minded teachers could form to develop ancillary elements to complete the course. For the ultimate course the teachers would need more than course materials—they need course software enabling students to interact with the content, supporting small group discussions, enabling testing and so on. These can be developed using tried-and-true techniques and tools of the open-source software movement. If thousands of people can develop the most sophisticated computer operating system in the world (Linux), they can certainly develop the tools for a psychology course. Indeed, there are already many well-known open-source software projects under way in the academic community, and there should be more. One of the most popular is Sakai. Built by educators for educators, Sakai facilitates collaboration in and across courses, research, projects, administrative processes, and multidisciplinary and multi-institution efforts. Creation of the software itself is a product of content co-innovation, and then the product in turn helps co-innovate content that can be taught to students.

Used properly, wikis and other social media should be used to co-innovate content directly with students too. Rather than simply being the recipients of the professor's knowledge, the students co-create knowledge with light supervision, which has been shown to be one of the most effective methods of learning. Matt Barton, a professor of English at St. Cloud State University in Minnesota, recently enlisted his students to help build a living, breathing resource on English rhetoric, including its history, uses, and meaning, on Wikipedia. "I could sit down and take days, weeks, even months to find all the terms," says Barton, "but with Wikipedia, I can start the list with three or four definitions and then kick back and let the community chip in a little." Barton figures the more eyes on his work the better, so he blogs about his work and, beyond students, he invites his peers to get involved too. "I might make a mistake that I'm not seeing, so having those people watching over it is a good thing." There is a growing contingent of scholars like Barton who see the value of a dynamic, evolving body of knowledge that students are active in shaping as part of the learning experience.

Level 3. The Collaborative Learning Connection

The digital world, which has trained young minds to inquire and collaborate, is challenging not only the lecture-driven teaching traditions of the university but the very notion of a walled-in institution that excludes large numbers of people. Why not allow a brilliant ninth-grade student to take first-year college math, without abandoning the social life of his or her high school? Why not encourage a foreign student majoring in math to take a high school English course? Why is the university the unit of measurement when it comes to branding a degree? In fact, in a networked world, why should a student have to assign his or her "enrollment" to a given institution, akin to declaring loyalty to some feudal fiefdom?

Luis M. Proenza, president of the University of Akron, asks exactly these questions, challenging the notion of the ivory tower itself as the fundamental unit of higher education. True, students can obviously learn from intellectuals around the world through books or via the Internet. Yet in a digital world, why shouldn't a student be able to take a course from a professor at another university? Proenza thinks colleges and universities should use the Internet to create a global center of excellence. In other words, an institution should choose its best courses and link them with the best at a handful of other institutions around the world to create an unquestionably best-in-class program for students. Students would get to learn from the world's greatest minds in their area of interest—either in the physical classroom or online. This global academy would also be open to anyone online.

In this vision, a student receives a custom learning experience from a dozen universities. The student enrolls in his or her primary college and is assigned a "knowledge facilitator," who works with the student to customize a learning experience, the journey, and outcomes. The student might enroll in the primary college in Oregon and register to take a behavioral psychology course from Stanford University and a medieval history course from Cambridge. For these students, the collective syllabi of the world form their menu for higher education. Yet the opportunity goes beyond simply mixing and matching courses. Next-generation faculty will create a context whereby students from around the world can participate in online discussions, forums, and wikis to discover, learn, and produce knowledge as a community of learners who are engaged directly in addressing some of the world's most pressing problems.

Of course, such open platforms could provide a means to address the needs of all learners, not just twenty-somethings. For today's knowledge workers, remaining truly competitive in fast-moving fields of research and innovation means constant retraining and retooling to begin and/or continue their working lives in a modern, dynamic, and technology-focused environment. The cost of building new continuing education programs from scratch could be prohibitively high, but new models of collaborative education can help bring greater efficiency and creativity to the efforts to help graduating students and aging employees update their skills.[24] Indeed, why not allow companies and governments to participate in this global network for higher learning too? Fees collected from commercial users could be used to subsidize ongoing development of the platform.

TIME FOR REINVENTION OR ATROPHY?

The combination of the new Web, the new generation of learners, the demands of the global knowledge economy, and the shock of the economic crisis is creating a perfect storm for the universities, and the storm warnings of change are everywhere. In 1997, none other than Peter Drucker predicted that big universities would be "relics" within thirty years.[25] Today, Drucker's seemingly hyperbolic predictions seem less shrill and even prescient.

Some universities and some faculty are more vulnerable than others. The great liberal arts colleges are doing a wonderful job of stimulating young minds, because with big endowments and small class sizes students can have more of a customized collaborative experience. If you're fortunate enough to get into Amherst College, Williams, or Swarthmore, you're in good hands. But the same cannot be said of many of the big universities that regard their prime role to be centers for research, with teaching as an inconvenient afterthought, and class sizes so large that the only way to "teach" is through lectures.

As the model of pedagogy is challenged, inevitably the revenue model of universities will be too. If all that the large research universities have to offer to students are lectures that students can get online for free, from other professors, why should those students pay the tuition fees, especially if third-party testers will provide certificates, diplomas, and even degrees? If institutions want to survive the arrival of free, university-level education online, they need to change the way professors and students interact on campus.

Many will argue: "But what about credentials? As long as the universities can grant degrees, their supremacy will never be challenged." This is myopic thinking. The value of a credential and even the prestige of a university are rooted in its effectiveness as a learning institution. If these institutions are shown to be inferior to alternative learning environments, their capacity to credential will surely diminish. How much longer will, say, a Harvard undergraduate degree, taught mostly through lectures by teaching assistants in large classes, be able to compete in status with the small class size of liberal arts colleges or the superior delivery systems that harness the new models of learning?

Others will argue: "What about the campus experience? That will never be replaced." Again, if campuses are seen as places where learning is inferior to other models or, worse, as places where learning is restricted and stifled, the role of the campus experience will be undermined as well. The university is too costly to survive as simply an extended summer camp.[26] Conversely, campuses that embrace the new models will become more effective learning environments and more desirable places. Even things as simple as online lectures do not undermine the value of on-campus education. Video lectures enhance education by allowing students to absorb course content online—whenever it is convenient for them—and then get together to tinker, invent new things, or discuss the material. The OCW experience has shown MIT that the real value of what it offers is not the lecture per se but rather the whole package—the content tied to the human learning experience on campus, plus the certification. Colleges and universities cannot survive on lectures alone.

How, then, can universities reinvent themselves, rather than atrophy? What are the steps to be taken?

Adopt Collaborative Learning as the Core Model of Pedagogy

Professors who want to remain relevant will have to abandon the traditional lecture and start listening to and conversing with students—shifting from a broadcast style to an interactive one. In doing so, faculty can free themselves to be curators of learning—encouraging students to collaborate among themselves and with others outside the university. They should encourage students to discover for themselves and learn a process of discovery and critical thinking instead of just memorizing the professor's store of information. Finally, they need to tailor their approach to their students' individual learning styles. The Internet and the new digital platforms for learning are critical to all of this, especially given the high student-faculty ratio in most universities. But most faculty do not have the resources to develop the courseware required. This must be co-innovated globally through new partnerships.

Build New Revenue and Collaboration Models Between Higher Education Institutions to Break Down the Silos Between Them

Right now, universities around the world are embracing, to varying degrees, levels one and two—course content exchange and co-innovation—of the Global Network for Higher Learning. But we need to move to the next level. To achieve a Global Network for Higher Learning where students can benefit from the capability of any university in the world, we will need a collaborative revenue model and a new structure of transfer pricing. Students would enroll with their "primary" university, which would handle the disbursement of their tuition fees depending on what other university courses they study. The value of, say, a second-year psychology course at Stanford would be determined by market forces, not some central bureaucracy.

Change Incentive Systems to Reward Teaching, Not Just Research

Why are universities judged by the number of students they exclude or by how much they spend? Why aren't they judged by how well they teach and at what price? If universities are to become institutions whose primary goal is the learning by students, not faculty, then the incentive systems will need to change. Tenure should be granted for teaching excellence and not just for a publishing record.

How can this be done? Student input is important. Web sites such as RateMyProfessors.com can provide helpful input. Though they are not simply a popularity contest, as some suggest, they also cannot serve as the only basis for rewarding professors. Peer review can provide helpful input as well, and administrators often have a view on teaching effectiveness. In addition, measures of economic success may make sense. If a professor at Stanford has an enormously popular psychology course that is subscribed to by thousands of students from around the world, shouldn't he or she share in the revenue created for Stanford? Such measures would be insufficient alone, however, since they reorient professors to be profit centers rather than learning curators. Ultimately, we will need more objective measures centered on students' learning performance.

Build the Infrastructure for Twenty-first Century Higher Education

While governments are investing in "shovel-ready infrastructure" to turn around the current economic crisis and global recession, a new kind of infrastructure is required to realize the University 2.0. Some of this is technological. Initiatives like the Wikiversity from the Wikimedia Foundation represent a good start in creating a national and global platform for all scholars and learners to build the content required. But we need more entrepreneurs building interactive courseware for all disciplines and categories of human knowledge. Governments could help by investing in networks to build the access and broadband capacity required to close the global digital divide. The world needs a "Digital Marshall Plan."

Governments should terminate their subsidization of academic journals in libraries and shift funding to building the digital infrastructure. David W. Lewis, dean of the University Library at Indiana University–Purdue University Indianapolis (IUPUI), argues that scholarly communication is a public good and, as such, requires subsidy. But because subsidies have been routed through libraries, corporate publishers figured out that science journals had inelastic prices and began to suck the subsidy out of the system. "The economics of the Internet turn all of this on its head," Lewis says. "The most effective way to use the subsidy is to support open access, which funds the infrastructure and gives away the works to everyone." Lewis argues that this will cause a battle: "There are many established institutions which get left out of this picture, libraries for one, and also much of what academic publishers do. These institutions are doing what established institutions always do—stay alive."

Universities in the United States typically lack the broadband infrastructure necessary for everywhere multimedia access and use. They also need to invest in building the applications and courseware for collaborative learning.

TOWARD UNIVERSITY 2.0

So why haven't these changes happened yet? Where is the University 2.0? "It's the legacy of established human and educational infrastructure," says Proenza. The analogy is not the newspaper business, which has been weakened by the distribution of knowledge on the Internet, he notes. "We're more like health care. We're challenged by obstructive, non-market-based business models. We're also burdened by a sense that doctor knows best, or professor knows best."

The industrial age model of education is hard to change. New paradigms cause dislocation, disruption, confusion, and uncertainty. They are nearly always received with coolness or hostility. Vested interests fight change. And leaders of old paradigms are often the last to embrace the new.

A powerful force to change the university is the students. And sparks are flying today. A huge generational clash is emerging in our institutions. The critiques of the university from fifteen years ago were ideas in waiting—waiting for the new Web and for a new generation of students who could effectively challenge the old model.

Changing the model of pedagogy and the model of knowledge production is crucial for the survival of the university. If students turn away from a traditional university education, this will erode the value of the credentials that universities award, along with the position of these institutions as centers of learning and research and as campuses where young people get a chance to "grow up." The Global Network for Higher Learning is not a pipe dream. Leading scholars are beginning to implement many aspects of it today. They know that universities and their faculties cannot continue to operate as separate ivory towers but must work toward collaborative learning and collaborative knowledge production. It's time!

9. SCIENCE 2.0: IGNITING KNOWLEDGE CREATION IN A NETWORKED WORLD

As an astronomy graduate student, Kevin Schawinski faced a rather sizable problem: he needed to sort through fifty thousand images of galaxies taken by a robotic telescope and classify each galaxy according to its formation. Even in the age of computer-aided science, each image needed to be closely scrutinized by hand to ensure the classifications were accurate. According to Schawinski, even the smartest and most powerful computers are prone to making mistakes that would never trip up a human.[1]

After spending a week doing nothing else, Schawinski was convinced there had to be a better way. Sitting in a pub one night in the summer of 2007, he complained to colleague Chris Lintott, an astronomer at Oxford, about how long it would take him to go through even a fraction of the images. It became clear he needed help. And Lintott had a suggestion. Why not do what Linus Torvalds did with Linux: toss the problem out to the world and see if anyone in the amateur astronomy community would be willing and able to pitch in.

It was an unorthodox notion, to be sure. But, if it worked, Schawinski and his colleagues could vastly accelerate the pace of scientific discovery in their field and investigate possibilities that most astronomers only dream of exploring. They might even end up challenging some of the core assumptions underlying today's scientific institutions—like the assumption that ordinary people can't participate meaningfully in scientific research, except as passive consumers of science journalism or as beneficiaries of scientific advances. So, with the aid of Lintott and several others, Schawinski cooked up a scheme whereby an army of armchair astronomers would help them sort through the millions of galactic images they had stored up in their databases. The result was Galaxy Zoo, a clever online citizen science project where anyone interested can peer at the wonders of outer space, while simultaneously helping advance an exciting new frontier in science.

The premise of Galaxy Zoo was simple. Users would be shown an image of a galaxy and asked two basic questions: Is the galaxy an elliptical galaxy (a type of galaxy with no dust or gas, but many stars) or a spiral galaxy (with rotating arms, like our own Milky Way galaxy); and, if it's a spiral, in which direction are the arms rotating? In order to participate, users would need to take a ten-minute tutorial teaching them the basics of galaxy morphology. But could it really work? Even Schawinski, now a postdoctoral associate in Yale's Department of Astronomy, had his doubts. "I thought there were maybe a couple of dozen hard-core amateur astronomers out there who might possibly be interested in this," he said.

He was wrong.

Thanks in part to a story by the BBC, where Lintott cohosts an astronomy program, thousands of users were busily classifying galaxy images within the first twenty-four hours of the site's launch—as many as 70,000 images per hour. Schawinski was caught totally off guard—the overload on the server storing the images literally melted a cable and nearly derailed the whole project. "We were overwhelmed," Schawinski says. "It was a complete surprise. Now, when people ask us for advice about how to do this, the first thing I say is prepare for success." Two and a half years later, merely claiming success may be an understatement. Galaxy Zoo is thriving, with more than 275,000 users who have made nearly 75 million classifications of one million different images—far beyond Schawinski's original 50,000. If Schawinski were still laboring on his own, it would have taken him 124 years to classify that many images!

But Galaxy Zoo is about more than just looking at pretty pictures of galaxies. The project has resulted in real scientific discoveries, with several papers already published using the data and a dozen or so more on the way. The Galaxy Zoo team—which includes astronomers from Yale and Johns Hopkins universities in the United States, and the University of Oxford and the University of Portsmouth in the United Kingdom—has often been surprised by the results. Bill Keel, an astronomy professor at the University of Alabama who studies overlapping galaxies, decided to ask Galaxy Zoo users to contact him if they came across an example of this rare phenomenon. Throughout his career, Keel had studied the dozen or so overlapping galaxies then known to astronomers. Within a day of posting his question on the Galaxy Zoo forum, he had more than one hundred responses from users who had indeed found such galaxies. Today, thousands have been identified.

To be sure, Galaxy Zoo is only possible because Schawinski and his colleagues have eschewed the usual inclination to keep their discoveries private until they are ready to publish. "We count community members as colleagues, even peers, and we're very open with everything that we find," Schawinski says. "We blog about our findings and we give people feedback. We've probably disclosed more preliminary science than is usual in our field, but our collaborators put in all this work, and they deserve to know." This kind of openness is not yet standard practice in the scientific community, but it is clearly growing, in no small part due to the success of Schawinski, Lintott, and their peers. Indeed, the promise of citizen science is that if you can make a task small enough and simple enough for someone to do at his or her leisure, you can actually aggregate a lot of talent and labor power. Even a massive task like analyzing millions of galaxies can be done by relative amateurs—and it can get done a whole lot faster, without compromising quality, so long as researchers are careful to maintain the rigorous standards of scientific inquiry. "We never thought of it as an outreach project, we never thought of it as an educational project. We always thought of it as, 'Please help us with our research, in a way that only you can.' And that really attracted people," he says.[2]

When asked where things might go next, Schawinski is emphatic about the possibilities. "Mass collaboration on the Internet is a powerful multiplier," Schawinski says. "It makes research possible that just wasn't possible before." In fact, the success of Galaxy Zoo has already inspired a growing portfolio of complementary projects. Zoo members can now hunt for supernovae or simulate cosmic mergers on their computers to help scientists better understand how galaxies form, merge, and evolve. There is a project to spot solar explosions and track them across space to Earth. If detected early enough, news of a brewing solar storm can be relayed to astronauts, giving them an early warning if dangerous radiation is headed their way. There is even a new umbrella organization called the Citizen Science Alliance that is nurturing new citizen science projects in disciplines ranging from data engineering to oceanography. As for the Zoo members, Schawinski says users are clamoring for these enhanced versions and relish the opportunity to contribute to new science. "They're our co-investigators—they're scientists, too," he says. "And they really care about the science they've contributed to."

UNLEASHING A BOLD NEW PARADIGM OF COLLABORATIVE SCIENCE

So, now here's the question: If ordinary citizens can help transform a modest PhD project into a path-breaking global science initiative overnight, what else could mass collaboration unleash in the scientific community? It turns out that citizen science is only one aspect of a much deeper transformation in science and invention that we describe in this chapter. The scientific revolution under way changes everything from the way scientists gather data and draw conclusions to the way companies harness new knowledge to drive economic and technological progress. Call it the new age of collaborative science. Just as the Enlightenment ushered in a new organizational model of knowledge creation, the new Web is helping to transform the realm of science into an increasingly open and collaborative endeavor that will accelerate scientific discovery and learning.

Much has already changed. The Internet is not just a low-cost medium for disseminating scientific information; it's the new platform for doing science, period. Thousands of scientific pioneers now rely less on "the paper" as the prime vehicle for scientific communication and more on tools such as blogs, wikis, and Web-enabled databanks. Rather than wait a year or more to crank out a traditional publication, they use Twitter and other social media to share day-to-day findings and observations with a global network of peers. Better and cheaper access to knowledge and scientific tools, in turn, is making the whole process of learning and economic change more efficient. Superior techniques are spreading faster. New technologies are being more widely deployed and improved. More minds are being trained in science, and more skills brought to bear on the time-urgent problems facing the world.

As these new forms of mass collaboration take root in the scientific community, there is an opportunity to completely rethink not just how we publish science, but how we apply science to drive innovation, solve problems, and create economic opportunity. Companies can scale and speed up their early-stage R&D activities dramatically, for example, by collaborating with scientific communities to aggregate and analyze precompetitive knowledge in the public domain. Even highly secretive pharmaceutical companies are getting involved. Novartis and GlaxoSmithKline demonstrate that even competitive rivals are seeing the benefits of collaborating on initiatives that will establish and grow a market for new products and services. Depending on the type of venture, firms can identify and act on discoveries more quickly, focus on their area of competence, facilitate mutual learning, and spread the costs and risks of research. Analogous opportunities exist for governments, major philanthropic donors, and private investors that all play important roles in funding science and applying the results.

If this plays out the way we predict, the new scientific paradigm holds a more than modest potential to rapidly improve human health, turn the tide on environmental damage, develop breakthrough technologies, and explore outer space—not to mention help companies grow wealth for shareholders. That's a bold statement. But there is growing evidence to support it. In partnership with other key institutions, scientific communities can embrace new wikinomics principles like openness, collaboration, and sharing to fundamentally change the world we live in.

COLLABORATIVE SCIENCE COMES OF AGE

Starting as early as the seventeenth century, we began to create, accumulate, and harness knowledge in new ways.[3] A rising class of engineers, mechanics, chemists, physicians, and natural philosophers began forming circles whose primary objective was to further scientific understanding of the world around them. They exchanged letters, met in Masonic lodges, attended coffeehouse lectures, debated in scientific academies, and, for the first time in history, they made knowledge about the natural world increasingly nonproprietary. In doing so, they not only advanced the basic science, they helped transfer the knowledge to entrepreneurs and early industrialists who could apply it to solve practical problems and build new industries. As time went on, Gutenberg's marvelous invention took the sharing of knowledge to a new level, and the interplay between open science and private enterprise provided a sustained basis for technological innovation and economic growth that characterized most of the twentieth century.

Today a new scientific paradigm of comparable significance is on the verge of ignition, inspired by deep transformations in the nature of scientific communications—transformations that will rival or even exceed those catalyzed by Gutenberg's press. The same technological forces that are turning the Web into a massive collaborative workspace are leading to a new generation of science that is fast, global, data intensive, and increasingly plays out on a scale that makes large and prestigious institutions like Harvard look like those seventeenth-century debating clubs. Indeed, just as other institutions are facing wrenching changes, there are some pretty big challenges for the scientific community to solve this decade. The amount of available data and information about everything is exploding, raising new challenges for researchers who must master increasingly advanced computing capabilities to manipulate and explore massive data sets. The models for funding science are antiquated and at times they perpetuate outmoded ways of doing and publishing research. Some scientists fear that commercial incentives are unduly influencing the direction of research and even eroding the culture of sharing that underpins the progress of science itself. At the same time, the recent controversy around climate data has led to louder calls for greater transparency in the way scientists manage data and report their findings. None of these challenges are insurmountable. In fact, the innovative approaches to science described in this chapter are already providing solutions in fields ranging from oceanography to organic chemistry. The question is whether the broader community of institutions underpinning science—including the funding bodies, universities, publishers, and scientists themselves—will willingly adapt to the changes or whether change will be forced upon them.

When Science and Innovation Go Global, What Happens to Funding Models?

Organizing the pursuit of knowledge in a peer-to-peer fashion is certainly nothing new in science. But recent research suggests that collaboration is exploding. One study conducted by the Santa Fe Institute (a private, not-for-profit research center) found that the average high-energy physicist now has around 173 collaborators. The same study found that the average number of authors per scientific paper has doubled and tripled in a number of fields. A growing number of papers have between 200 and 500 authors, and the highest-ranking paper in the study had an astonishing 1,681 authors.[4] It's the same story in many fields. Some 3,000 climate scientists from over 130 countries are publishing over 2,000 papers a year, investigating the science and societal implications of global climate disruption. Projects like the Earth Systems Grid combine supercomputing facilities with data, a wide range of sources including ground- and satellite-based sensors, computer-generated simulations, and thousands of independent scientists uploading their files from around the world.

Over at the California Institute for Telecommunications and Information Technology (Calit2), a faculty of three hundred researchers is turning the institution into a laboratory for new models of collaborative science. "We simply cannot do a project that doesn't become global overnight," says its founding director, Larry Smarr. A veteran astronomer himself, Smarr remembers applying for research grants from the U.S. government in the 1980s and early 1990s and waiting what seemed like eons to see whether his project would be approved. "The gap between your idea and the money flow was at least a year, and maybe even two years in some cases," he said. That gap is narrowing as funders speed up their processes. But the grant-driven, rank-ridden approach to funding research is still a pretty poor match for an era when large-scale collaborations are an increasingly routine feature of how science gets done.

In a typical funding scenario, government agencies put out an RFP (request for proposals) and scientists compete rather than collaborate to win it. They typically go off to do their research privately, and when it's done they report the results back to their funders, who eventually publish it for others. It's closed. It's anticollaborative. And it's antithetical to the demands of the global knowledge economy. Rather than turn inward, or focus narrowly on national goals, funding bodies should open up and create a truly open market for science funding—one that rewards only the most qualified candidates and doesn't pay heed to passports, seniority, or star status. Sure, this arguably undermines one of the key reasons why many funding bodies exist: to promote national science objectives and national research institutions. But why, in today's world, should we strive to contain science within geographic boundaries? Shouldn't funding bodies maximize taxpayer dollars by seeking out scientific ideas and talent on a global basis? And shouldn't the national priority be to create a large pool of highly trained scientists that can compete globally for funding with the best colleagues and the best institutions?

While funders have been slow to contemplate these questions, smart researchers are racing ahead: tapping a global pool of resources and exploiting creative new ways to finance and disseminate their research. "You're totally visible to a global community now in a way that we just had no idea about even five or eight years ago," says Smarr. "I don't have to limit myself to the federal government . . . I can look globally. And if I find somebody who's working in a similar field to mine who would love to collaborate with me and they've got resources already in hand, literally within a few weeks you could be doing a project." Harvard neurologist Seward Rutkove also has firsthand experience tapping the Web for opportunities. In 2009 he competed in an InnoCentive challenge seeking a way to track the progression of amyotrophic lateral sclerosis (ALS), or Lou Gehrig's disease, because he was researching that topic already. He won $50,000 for presenting his solution and is now in the running to collect a $1 million prize for demonstrating the ability to ward off an ALS-like disease in mice.

Some major funding agencies have taken note of these innovations and adopted a less parochial approach. The National Science Foundation's nanoHUB provides Web-based resources for research, education, and collaboration to a global community of over 100,000 nanotechnology researchers stationed not just in the United States, but in 170 countries around the globe. In the meantime, the U.S. government and many other research funders in Europe have backed the idea of a global scientific data commons by requiring scientists who receive public funding to place a copy of their paper (and often the data too) in a free, publicly accessible database—a move that most scientists, if not the publishers, are enthusiastically backing.

If there is another thing that most scientists agree on, it's that the globalization of science is accelerating the pace of discovery. And we're not just talking about a nominal percentage increase; in some fields we're talking orders of magnitude. Leading scientific observers are predicting more change in the next fifty years of science than in the last four hundred years combined.[5] Even the definition of what counts as discovery is expanding. In the early nineties, for example, a PhD candidate in biosciences could earn his or her doctorate by sequencing one gene. Then Craig Venter completed the first full sequencing genome in 2000, and one could no longer get a PhD for doing a gene anymore; you had to do a genome. And today, unless you're doing a population of genomes (for example, by looking at an entire species, rather than an individual member of that species), it's not very interesting.

Not only is everything speeding up, this globalization of science is rearranging the geography of research and innovation itself. In the twentieth century, the United States, Europe, and Japan could be reasonably confident in thinking that if new and important ideas were going to be invented they would be invented on their shores. In the twenty-first century, citizens of the United States and Europe will consider themselves fortunate if they produce maybe one of every four or five major inventions. "Today, innovation has gone absolutely global," says Smarr. "It's no longer sufficient, if you're a researcher, to know what somebody in your country is doing. You have to have a global watch."

Competition may be ratcheting up, but on the flip side there will be more smart minds focused on problem solving. And with those minds increasingly connected by the global fabric of the Web, we can boost the productivity of the scientific community. "As the productivity of the globe goes up," says Smarr, "the time from posing a question to getting the answer goes down." That's good news in a world with no shortage of time-urgent situations, each needing its own unique infusion of new thought, new technologies, and new capabilities.

As Science Becomes Integral to Public Policy Making, Can We Increase Scientific Transparency and Improve the Capacity of Citizens to Participate?

Science and public policy have long gone hand-in-hand. From decisions about whether to regulate a new technology, to the ongoing need to assess the impacts of urban development on the local ecology, objective scientific analysis is often central to the formulation of effective public policies. As the intermingling of science and public policy intensifies in an era of new global risks, questions about how scientists relate to the public and how the public relates to science are becoming critical. Nothing illustrates the challenges better than the recent "climategate" scandal in which a large stash of e-mails from and to investigators at the Climatic Research Unit of the University of East Anglia provided more than enough evidence for concern about the way some climate science is done.

The science discussed in the e-mails is mostly from one small area of climate research—the taking of raw temperature data from thermometers, satellites, and proxy measures of historical climate such as tree rings and turning it into usable information on temperature trends. Under director Phil Jones's management, the CRU assembled the most comprehensive thermometer data record in the world, much of it under contract to the U.S. Department of Energy. One result was the creation of iconic graphs like the famous "hockey stick," first published twelve years ago and one of climate science's most famous and controversial products. It shows a long period of natural stable temperatures followed by a sharp, exceptional warming in the late twentieth century.

Within academia, the reliability of the CRU's temperature data is widely accepted. But various industrious bloggers are not so convinced, and they accuse climate scientists of cherry-picking their data. One such skeptic is Steve McIntyre, a Canadian squash-playing blogger and data-obsessive in his sixties who recently retired from the mining industry and now spends most of his time trying to publicly debunk and deconstruct the science of climate change. It started in 2002 with an e-mail from his home in Toronto to Phil Jones, asking for some weather station data. Initially the exchanges, as revealed on McIntyre's blog, Climate Audit, were civilized. But as the years passed, and his data demands grew greater, relations quickly soured. McIntyre clearly doubted the statistical techniques being employed by the climatologists, and felt that, as a trained mathematician, he could do better, despite his ignorance of climate science. Although McIntyre rarely questions the "edifice" of climate science himself, his analysis provided fodder for a more excitable and less fastidious network of fans who point to his findings as evidence that all climate science is a hoax of epic proportions. So when McIntyre tried to access their raw data and computer programs using freedom of information laws, the researchers resisted, fearing that the information would be distorted. The situation came to a head when the e-mail records of prominent climate scientists were stolen and then published, exposing years of heated and often unfortunate exchanges between climate researchers and the bloggers who were hounding them.

Depending on your perspective there are two competing analyses of what "climategate" means, says Fred Pearce, an environment writer who led a major investigation into the controversy on behalf of the *The Guardian*.[6] Climate scientists tend to see it as the mob storming the lab—the story of a malicious attempt to disrupt, cross-question, belittle, and trash the work of mainstream scientists. Their critics see it as democracy in action—the outcome of an entirely laudable effort by amateur scientists and others outside the scientific mainstream to gain access to the complex data sets behind some of the climate scientists' conclusions and to subject them to their own analysis. While there is no reason, in our minds, to question the integrity of the world's climate scientists, there is evidently some truth in both narratives. Pearce's investigation found evidence of slipshod use of data and apparent efforts to cover that up. It also found persistent efforts to censor work by climatic skeptics regarded as hostile—especially those outside the scientific priesthood of peer review—or those able to generate headlines in media outlets thought unfriendly, like FOX News.

To be fair, the scientists claim they were under intense and prolonged attack from politically and commercially motivated people who wanted to prevent them from doing their science and trash their work. In turn, they accuse their attackers of conflating McIntyre's legitimate technical criticism of their methods with "unsupported, unjustified and unverified accusations of scientific mal-conduct that confused the public." They were partially vindicated when the U.S. House of Representatives' Science and Technology Committee concluded in March 2010 that there was, in fact, no evidence to support charges that the University of East Anglia's Climatic Research Unit or its director, Phil Jones, had tampered with data or perverted the peer review process to exaggerate the threat of global warming. Moreover, the committee noted that nothing in the more than one thousand stolen e-mails challenged the scientific consensus that "global warming is happening and that it is induced by human activity."

Still, that doesn't mean that Jones and colleagues are off the hook. The CRU's habit of keeping much of its data, methodology, and computer codes secret is clearly counterproductive. By opening up the temperature databases to independent analysis and interpretation, climate scientists could help restore the credibility of land-surface records and demonstrate an openness on the part of climate science that has not always been evident in the past. Indeed, if the scientific community wants to maintain credibility in the eyes of the public, it will no longer be sufficient for scientists to speak only to each other. They must engage with the rest of the world, even if it feels like an irritating distraction at times. Scientists need make no intellectual concessions, but they have a duty to understand the context in which they operate. It is no longer acceptable for climate researchers to wall themselves off and leave the defense of climate science and climate policy to politicians, environmental groups, and the media. Other institutions are embracing transparency. It's only natural that scientists should embrace transparency too.

When We Get Real-Time Data About Everything, How Will Scientists Make Sense of It?

The battle over climate data is just a harbinger of things to come. Increasingly powerful and pervasive computing capabilities are driving quantum leaps in the amount of data available to scientists and other interested parties. When scientists are equipped with the right tools and the right training, this data will eventually revolutionize our ability to model the world and all of its systems, giving us new insights into natural phenomena and the ability to forecast trends like climate change with greater accuracy. This level of data granularity will usher in some big changes, including a growing reliance on computers. It is also sure to stir up new controversies as societies wrestle with the social implications of a world with ubiquitous connectivity, where every minute movement or trivial utterance can be detected and recorded for subsequent analysis.

Just think about it. When the devices we use to capture and process data are sparsely distributed and intermittently connected, we get an incomplete, and often outdated, snapshot of the real world. But distribute billions and perhaps trillions of connected sensors around the planet—just as we are doing today—and virtually every animate and inanimate object on Earth could be generating data, including our homes, our cars, our natural and man-made environments, and yes, even our bodies. Although our bodies are not connected to the Internet today, they will be, as biochips embedded in patients with chronic conditions report their vitals back to a central database that is monitored remotely by physicians. Our cars aren't sharing their data, but they will be too, including their speed, fuel performance, location, and ultimate destination with an incredibly high degree of accuracy. And while our physical environment (both natural and man-made) may be sparsely connected now, soon we'll have global insight into water quality, vegetation, temperature variations, wind speeds, and much, much more, at the click of a button.

It's not just that the sensors are getting smaller and faster, as Moore's law predicts they will. The absolute number of sensors is exploding as more and more applications emerge. "Two things are expanding exponentially," says Smarr, "it's a double exponential in terms of the amount of data being generated." That data will provide the raw materials for a much more real-time and granular view of the world. At the same time, it will revolutionize the practice of science and even alter the basic skill set required to enter the field.

Of course, the real challenge is not to collect the data; it's to analyze and make sense of it. And not just each individual data stream in isolation, but the larger emergent patterns arising out of the cacophony of information we are constantly assembling. Needless to say, this is an increasingly big challenge for scientists. "We already have orders of magnitude more data than before," says Euan Adie, who works in the online division of Nature Publishing Group. "It's not like one person can collect the data, analyze it and then exhaust all the possibilities with it." As we saw with the Galaxy Zoo project, increasingly powerful telescopes like the Sloan Digital Sky Survey and the data they generate have fundamentally changed astronomy. A decade ago, astronomy was still largely about groups keeping observational data proprietary and publishing individual results. Now it is organized around large data sets, with data being shared, coded, and made accessible to the whole community. In the process, astronomers went from having dozens and hundreds and thousands of galaxies to handling hundreds of thousands and now millions.[7]

The need for more computer-aided research and new models of collaboration will only increase as the Internet extends its reach into every last corner of the planet, including some places where you might least expect. Take the vast and largely unexplored swath of ocean floor lying off the west coast of Canada. The Venus Coastal Observatory, a project run by the University of Victoria, provides oceanographers with a continuous stream of undersea data once accessible only through costly marine expeditions. When its sister facility Neptune Canada launches in the summer of 2010, the observatories' eight nodes will provide ocean scientists with an astonishing wealth of information. "We're changing by orders of magnitude the sampling ability we have for the oceans," says Benoît Pirenne, associate director of Neptune Canada.[8] To cope with the flood of data, researchers using Neptune's Oceans 2.0 platform can tag everything from images to data feeds to video streams from undersea cameras, identifying sightings of little-known organisms or examples of rare phenomena. Wikis provide a shared space for group learning, discussion, and collaboration, while a Facebook-like social networking application helps connect researchers working on similar problems.

Meanwhile, over at the European Bioinformatics Institute, scientists are using Web services to revolutionize the way they extract and interpret data from different sources, and to create entirely new data services. Imagine, for example, you wanted to find out everything there is to know about a species, from its taxonomy and genetic sequence to its geographical distribution. Now imagine you had the power to weave together all the latest data on that species from all of the world's biological databases with just one click. It's not far-fetched. That power is here, today. Projects like these have inspired researchers in many fields to emulate the changes that are already sweeping disciplines such as bioinformatics and high-energy physics. Having said that, there will be some difficult adjustments and issues such as privacy and national security to confront along the way. "We're going from a data poor to a data rich world," says Smarr. "And there's a lag whenever an exponential change like this transforms the impossible into the routine." People aren't necessarily good at thinking about exponential changes, he argues, and as a result, it seems scientists are underinvesting in the analysis and visualization tools we need to handle it.

Fortunately there are trailblazers to show us the way in a world where we have data about anything and everything. And in scientific pioneers like Neptune, Galaxy Zoo, and Calit2 we are seeing a new kind of analysis, a new kind of science, and a whole new kind of organization come into being. The question now is whether the rest of the scientific world is ready. For a glimpse of what's coming next, let's revisit the life sciences industry, where the nascent models of open collaboration have now come of age.

OPENING UP THE NEW SCIENCE OF LIFE

Scientists have long suspected that our genes determine things like what we look like, our intelligence, how well we fight infection, and even how we behave. Armed with a fully sequenced genome, scientists are now convinced that these microscopic spirals of DNA amount to something like an operating system for humans. Learning how to "program" this operating system could hold the key to eliminating dreadful diseases such as Alzheimer's, diabetes, and cancer. Applications of this research in fields such as agriculture and ecology could help us end world hunger and take better care of the planet. But while we've sequenced the genome, progress in exploiting this knowledge has so far been relatively slow. In the medical world, effective and well-tolerated treatments for many forms of cancer, as well as chronic diseases and disorders such as Alzheimer's, Parkinson's, and schizophrenia, are still lacking. There has been almost no research on tropical diseases such as malaria and typhoid, the burden of which falls almost entirely on the world's poorest populations. In fact, only 1 percent of newly developed drugs will help the millions of people in Africa who die annually from these diseases. "The rapid pace of scientific discovery, especially following the sequencing of the human genome, has led many to believe that a pharmaceutical renaissance is just around the corner," argues Gigi Hirsch, executive director for MIT's Center for Biomedical Innovation. "It is perhaps the most frustrating fact of the industry that despite an enormous increase in R&D investment, and historical advances in technology through genomics, automation and computation, the number of new drugs produced each year remains at the same level that existed over 40 years ago (about 20 per year)."[9] Indeed, as increased research spending collides with pressure to contain health care costs, the factors that affect the efficiency of biomedical research and drug discovery have rightfully come under scrutiny. The promise of biomedical research to relieve human suffering and create wealth has never been higher. But the ability of science and industry to deliver on this promise depends critically on the ability of both institutions to control costs, marshal resources effectively, and manage an increasingly complex knowledge base.

Dr. Frank Douglas, former executive vice president and chief scientific officer of Aventis, agrees that there are many concerns to address. "The productivity of large pharmaceutical innovation has decreased," says Douglas. "We lack the ability to properly predict the side effects of new compounds, and we don't have good ways to monitor and assess them once they are in the market. Pricing models have become untenable. So has the 'blockbuster' mentality. Across the board, a lot of old models really need to be examined."[10] But what's the underlying source of this malaise? Extremely long product development times, and the high cost of R&D, argues Hirsch, have led to a very competitive industry culture, with little interest in cooperative ventures and a bias toward fiercely protecting its intellectual property. These inefficient practices have become deeply ingrained by a highly risk-averse and legalistic corporate culture, often at the expense of opportunities to codevelop early-stage technology tools, establish data standards, share disease target information, or pursue other forms of collaboration that could lift the productivity of the entire industry.

Having witnessed what Linux has done for software production, it's worth asking whether a flurry of open-source activity could unleash a similar revolution in the life sciences? What if the drug discovery process, for example, was opened up so that anyone could participate, modify the output, or improve it, provided they agree to share their modifications under the same terms? Could the collective intelligence of the life sciences community be harnessed to enable a more coordinated and comprehensive attack on the intractable diseases that have so far stymied the industry? Could opening up the expensive and time-consuming clinical trials process to a broader community of researchers lower the cost of drug development to the point where the resulting medicines are within reach of the world's poor? Back in 2006, when we first published *Wikinomics*, open-source biology was mostly a beta concept supported by a small number of visionaries who understood and were passionate about the open-source opportunity in medicine. Today, the concept has become reality as a growing number of scientists and companies are seeing openness and collaboration as key to unlocking the human genome's full potential. To see where it's going, we caught up with Jean-Claude Bradley, who heads up an antimalarial research program at Drexel University and just happens to be one of the pioneers among a growing number of scientists that aren't all that averse to getting naked in front of their colleagues. In fact, he prefers it!

Getting Naked

Bradley is not really a nudist. He's a synthetic organic chemist by training. His specialty is making organic compounds, particularly those that can help prevent or cure malaria, an affliction that affects 350 million people annually and kills over one million, most of them young children living south of the Sahara desert in Africa. But Bradley has a problem. A typical new drug takes ten to fifteen years and an average of $800 million to develop. It requires access to expensive laboratory instruments and a whole lot more expertise than can be reasonably possessed by one person.

Commercializing biological inventions means years of painstaking clinical trials and a healthy dose of regulatory know-how to reach that point. So to create the ultimate cure for malaria he depends on a large network of colleagues with complementary skills and knowledge. "Collaboration just makes sense," says Bradley. "Our research is done openly so really anyone can just drop in and join the collaboration or join the conversation, and that's pretty exciting." Bradley has been using wikis as a lab notebook since 2005. So do all of the students in his lab, who post their experiments online in as close to real time as possible. "We make all of our research open to the world," he says. "We're doing open science, and we're doing it in real time." Instead of a life stream on Facebook or Twitter, imagine a scientific activity sort of stream documenting day-to-day research findings. "Every time you edit a record in the database, update some details about a protein on the Web, or write a comment on a paper, it appears in your scientific stream of activity," says Bradley. All of his colleagues in the UsefulChem network are doing the same. So it all adds up to a real-time storm of information about the progress they're making. Bradley says collaboration is speeding up the metabolism of his research. "Conventional publishing can take months or years, but when my colleagues started blogging," he says, "I'd find out pertinent information within days. And now with things like Friend Feed and Twitter, it's down to hours and minutes after relevant things happen in my field."

A handful of smart companies are getting naked too. In 2008, for example, GlaxoSmithKline (GSK) pledged to provide cheaper medicines in the developing world and help bolster research on neglected diseases by placing any relevant chemicals or processes into a "patent pool" so they can be explored by other researchers. If other companies were to follow suit, a patent pool for neglected diseases would provide a significant boost to researchers who have been working on treatments for diseases such as TB, malaria, and river blindness. In the near future, intelligent machines will accelerate biomedical research even further. Bradley's team already translates their "human readable" logs into a machine-readable format. "We really want to get to the point where machines can design experiments, can execute experiments, and can analyze them," he says. Soon it could become difficult to tell if you're interacting with a machine or a human. "We're not actually that far from that point," he says. "And I think that's when things will really accelerate." Now just imagine how long it will take to disseminate new findings when machines are talking to other machines rather than humans. "I think we're now talking milliseconds instead of hours," he says.

Synthetic Humans and the New Building Blocks of Life

From our conversations to date, we are convinced that we have only seen the tip of the iceberg when it comes to exploiting the new sciences of life. It turns out that scientific openness is expanding way beyond papers and even data to the very building blocks of life itself, all thanks to a burgeoning field called synthetic biology that applies engineering principles to building new biological systems ranging from food to fuel to medicine.

That's a pretty big idea, so here are the basics. Imagine that all of life's complexity could be boiled down to a library of interoperating components. The basic fundamentals of life, all indexed, open to the public, and waiting to be reassembled into new creations in the same fashion as one might twist and scramble a Rubik's cube. Call it the open-source library of life. Now picture an Olympic-size gymnasium full of lab students building new organisms and life forms. It's not a science fiction movie; it's for real. And every year since 2004, thousands of lab students from around the world gather at MIT for the International Genetically Engineered Machines Competition. Once there, they spend the entire summer building biological systems from standardized parts and then operating them in living cells. In 2009, a team from Cambridge University claimed honors, and the year before it was a team from Slovenia. After the competition, the students' organic fabrications are made available to the public free of charge via a registry hosted by MIT. Think of it as the open science equivalent of Amazon; just replace books with biological machines. Over eight thousand open-source "BioBricks" (a concept invented by MIT scientist Tom Knight) are already hosted there. There's even an iPhone app called BioBricks Studio Mobile, just in case you get the urge to doodle a new molecular creature while riding the subway.

To be sure, the prospect of engineering life on your iPhone raises all kinds of provocative questions and some intriguing possibilities. What if engineering a new, tastier form of spinach to suit your kid's unique palate was as easy as assembling a Lego kit? What if every household could produce a lifetime's worth of clean biofuel from a test tube housed in your utility closet? If creating synthetic humans turns out to be feasible, will society seize the possibility? Or will we recoil from genetic manipulation, for fear of unleashing something much worse than Frankenstein's monster? For some, radically reengineering life is the stuff of nightmares. For others, it's the key to a sustainable future on Earth. But before getting carried away, you might be asking whether any of this is remotely possible. After all, scientists have been engineering the genetic makeup of life for thirty to forty years. Aren't we still a very long way from exercising the kind of futuristic capabilities just described?

All that's about to change, insists Jason Kelly, cofounder of OpenWetWare and a new company called Ginkgo BioWorks that has been set up to capitalize on the field of synthetic biology. "A lot of what has been done to date is really just tinkering," he says. "You sort of get in there and you hack around. You make some changes that are really unpredictable. You don't know what's going to happen." Kelly argues that what genetic engineering really needs now is a set of standardized, interchangeable parts with standardized interfaces for assembly. A bit like Lego bricks, these components would be engineered so that the whole library of complementary BioBricks could fit seamlessly together in a more complex life form, perhaps something as complex as a human. If successful, it will pave the way for advances in medicine, agriculture, and energy we scarcely anticipate today. "Biology is a ridiculously powerful technology," says Kelly, "it's just we've been really horrible at engineering."

If this sounds a little like computer science 101, it's intentional. Kelly thinks software programming is precisely the right metaphor for this new form of biological engineering. "You need to have abstraction like you have in computer science where you build things out of submodules that act predictably. From there, you make larger systems and ever larger systems based on a whole set of interoperating components." The truth is that you no longer need to even step into a lab to engineer a new life form. "We're doing what many other engineering fields have long done: decoupling the design and the fabrication," says Kelly. Just as you don't have to be a blacksmith to design and fabricate a car frame, scientists can design life forms on their computer screens and have the real thing assembled in a lab halfway around the world. Kelly even envisions new business models that are reminiscent of those used in open source. In the same way that IBM and Red Hat build proprietary value on top of Linux, Ginkgo BioWorks and other companies will build new applications and services on top of MIT's Registry of Biological Parts. Now how's that for wiki science?

SCIENTIFIC PUBLISHING IN AN ERA OF MASS INVENTION

For scientists on the cutting edge, there remains one fundamental frustration in their day-to-day work. The new model of collaborative science doesn't gel with one of the research world's core institutions: the scientific paper and the broader publishing industry that has long maintained a central role (some might say dominant) in the world of scholarly communication. This model dates back to Gutenberg. It's a one-way stand-alone document that is expensive to produce and update, while the end product spends most of its time collecting dust on library shelves. The inertia is leading more and more scientists to emulate the open notebook methods of Jean-Claude Bradley and his UsefulChem network. If trends continue, the scientific publishing industry as we know it is in for a pretty big shakeup. Will it still have a role to play in mediating the flow of scientific information around a networked globe? Perhaps. But it's not a stretch to predict that within a decade or less all of the world's scientific data and research will at last be available to every single researcher—gratis—without prejudice or burden. Like the newspapers, publishers can either adapt to new models or risk irrelevance. Ultimately, it's a matter of when, and not if, open access becomes the dominant paradigm.

Unrealistic, you say? Not really, when you consider that conventional scientific publishing is both slow and expensive for users, and that these issues, in turn, are increasingly big problems in science. Visit any campus today and you'll hear ever-louder vocal cries for the old paradigm to be swept aside. Until quite recently, frustrated authors could find their cutting-edge discoveries less cutting-edge after a lumbering review process delayed final publication by up to a year, and in some cases longer.[11] Publishers have done much to speed up the publication process in response to increased competition on the Web and a growing trend toward self-publication. Even still, the traditional approach to publishing "papers" is arguably too rigid and slow to keep pace with today's high-octane world of science. An even bigger problem than speed is access, and the ever-increasing subscription fees that have made valuable research less accessible. The vast majority of published research today is still only available to paid subscribers, despite the availability of much cheaper electronic publishing methods. Though an unlimited number of additional readers could access digital copies of research at virtually no additional cost, publishers hold back for fear of creating a Napster-like phenomenon.[12]

As the scientific endeavor swells in scale and speed, a growing number of participants in the scientific ecosystem are questioning whether the antiquated journal system is adequate to satisfy their needs. New communication technologies render paper-based publishing obsolete. The traditional peer-reviewed journal system is already being augmented, if not superseded, by increasing amounts of peer-to-peer collaboration. Best of all, there is an alternative. It's called open access—and a growing number of scientists are signing up.

Why Open Access Makes Sense

Open access is to science what open source is to software. Rather than scientific insights and data being deposited behind a firewall, it's made freely available to whoever cares to make use of it. The poster child for open access is the Public Library of Science. Having started from scratch in 2000, it's now one of the world's most respected publishers of medical research, period. To get a better understanding of how it works, we sat down with Peter Binfield, the publisher of *PLoS ONE*, one of seven journals published under the PLoS brand. Binfield has fifteen years of experience in academic publishing and came to the Public Library of Science from a "very big corporate environment" where he ran multimillion-dollar programs, with two to three hundred journals under his supervision at a time. His reservations about the closed-access model are not driven by ideology, or distaste for traditional publishing, but from an inside view of how the subscription model is harmful to the way we disseminate science.

In most journals, an editorial staff makes decisions about which content is interesting enough for their audience. In other words, a small number of people filter everything at the front end. Editors typically assume that the only researchers who have interest in that content are already sitting in well-funded libraries, so their subscription fees are not much of a barrier. PLoS couldn't be more different. Rather than charge readers, authors pay $1,300 to have their articles published. These costs are usually built into research grants, and it's ultimately a small price to pay to make your research available to the world through a well-respected scientific publication. Unlike traditional journals, Binfield publishes everything that is scientifically sound. There's no need to prejudge audience or the eventual applications of the research. Readers can judge for themselves which research is worth paying attention to. And using the site's new rating system, they can participate in promoting the research they like.

Whereas most journals jealously guard their content, *PLoS ONE* content is freely available for anyone to repackage and reuse. Entrepreneurs could even sell it commercially, and make a profit, Binfield says. Arguably a whole new class of services could be built on top of PLoS to filter, annotate, and apply the growing body of public knowledge that scientists generate. In other words, journals like *PLoS ONE* could become platforms for innovation in the same way the iPhone is a platform for over 100,000 third-party apps.

In some cases, the ultimate applications for research may not even be known until some time in the future. "You never know what could turn out to be valuable down the line," says Binfield. "In years to come when there's better data discovery tools and better text mining, somebody or some machine somewhere will pull out the one data point or one insight in a paper that may have been disregarded when it was first published."

But the key point for Binfield is that open-access content is inherently more valuable and more useful than subscription content because it reaches a broader audience—the same audience that traditional science publishers assume is irrelevant. "There's an entire 99 percent of the rest of the world that might have an interest in that content that can never access it in a subscription model," he says. "But with a wiki model you've got a chance to get some useful insights out of those 99.9 percent of people that you couldn't have got otherwise."

From Open Access to the Singularity

The power of collaborative science will only grow as open-access publishing goes beyond journals and research papers to include growing access to science's raw materials: massive databases filled with information about everything from genomics to species extinction. After all, science and commerce depend upon the ability to observe, learn from, and test the work of others. Without effective access to data and source materials, the scientific enterprise becomes impossible. Open up access to all of the world's knowledge, on the other hand, and we will help deepen and broaden the progress of science, giving everyone from knowledge-thirsty students to innovation-hungry entrepreneurs the opportunity to tap new insights and contribute their own.

Just ask the scientists involved in OpenWetWare, an MIT project designed to share expertise, information, and ideas in biology. Twenty labs at different institutions around the world already use the wiki-based site to swap data, standardize research protocols, and even share materials and equipment. Researchers speculate that the site could provide a hub for experimenting with more dynamic ways to publish and evaluate scientific work. Labs plan to generate RSS feeds that stream results as they happen, and use wikis to collaboratively author/modify reports. Others have suggested adopting an Amazon-style reader review function that would make the peer review process quicker and more transparent.

As large-scale scientific collaborations become the norm, scientists will rely increasingly on distributed methods of collecting data, verifying discoveries, and testing hypotheses not only to speed things up, but to improve the veracity of scientific knowledge itself. This will impose a new level of transparency on scientists. But practitioners like Jean-Claude Bradley think more transparency is long overdue. "My view is that you're far better actually supplying all of the raw data, including the stuff that didn't work, so that people can evaluate for themselves exactly what kind of claims you can make against your data set," he says. The hope is that increasingly rapid, iterative, and open-access publishing can engage a much greater proportion of the scientific community in the peer-review process. Rather than rely on a handful of anonymous referees, a self-organized community of participants can vet results continuously. This, in turn, will allow new knowledge to flow more quickly into practical uses and enterprises.

Binfield even goes so far as to suggest that open access is underpinning the kind of exponential advances that will trigger the singularity, Ray Kurzweil's theory that human civilization will soon be superseded by advanced forms of artificial intelligence and a new race of cybernetic beings. "Open access will drive exponential leaps in scientific understanding," he says.

Time will tell whether Kurzweil's singularity plays out as he predicts. But one thing seems certain. When fully assembled, open-access libraries will provide unparalleled access to humanity's stock of knowledge. New science results that might have been available only to deep-pocketed subscribers will now be widely and freely available for education and research. Older resources that might otherwise have wallowed in dusty archives will be given new life and new audiences in digitized formats. Perhaps at long last we will fulfill the true promise of the Great Alexandrian library, but with a twenty-first-century twist: assembling the sum of mankind's knowledge on the Web and sharing it for the betterment of science, the arts, wealth, and the economy.

10. COLLABORATIVE HEALTH CARE

In 1997, Stephen Heywood, a twenty-eight-year-old custom home builder from Palo Alto, California, noticed that he couldn't turn a key with his right hand. A year later, he was diagnosed with Lou Gehrig's disease (also called amyotrophic lateral sclerosis, or ALS). It's a neurodegenerative disease that paralyzes and eventually kills its victim. As soon as the diagnosis was made, Stephen's brothers, James and Ben, both MIT engineers, struggled to learn as much as they could about the disease. But they were stunned by the paucity of information available to them, even as members of the academic community with broad access to information resources. With thirty thousand cases in the United States, ALS is not as common as the better-known multiple sclerosis, but it's just as debilitating. Yet ALS patients had no efficient way to share information about their disease with one another, or with their doctors, so both diagnosis and treatment were delayed. What's more, it was extraordinarily hard to find patients just like Stephen, who by his thirty-fifth birthday was breathing on a ventilator and confined to a wheelchair.[1]

What these patients needed, the Heywood brothers thought, was good data on their condition and treatment options. In particular, they needed answers to two questions: Given my status, what's the best outcome I could hope for? And how do I achieve it? Doctors couldn't help, because they saw so few patients with these afflictions. Yet if they could gather people with rare conditions together, they'd form a community, and if they aggregated information from that community, they could provide meaningful clinical data to patients and their doctors. This would help both doctors and patients make more informed treatment decisions.

Stephen died in 2006 when his ventilator disconnected in the middle of the night. Yet the Heywood brothers were determined to improve the lives of thousands of people with ALS—by forming a new online community. Two years before Stephen's death, in 2004, with the help of longtime friend Jeff Cole they launched PatientsLikeMe, a Web-based community of patients with rare or life-altering disease states such as ALS, progressive supranuclear palsy, corticobasal degeneration.[2]

PatientsLikeMe took off and has become one of the Web's most vibrant health care communities. Its members—sixty thousand and growing—share personal details of their medical history with fellow members. The data they contribute is aggregated to track patterns and responses to various reported treatments. "People think we are a social networking site," says cofounder Ben Heywood. "But we're an open medical framework. This is a large-scale research project."

Whereas most health care sites fervently guard their patients' data, the Heywoods believe sharing health care experiences and outcomes is good, and perhaps even integral, to speeding up the pace of research and fixing a broken health care system. Why? Because when patients share real-world data, collaboration on a global scale becomes possible. The health care system becomes more open and this in turn improves outcomes for patients, doctors, and drug makers. New treatments can be evaluated and brought to market more quickly. Patients can learn about what's working for other patients like them and, in consultation with their doctors, make adjustments to their own treatment plans.

An example of this is David Knowles, a fifty-nine-year-old property manager who lives in the U.S. Virgin Islands, who has struggled with multiple sclerosis (MS) for ten years. Knowles was searching online for information about a new treatment for MS when he came across PatientsLikeMe. "There's no other site with data like PLM has," says Knowles. "You can click on a symptom and say, 'Well, there's 850 people with this symptom, and this is what they're using to treat it.'"[3] Knowles wanted to know more about a drug called Tysabri and its possible side effects, which included brain infection, anxiety, and fatigue. One of his doctors was suggesting the drug, and PatientsLikeMe offered data from hundreds of patients taking Tysabri. After reviewing their results, Knowles decided that in his case the risks outweighed the rewards, and he went to his doctor with a list of other treatments he wanted to explore instead. "I feel like I'm in charge of my medical care now," he says. "Of course, I still listen to my neurologists, but now it's more of a team approach."[4]

THE OLD MODEL: PATIENTS ARE PASSIVE RECIPIENTS OF HEALTH CARE

Knowles represents the future of medicine, with engaged patients becoming prosumers of wellness rather than passive consumers. This is a dramatic change from current practice. For centuries the medical industry operated under the following proposition: doctors are smart because they have education and hands-on experience, and patients are medically dumb because they have no relevant knowledge. Doctors would wait in their office or hospital for sick people to come to them, and doctors would treat them and tell them what to do, one-on-one, face-to-face. If patients didn't like what their doctor told them, they could shop around for a second or third opinion (if they could afford it) by getting appointments with additional doctors. But these subsequent doctors would still subscribe to the same basic model of health care: Patients are passive recipients of medical care with little or no role to play in deciding their own treatment plans should they get sick. Patients are isolated from one another and rarely communicate or share knowledge. Health care occurs primarily when you enter the health care system. For many years, this was the only model possible.

But this system has grown increasingly expensive and ineffective, particularly in the United States. In 2007, Americans paid $7,290 per capita for health care, 87 percent more than say, Canada.[5] The factors that contribute to inefficiency and rising costs are complex and subject to ongoing debate. They include, among other things, the fact that preventable chronic diseases are rampant, health insurance is typically irrational and subject to abuses, prices are opaque, patients are uninformed consumers, and up to a third of health care spending is wasted.[6] To top it off, in America's decentralized system, these spiraling health care costs are seen as nobody's fault, and nobody's responsibility.

These issues are further compounded by the fact that, in America at least, an individual's encounter with health care can be financially devastating. According to a study by Harvard researchers, medical problems caused 62 percent of all personal bankruptcies filed in the United States in 2007. Even more surprising was that 78 percent of those filers had medical insurance at the start of their illness, including 60.3 percent who had private coverage, not Medicare or Medicaid. Medically bankrupt families with private insurance reported average out-of-pocket medical bills totaling $17,749, while the uninsured's bills averaged $26,971.[7] To make matters worse, the vast majority of these health care dollars (approximately 77 percent) were spent during the last year of a patient's life as they tried to prevent the inevitable.[8]

These high costs and inefficiencies, in turn, translate into poor outcomes for patients. Despite the United States being the world's largest spender on health care,[9] 35 countries have longer life expectancies[10] and 179 countries have lower infant-mortality rates.[11] The health care performance of the

United States, a global superpower, ranks 37 out of 191 countries.[12] Unbelievably, the health care system itself is the third largest killer of people in the United States after heart disease and cancer.[13] According to research studies, a total of 225,000 Americans per year die as a result of their medical treatments.[14] This comprises 12,000 deaths due to unnecessary surgery; 7,000 deaths due to medication errors in hospitals; 20,000 deaths due to other errors in hospitals; 80,000 deaths due to infections in hospitals; and 106,000 deaths due to adverse side effects from the drugs patients have been prescribed.

And these are just the deaths resulting from the current system. There are hundreds of thousands more Americans whose ailments or disabilities are actually made worse by the system rather than being helped. This includes patients who are misdiagnosed, and those who are diagnosed correctly but given inappropriate treatment.

We also know that loneliness and isolation can be a medical risk factor. Lonely people get sicker than the population as a whole.[15] They suffer from a wide variety of ailments, ranging from colds to heart attacks. Lonely people with HIV respond less well to antiretroviral drugs. People who are lonely in their old age are twice as likely to develop Alzheimer's as other seniors who are socially active. Socially isolated women have a greater risk of dying once they have been diagnosed with breast cancer.[16] Doctors often encourage patients to join a support group for their health problems, and many do. A support group can do wonders for a patient by providing advice and friendship. But often there are too few people in a community or city suffering from the same ailment to sustain a group. Of course, the online health care communities can't replace the need for face-to-face contact and support, but they can increase one's likelihood of connecting with people with similar conditions.

Possibly the greatest failure of the current health care system is that it clearly doesn't engage a large part of the population. And when we don't think about our health we get unhealthy. Close to two thirds (63.1 percent) of adult Americans are becoming overweight or obese, exercising less, and eating unhealthy foods.[17] Compared with healthy-weight people, overweight and obese people have particularly unhealthy lifestyles—lifestyles that contribute to the skyrocketing rates of preventable diseases like diabetes and heart conditions, which are among the most costly public health afflictions.[18] A population truly engaged in the issue of wellness would not act so recklessly with respect to its own well-being.

Moreover, the model where the patient is an isolated recipient of health care is not designed to capture knowledge. A doctor successfully dealing with a patient one-on-one in a small treatment room has no way to record the process of diagnosis and treatment in a manner that would be educational to others. But this is something the system should be capable of doing. Rather than handwritten data housed silently in filing cabinets, the data should be captured in a way that could tell us quickly if a treatment is exceptionally beneficial or, conversely, is doing harm. Such information could be used to teach new doctors, or enable researchers to investigate new approaches to medicine. Moreover, if patients were collaborating among themselves, sharing experiences and learning from one another on a mass scale, this would create an almost infinitely large database that could feed science and the advancement of medicine.

THE ADVENT OF COLLABORATIVE HEALTH CARE

Since the mid-1990s, the Internet has housed a large and ever-expanding volume of information on the universe of medical experiences, conditions, and treatments. Most of this information was generalized, and didn't readily convert into practical information for a typical patient. And some of it was quackery. The notion of patients flocking to Dr. Google before they went to their own doctor drew the ire of the medical community. As recently as 2001, the American Medical Association told the public to "trust your physician, not a chat room," stating that the information found online puts "lives at risk."[19]

But Web 2.0 puts the informed patient into a new context. It enables a new model of medicine we call "collaborative health care," which would give us a system that is cheaper, safer, and better than what we have today. A key component of collaborative health care is user-generated medical content, just as new user-generated content is becoming an essential component in fields such as journalism, learning, science, and the media. In the health care industry, Web 2.0 enables people to self-organize, contribute to the total sum of knowledge, share information, support each other, and become active in managing their own health. Much like the students we discussed in chapter 8—students who collaborate and learn from online material with teachers providing guidance—citizens will be active in researching their own health, sharing their experience with others, and consulting with health care professionals when necessary. Because patients are engaged, they are managing their own health more effectively, reducing costs and improving outcomes. This is already happening on a small scale, and as you will see in this chapter, we're only scratching the surface of the possibilities.

Collaborative health care goes beyond the current catchphrase of health care being "patientcentric." Not only is collaborative health care focused on the patient; the patient co-creates health care and wellness, producing an outcome that is more evidence-based and cost-effective. Though pockets of resistance to this idea are to be expected, interviews we conducted suggest that a growing number of doctors and health care professionals agree that collaborative health is the way forward. "Public voice gives explicit power to patients," says Paul Hodgkin, a Sheffield-based GP and founder of Patient Opinion, an independent site that gathers patient feedback for the UK's National Health Service (NHS). "And becoming a real co-creator of change in the system empowers people in their struggle with illness," he adds. "Wherever patients are struggling with the helplessness, lack of control, and fear associated with illness, they can find tools that help them to feel more informed, included, and valued."

We should note that the vision of collaborative health care we outline transcends the contentious health care legislation passed by the U.S. Congress earlier this year. The legislation that was passed by the Democrats, and the alternatives put forward by the Republicans, embraced the old model of medicine and did nothing to rebalance the doctor-patient relationship or encourage patients to claim a much more active role in defining and managing their health care needs. Collaborative health care also transcends different models of health care provision, pertaining as much to the public health systems as it does to the private system preferred in the United States.

In short, collaborative health care is an idea whose time has come, and in our view, all key participants in health care systems globally should embrace its four key elements. First, all people involved in the system of health care, including the patients, should use the Web as a platform to share information, deliver care, and build communities such as PatientsLikeMe. This would make patients more engaged in managing their own health, and therefore more committed to being healthier. Patients could also self-organize and find others that share their medical interests and goals.

Second, when born, everyone should get a Web site: their Personal Health Page (PHP). The patient would own and control his or her own data, but health care professionals would access it as required with appropriate levels of privacy and security. This PHP would be the patient's personal window into his or her own health. It would also provide the basis for one's participation in a broader health social network.

Third, extensive digital collaboration among all stakeholders would generate vast amounts of data that should be mined for new medical insights and become part of the knowledge base for science, health, and medicine. Rather than relying on past practices or following a doctor's intuition (a potential source of mistakes when confronting novel situations), collaborative medicine would bring the best scientific knowledge into the health care process for each individual patient. Health care professionals would build on this foundation of information to collaborate, learn, and teach.

Finally, physicians and other medical personnel should become more engaged online and work toward achieving a new balance between professionals and patients. Collaborative health care would not undermine the health care professional's role but enhance it with the support of smart Web-based applications and self-monitoring tools for patients.

This vision may sound far-reaching, but it's not a utopian dream. The contours of the new model are emerging, and in many countries, collaborative health care is already having a material impact on human health and wellness. To see how, let's take a closer look at each key aspect of collaborative health care in turn.

1. Collaboration and Community

The PatientsLikeMe Web site that was developed in response to Stephen Heywood's ailment is a great example of the new collaborative health communities that are flourishing on the Web. The community has expanded from rare diseases to include many "prevalent disease communities"—everything from HIV/AIDS to depression. Patients participate in the site at no charge. The company running the community recoups its costs by rendering members' data anonymous and selling it to third parties such as pharmaceutical companies, insurance companies, and financial services firms.

The value of the community lies not only in the support patients receive, but also in the shared information. Everyone who joins has to share their medical stories—what drugs they're on, what condition they're in—and their status is tracked over time. They can use a username to disguise their identity, but many people don't feel the need to hide. Up to one fifth of all users opt to go public, adding pictures and other details to their personal profiles. In the ALS community, an astounding 70 percent go public.

In return for sharing details about their health, they can see where they fit into the picture, which helps them answer the question: What outcome can I achieve that's normal for someone like me? "Just knowing puts you more in control," says cofounder Ben Heywood. "We're helping them track and share in the community, so that everyone can learn." The user-generated information on the site—packaged into convenient charts and graphs—is improving lives. "We had an MS patient join about three months ago," Heywood told us. "He's taking Baclofen for his leg spasticity, and he's been taking 10 mg for 10 years. So 10 years ago his doctor said you can't take it at a higher dose, it's going to mess up your system; it's too toxic." When the MS patient clicked onto Baclofen on the PatientsLikeMe site, he was in for a surprise. Patients like him were currently taking much higher doses. "So here he now gets to go back to his doctor and open up that dialogue again," Ben said. "And this is a problem he had for 10 years, and it's only because he didn't get the right information."[20] Because of PatientsLikeMe, the patient was able to contribute information to the decision-making process and achieve a different outcome.

Some progressive health care providers are starting to use online health communities to encourage healthier lifestyle choices, such as improved diet. These tools are convenient, low cost, and able to use audio and video to increase patient engagement and appeal to a range of learning styles.[21] The content can be tailored to an individual's needs, preferences, and psychological characteristics. The use of tailored programs to change behavior is generally more effective than using those with untailored content, in part because they are perceived as more personally relevant and salient.[22] Research also shows that, as a general rule, the more time a patient spends in a community site, the more committed they are to the program and the more their treatment outcomes will improve.[23] This should be no surprise; it's simply human nature. People like to be part of a good thing. The popular Nike+ program connected Nike sneakers to an iPod and let runners record the details of their jogging. But in an inspired move, Nike let runners upload their data to a Nike community so they could compete with one another. Once new runners had uploaded their data five times or so, their commitment to the sport soared.

Online communities can offer many advantages. The Center for Connected Health conducted an online survey of 260 individuals recruited from five online psoriasis support groups. Nearly half of all study participants reported improvements in their quality of life (49.5 percent) and psoriasis severity (41 percent) since joining the site. The key reasons the patients had joined the community included availability of resources (95 percent), convenience (94 percent), and the lack of embarrassment when dealing with personal issues (91 percent).[24]

"Increased access and use of the Internet is allowing patients to gather information and interact in new and beneficial ways, and is helping to empower people to improve their health and quality of life," says Joseph C.

Kvedar, MD, director of the Center for Connected Health and associate professor of dermatology at Harvard Medical School. "This is the first such study to evaluate online support groups in the field of dermatology, and offers rich insight into the potential benefits and usage of virtual patient communities."[25]

Another online health community called WeAre.Us was founded in 2007 to create micro social networks for patients with rare diseases—such as WeAreLupus.org (lupus); WeAreEndo.org (endometriosis); WeAreFibro.org (fibromyalgia), WeAreHD.org (Huntington's disease); and WeAreCrohns .org (Crohn's disease). Each site gathers together patients, families, and other related stakeholders and facilitates a conversation, which in turn creates a foundation of peer-to-peer support. Patients love it. "I no longer feel alone," said Becky, a WeAreCrohns.org member. "It is comforting to be able to talk to people who are exactly like me," said Wendy about WeAreEndo.org.[26]

WeAre.Us is more than just an online support group. The site uses data mining and collective intelligence to help advise and inform patients about which actions (medical and lifestyle) may have the most impact on their condition. WeAre.Us collects and aggregates daily tracking data from individuals and then shares recommendations based on group information. "WeAre.Us uses social media and proprietary technologies to connect individuals who were once disconnected, but share a commonality—and the result is community where before there was none," said Dr. Declan Doogan, MD, president of WeAre.Us. "This is the start of a larger movement that will bring disparate patient communities together." We hope so.

Ad hoc medical communities can be terrific in a crisis. The 2009 H1N1 flu pandemic generated many helpful communities. FluWikie.com, for example, is an up-to-the-minute and complete resource on everything to do with a potential flu pandemic, and it offers comprehensive information for people around the world. It is the kind of resource that would be difficult for, say, a government agency to put together on its own.

Information, experiences, and opinions generated by users can be combined with the power of data aggregation to produce data that is often of clinical significance for both health care providers and patients. Connecting patients to patient communities and physicians to physician communities allows for a more powerful analysis of data and treatment innovation because exchanges are happening in real time and are unhindered by the constraints of clinical research trials, journal article publication, and medical conferences—all of which take time and resources to produce.

Using a completely different strategy, many patients are taking collective action to change the behavior of the health care system and its players.

RateMDs.com gathers feedback on physicians from their patients, and has rated 160,000 doctors so far, with over 400 new ratings being added daily.[27] Comments are anonymous, which means the information generated can be suspect.[28] MDJunction.com is aiming higher. In addition to giving you the chance to rate a doctor, it allows you to search for a doctor by specialty, and to join a variety of discussion forums and support groups. Doctors can add their own profiles to the site. Needless to say, many physicians don't welcome this new public report card. Some sites have been threatened with libel action. Nevertheless, some national health care services have decided to follow suit. For example, the UK's NHS recently set up a patient feedback site called NHS Choices where patients can rate different hospitals and provide feedback on their experiences. Subsequent visitors benefit from the wealth of patient knowledge as it accumulates and the NHS gains the ability to easily compare ratings across NHS facilities and thus pinpoint weaknesses in the system.

2. Everyone Gets a Personal Health Page

Health care in America—and most of the world—suffers from a reluctance to engage in fresh thinking and new delivery platforms based on new technologies. Yes, we see new medical technologies used in laboratories to cure and treat diseases. Yes, we see new technologies changing business and social interactions. What we don't see is the crossover of these two worlds: new technologies to change the business and social interactions of the medical world. We don't see the appropriate uptake of the benefits of Web 2.0 innovations by the medical establishment.

Imagine a scenario like this: At birth you receive a URL for your Personal Health Page. Unlike today, you would have full access to your own health record and transparency in everything to do with your care. Incredibly, many jurisdictions and doctors insist that health records are the property of the doctor, not the patient. This would change. The dismal reality is that today, less than 1 percent of patients have online access to their own medical records.[29] This contrasts sharply with the near universal availability of online personal financial records from banks, brokers, and information management firms.

Your Personal Health Page would empower you with an online record of your health-related information filed by the different places you've received health care services, such as clinics, hospitals, and diagnostic labs.[30] Obviously, your medical information would remain secure and confidential. You would determine how and when other people, such as medical personnel, pharmacists, and insurers, could access the information. For example, you might allow researchers to access your information only after it has been stripped of any personally identifying data.

The Personal Health Page would be the new foundation of collaborative health care. It would be the window into wellness for everyone, providing them with access to their own information, the world of medical knowledge, and other people who care about their health, such as spouses or siblings. Your PHP would include more than just your medical facts. It would supply you with useful information to help put the data into context and give meaning to the numbers. You would use your PHP to subscribe to a number of medical information services to keep you up-to-date on health and medical advancements in areas of interest to you. It would encourage you to get engaged with your own health care.

The Personal Health Page would also transform the collaboration between practitioners and patients—providing an opportunity to fully engage patients during all phases of health care by making the prevention and treatment process transparent. Having instant access and instant feedback through their PHP would enhance patients' sense of control over their health, and provide them with a better way to track nonacute concerns that can be otherwise ignored or overlooked. Caregivers could also be provided with complete information to be able to track and manage their patient's progress.

As Dr. Michael Evans explained to us: "People are not so personally engaged in their health, and health, at the end of the day, is about behavior change. When we look at chronic diseases, it's the same sort of four or five behaviors that predict all the chronic diseases with some genetic overlay. And when we look at why people die early, you know, 40 percent of the pie is behavior change. And so it's one hundred decisions we make a day. Giving patients generic information has little impact. We need personalization. Just as we see mass customization in other industries, that's what needs to happen in medicine."

Part of this personalization would involve the Personal Health Page also being your home page for a Social Health Network—much like a Facebook for health. Your personal information would be downloaded directly from Facebook or other social networks. More than just a private medical record, your Personal Health Page would help you reach out to other organizations such as Weight Watchers or a local health club that could contribute to your well-being. Imagine that you log on to your PHP and the first thing is a news feed giving you updates from others in your community who share a common illness, or from your medical professionals with the latest information about a new medical discovery. You are able to create communities or join medical "causes" much like with a generic social network. There are myriad applications that you could use to measure your health, do a prediagnosis of a sick child, or test for possible drug interactions. Scheduling tools help you plan a visit to your family doctor, and your doctor can request information from you about how a certain treatment is working. The possibilities are almost limitless. Giving patients access to personal health records gives them the opportunity to identify and update missing or incorrect information.[31] With its ability to aggregate patient medical information across providers, the Personal Health Page could also help provide the holistic view of a patient's medical history that would help reduce health care errors.[32] No wonder the National Committee on Vital and Health Statistics and the Institute of Medicine have identified the personal health record as a key dimension of a national health information infrastructure. National health information technology coordinator David Brailer says, "We want to see every person have access to a personal health record and be able to communicate with their clinician using it."[33]

3. Collaboration Generates Massive Data for Medicine

Scientists such as Larry Smarr at the California Institute of Technology have argued convincingly that we will never have a true science of medicine until we have much more granular data about the health of large populations over long periods of time. Researchers would not need to know your name or your home address, says Smarr, "but I would very much like to know what your glucose level is, what's your lipid profile, and maybe even things about your lifestyle: are you obese, or are you a smoker?" Once in possession of this level of medical data about hundreds of millions of individuals, researchers could begin to reliably inform the parts of medical practice that are subject to scientific methods. For example, evidence-based medicine could be used to choose the methods to ensure the best prediction of treatment outcomes. As more and more data is created, treatment decisions would become more personalized and evidence based—that is, based on data pertaining to individual patients rather than a larger group (or, more likely and probably worse, just based on the physician's typical way of doing things). "In addition, data collected over the continuum will be analyzed, and from that analysis, care processes as well as drug and procedure discovery will become more efficient and effective," says Matthew Holt of consulting company Health 2.0.[34]

For some, the promise of evidence-based medicine is still a distant reality. But thanks to new medical technologies and new communities such as PatientsLikeMe, the raw materials for a new science of medicine are quickly accumulating. "We're focusing on collecting real-world patient-recorded outcome measures over time," says Ben Heywood. He calls it "patient informatics." The fact that the community is tracking far more people with specific rare diseases than the number of cases typically featured in a scientific study makes online communities an attractive asset for researchers seeking subjects for new clinical trials that test new treatments. In fact, PatientsLikeMe has already provided reality checks for a number of clinical studies. For example, when a recent study suggested that lithium could slow down the progress of ALS, PatientsLikeMe checked out what its own patients were saying about the drug. The clinical study was based on observations of a handful of patients. But by then, some four hundred ALS patients in the community were taking lithium. Their verdict: lithium didn't work.

Some in the medical establishment remain skeptical. At a recent Health 2.0 conference, for example, a Pfizer representative said that research of the sort done by PatientsLikeMe was of little value to the company because it didn't meet the strict standards for clinical testing as set out by the U.S. Food and Drug Administration. That may be true in the short term, but the longer-term trend is clear. The data collected through online health communities, with patient permission, will be an important driver of evidence-based medicine. What makes PatientsLikeMe effective is the way it aggregates the data from its members, both for the benefit of individual members and for scientific research. "We're a very unique thing. We're thirty people working very hard to change the lives of the fifty thousand people we have now," says Ben. "We have very high expectations, but it's a long journey."

In addition to PatientsLikeMe, other significant players like Google are lining up to develop valuable new services using medical data. Google is, of course, the most powerful aggregator of information on the planet. Each week, millions of users around the world search for health information online, and not surprisingly there are more flu-related searches during flu season, more allergy-related searches during allergy season, and more sunburn-related searches during the summer. Google aggregates this information and presents it in real time on the Google Web site. It turns out that Google's user-based intelligence can predict flu before the doctors can. "We have found a close relationship between how many people search for flu-related topics and how many people actually have flu symptoms," Google reports. "Of course, not every person who searches for 'flu' is actually sick, but a pattern emerges when all the flu-related search queries are added together. We compared our query counts with traditional flu surveillance systems and found that many search queries tend to be popular exactly when flu season is happening. By counting how often we see these search queries, we can estimate how much flu is circulating in different countries and regions around the world."

The Google numbers show that over the years, the level of searches for flu tends to predict flu activity before medical practitioners do.[35] This is a form of de facto mass collaboration as people, through their behavior, provide information that can be aggregated and made helpful to all.

Part of the new collaborative health care is the exploitation of an array of data-generating Web-enabled devices now coming to market. Users will wear or plug into small devices that capture exact information about a wide range of data points such as weight, blood sugar levels, blood pressure, or sleep patterns. If patients so choose, this information can be shared on an ongoing basis via the Internet with their doctors or caregivers. This information continuum is much more useful than the readings taken during a visit to the doctor every one or two years. Studies show that constant attention to key indices can help motivate people to change their behavior. People who weigh themselves daily are more successful at weight loss and maintenance than those who weigh in weekly.[36] People on the Weight Watchers diet who attend meetings and use digital tools, such as an iPhone app, to follow their points are 50 percent more successful in reaching their weight loss goals than those who don't.[37] These new technologies are having real-world effects.

The current system of record maintenance is costly and inefficient. A 2003 article in the *New England Journal of Medicine* studied the impact of health care administration costs on physicians in the United States. Physicians spent an average of 13.5 percent of their time on administrative tasks, valued at $15.5 billion. Other overhead costs (clinical and clerical staff, rent, office expenses, accounting and legal fees) came to 28.4 percent of physicians' gross income, approximately $57.1 billion.[38]

Some of this is unavoidable. In the United States' largely private insurance system (a system of multiple providers offering multiple insurances), the multiplicity of payers causes, by definition, duplications and high overhead costs. Yet the health care industry has done little to modernize the management and transmission of data. It's inefficient in part because health care providers and patients lack an efficient data management system. It is estimated that less than 25 percent of American doctors have an electronic record-keeping system.[39] Despite the need for intense, frequent, and current information exchange, the health care industry is one of the least automated. In fact, the average company outside the health industry spends seven times as much as U.S. health care companies on information technology, and companies in some wealthier industries like banking spend up to twenty times as much.[40] If we are going to get to a new science of medicine soon, this reality will need to change.

4. Engagement of Medical Professionals

Collaborative health makes the health care system more productive. If doctors had a wealth of established medical information at their fingertips in a collaborative health care system, it could also help them make better diagnoses and prescribe better treatments. They would make more decisions based on hard evidence about the individual patient, instead of relying solely on experience.

Just as we saw in the chapter on collaborative education, with teachers working with empowered students, the Internet-enabled patient doesn't pose a challenge to the credentials or utility of the health care professional. Medicine is an increasingly challenging occupation. Medical knowledge is growing at an exponential rate, and doctors must scramble to keep pace. If the patient's visit to a doctor seems isolating to the patient, think of it from the doctor's point of view. As family physicians, they process patients through their sausage factory practice as quickly as they can. They can't afford to do otherwise. They want nothing more than to engage patients in a sustained dialogue that motivates them to pay more attention to their health. But the tools for such a patient-doctor dialogue are just now coming to fruition and have not yet been widely adopted by doctors.

Consequently, the doctors' effectiveness is not as strong as it could be. They're stymied by their inability to collaborate with patients, and their opportunity to collaborate with their peers is limited to the occasional conference (although that is finally changing as a result of communities like Sermo, described below). Doctors need the wikinomics tonic just as much as the patients. They want to enable their patients to strive for a healthier lifestyle. Instead of a patient showing up intermittently with a sheepish attitude as to why they didn't follow their doctor's advice, collaborative health care makes for a more frequent and productive dialogue. The relationship shifts from reactive to proactive. And doctors welcome the opportunity to brainstorm with their colleagues.

Consider Dr. Jay Parkinson, who took a long look at the state of primary care in the United States after completing residencies in pediatrics and preventive medicine, plus a master's in public health from Johns Hopkins. He didn't like what he saw. The system, built on the model of a high-volume, eight-minute-appointment, sixty-hour workweek, actually discouraged young doctors from entering primary care. The result was a shortage of primary-care physicians in the United States, which only made the life of the physician worse.[41] "This is a $2.4 trillion industry run on handwritten notes," Parkinson lamented. "We're using 3,000-year-old tools to deliver healthcare in the richest country on the planet."[42] At thirty-three, Jay Parkinson was faced with $240,000 in medical school debt, but he knew he could leverage online social networking tools to give patients better and more personalized care. So he launched a virtual practice called Hello Health. Patients book appointments by viewing Parkinson's Google Calendar. Parkinson gets an alert on his iPhone and receives payment via PayPal, not insurance. He even makes house calls. Three hundred patients signed up in the first three months.[43] Hello Health gives patients access to videos of virtual visits with Parkinson and links to test results. After visits, patients can rate their experience, add comments, and share them on Hello Health's social network. And when patients need a prescription, Hello Health sends patients a text message revealing which pharmacy in their neighborhood offers the lowest price for their medication.[44] By making the information easier to find, sites like Hello Health will not only change how people interact with their doctors, but also raise their expectations. "I think if you're a physician in the future and you aren't communicating online, and you don't have a Web site, you're going to be an anachronism," says Unity Stoakes, cofounder of OrganizedWisdom.

While Hello Health suggests a new model for doctor-patient interaction, Sermo is a physician-based online community that allows doctors to exchange information and clinical experience with peers. Founded in 2006 by a young surgeon, Daniel Palestrant, Sermo provides an online social nexus to augment and sometimes replace the conversations that used to take place at the bedside or at the nurse's station or in the doctors' lounges, and even on the golf courses that doctors frequent. As in many fields, and like so many parts of society, medicine has become more and more time-pressured, says Palestrant. The result is that doctors have become more isolated and lack the time and tools to share information and experiences with one another. By providing a platform for these interactions, Sermo not only fulfills a social need, it also helps doctors improve their practice. And with over 15 percent of all physicians in the United States now using Sermo, it is clear that doctors of all descriptions are finding this valuable.[45]

In the online environment, younger physicians have equal footing with veteran physicians and credentials in this community are measured by contributions and opinions rather than years of experience, shattering some of the barriers that come with traditional models of seniority and reputation in the medical community. The doctors elect to participate and choose how much information about themselves they want to share, understanding that this information is shared with health care companies as a means of improving service and products. Part of the popularity of Sermo is that they allow the community to self-govern and do not step in during disputes. With a loyal community, Sermo has found that issues are resolved respectfully and within the community.

The ultimate promise of communities like Sermo, however, is that near instantaneous access to a large number of experts in their peer community will help improve patient care. "Doctors can log in with a clinical picture or a question and then within a few hours have dozens of physicians weighing in on what they think is the right answer," says Palestrant. Indeed, Palestrant told us the story of a physician who was asked to examine a mole growing on the arm of one of his guests at a recent dinner party. It was not his specialty, but the physician whipped out an iPhone and posted a picture of the mole on Sermo. By the time dessert was served, a dozen dermatologists had replied with a diagnosis! Talk about collaborative health care in action.

ARE CITIZENS READY FOR COLLABORATIVE HEALTH CARE?

In the era of Twitter, Facebook, and LinkedIn, it's clear that most of us like to connect with others, share our thoughts, and brainstorm. But are ordinary citizens truly equipped to take on the larger role in managing their own health that we are prescribing here? Paul Hodgkin offers a cautionary note. "All health care systems," he argues, "are bedeviled by the problem that sick people have historically made poor shoppers. Patients are consistently disadvantaged by having less knowledge, less power, and more vulnerability than other players in the health care system."

These asymmetries of information and power, says Hodgkin, are one of the main reasons why health care systems around the world have been slow to change, despite increasing evidence and agreement that the conventional approach is broken. "The system has been stuck," he says, "because the costs of reversing the fundamental asymmetries of information, power, and vulnerability have been too high to change. The direct result on patients is that their views are consistently undervalued and, combined with the dependence inherent in illness, they are systematically, if unintentionally, disempowered." Thanks to the innovations described in this chapter, however, these asymmetries are lessening, although not disappearing completely, as researchers such as James Robinson and Paul Ginsburg point out.

"One must acknowledge that consumers often need support if their choices are to promote their well-being and constraint when they are spending other people's money," they write in the journal *Health Affairs*.[46] "Healthcare is complex at best and not infrequently rife with nontransparent, anticompetitive, and even fraudulent behavior on the part of the many self-interested agents." To address these imbalances, they argue that consumers need professionals to create meaningful products and processes from which they can choose—bundles of products and services that can be measured, priced, purchased, and used not only by the highly educated and motivated individual, but by those who are sick and scared, of only modest means and financial sophistication.

The upside, according to Robinson and Ginsburg, is that informed consumers faced with appropriate incentives make better choices for their own health than third parties like HMOs and insurance companies. But with that observation comes an appropriate qualification. "For choice to be meaningful," they conclude, "it has to be choice among meaningful options, and meaningful options need to be designed, built, and managed."

Collaborative communities and personal health records provide the ideal environment for this type of dialogue between patients and professionals. Indeed, they can not only make patients more informed, but also help speed up diagnosis and help individuals find the best treatment, argues John de Souza, the president and CEO of MedHelp, another popular online health care community. "When you look at people going through diagnostic odysseys of not knowing what they have, and trying to find out . . . going through multiple treatments, a lot of that could be short-circuited by the ability to reach an expert who knows what you have very quickly," he says. This is critical to controlling costs too. As de Souza points out, many people live in places where they don't have immediate access to experts who are knowledgeable about their conditions. "They get shuffled around the system," he says, and that inflates the cost of treating a patient. But in communities like MedHelp, patients can get access to top doctors from the top hospitals, including a wide variety of specialists. They can also use Web-based applications and self-monitoring tools to track the progression of their medical conditions.

MedHelp members, for example, are able to track over 1,500 symptoms and treatments on a daily basis.[47] The ability to document and share this information with their doctors has led to better communication and more active patient engagement, according to de Souza. MedHelp trackers are available on the Web site and on mobile phones through Web-based browsers and iPhone apps covering both general health conditions, such as weight loss and allergies, and very specific disorders, such as infertility and diabetes.

Broadening these practices could dramatically curtail the cost of treating chronic conditions by keeping the chronically ill out of the hospital. Indeed, several pilot studies aimed at reducing the cost of chronic care confirm that self-monitoring technology does help patients better manage their illnesses, which in turn decreases emergency department trips, unnecessary doctor's office appointments, and costly home nurse visits. Marco Smit of Health 2.0 LLC gives a simple example of the type of savings that could be achieved. Older patients are often on a variety of medications, and on their own will tend to go off their drug prescriptions after a couple of refills. They worry about the cost or start forgetting. The result: they become ill and end up back in the doctor's office or rushed to the hospital. "If we could engage these patients and keep them on their prescribed regimen," says Smit, "the health care system could save $290 billion annually."

While communities like MedHelp are growing, online versions of personal health records and self-monitoring tools are still too rare. While 40 percent of Americans claim they keep a personal health record, only 2 percent keep it online.[48] Back in the dot-com era, more than 120 online personal health record tools hit the market, but only 16 remain. The true value of personal health records wasn't widely understood or communicated, with Americans worried about their personal information being compromised or used against them by employers or insurers. The large number of competing proposals also made it challenging to select and maintain one personal health record tool. Compared to Europeans, who are largely supportive of interoperable medical databases, Americans are split, with 48 percent saying that the benefits outweigh the privacy risks and 47 percent saying the opposite. Twenty-five percent of the population will not use an online health record due to online privacy issues.[49, 50]

Despite these concerns, 71 percent of Americans surveyed believe having an online health record will clarify doctors' instructions, and 65 percent believe online health records will prevent medical mistakes.[51] Thirty-five percent of respondents would use an online public health record today, and over 60 percent indicate interest in using one in the future, with demand higher among those with chronic diseases.[52] We see a generational difference in the demand for online health records: 70 percent of people aged twenty-five to thirty-four want online access to their health information, as opposed to just 35 percent of those aged sixty-five and over.[53] This is no surprise, as youth instinctively turn to the Internet when they are in need of information and find online collaboration to be an ingrained part of the youthful personality.

That said, there are early adopters of collaborative health care in all age brackets. Consider Dave deBronkart, a sixty-year-old high-tech professional.

He survived a serious bout of cancer that had spread from his kidney to his lung. He survived partly because an online community led him to a course of aggressive and expensive treatment. Now, deBronkart operates a blog called "The New Life of e-Patient Dave," with the "e" standing for "empowered and engaged, equipped and enabled."[54] Through his blog, deBronkart champions patients' rights to have access to their personal health records along with affordable and effective health care coverage. So while cautions about information asymmetries are duly noted, it's in people like deBronkart and communities like MedHelp, PatientsLikeMe, and Sermo that we see the contours of a new model of health care emerging.

WILL VESTED INTERESTS KILL COLLABORATIVE HEALTH CARE?

As the battle over President Obama's health care reform plan showed, there are deep vested interests in the status quo, profound differences in perspective, and almost incomprehensible challenges in reforming the structure of the system. The one thing the opponents agree on: the American health care system is broken.

But won't big health care companies fight against collaborative health care? After all, the patient record is what locks a consumer into their service. True, there are vested interests, but the advantages of opening up are so compelling it is likely that everyone will have to comply. In fact, the national association of the American private health insurance companies says it supports the principle of data records being portable, so if a patient wanted to switch from one insurance company to another it would be a simple process. "Say you were a member of Aetna and you want to go to Blue Cross, or if you're at one hospital and decide you want to switch to another hospital across the street. Both of these are basically impossible, even though everyone says they want to do it," says Matthew Holt of consulting group Health 2.0. He argues that mass consumer demand or legislation can also be powerful forces causing private industry to change.

To be sure, there are numerous other challenges. Setting aside the vanguard identified in this chapter, it will take time for some doctors to adjust to this notion. The digital divide is a problem. Unfortunately, not everyone has equal access to the Internet, and measures need to be taken that ensure low-income citizens have good access to broadband service in areas such as local community centers, drop-in centers, and libraries.

Fear of litigation may discourage some physicians from committing their thoughts to paper on a medical record they don't control, especially in the current litigious environment in the United States. Doctors and other professionals are going to need assurance that their informal communications with patients won't be held against them, excepting cases of true malpractice.

Privacy is another huge concern. Any move toward collaborative health care must ensure the protection of records. This will also be difficult to achieve because patients will have control over what they release to the public, and as illustrated on social networks, individuals don't always have the best judgment.

However, all these obstacles fall into the category of "implementation challenges" as opposed to the category "reasons not to do this." The issues that surround new technologies—questions about privacy and safety of records, for instance—are issues that we can deal with through legislation, intelligent engineering, and common sense. The time for collaborative health care is now. It embraces all of the principles of wikinomics and can be supported regardless of one's political views. By harnessing self-organization and participation, it enables people to collaborate and innovate. It is open and founded on transparency and the sharing of intellectual property. It is rooted in the notion of interdependence and the deep interconnections and common interest we all have in improving human health.

But we can't take the budding platforms we have to full flower unless the establishment capitulates and says, "Yes, we need to change. We need greater transparency, better patient access to information, and a more cost-effective system." Without the buy-in of the biggest players—namely government and insurers—we won't be able to maximize these technologies. And more people will get needlessly sick. Harness these new capabilities, on the other hand, and the medical establishment can join with patients and other stakeholders in making collaborative health care a reality for everyone.

V

TURNING THE MEDIA INSIDE OUT

11. THE DEMISE OF THE NEWSPAPER AND THE RISE OF THE NEW NEWS

Arianna Huffington is a newspaper magnate—but she's no Citizen Kane. The casual observer might take her at face value when she says she is creating an "Internet Newspaper" and hopes it will be the biggest one in the world. At a glance, her *Huffington Post* sure looks the part—an online newspaper read by more than 20 million people per month and growing at a rate of 50 percent per year.[1] Huffington is the site's cofounder and editor-in-chief, and she says it is one of the most widely read, linked to, and frequently cited media brands on the Internet.

But the HuffPo, as it's called, is not just paving the journalistic cow path, or simply turning atoms into bits. Rather, it represents a new model of content production built on a new species of community. It has a small paid staff of 150, and relies on more than 3,000 contributors[2] to produce content on every conceivable topic. It has another 12,000 "citizen journalists" who are its "eyes and ears." Its readers also produce much of the HuffPo's content to the tune of over 2 million contributions per month.[3] Huffington says her readers' engagement and insights are essential to HuffPo's value. Huffington cofounder Jonah Peretti believes that the news model is no longer a passive relationship of news handed down but "a shared enterprise between its producer and its consumer."[4] Huffington's goals are clear: "We want to be the Internet newspaper, covering everything and catering to every interest," she says, but "driving this from our distinct editorial viewpoint." Huffington describes herself as a "progressive populist." Where traditional print newspapers are atrophying and dying, her model seems to be thriving. HuffPo has become an influential player in American business, political, and social life.

Huffington says the "real motivation for me to set up the site was that I could see that the world of news was going online but there were so many interesting voices, so many people I knew and respected that were not online and not available to the world." She describes how Arthur Schlesinger, Jr., a

Pulitzer Prize–winning author and well-known Democrat, was the first person she invited to blog and his response was "What's a blog?" When Schlesinger confessed he could barely use a computer, she told him not to worry because he could fax in his blog. "I wanted the voice there. I didn't care how it got there—fax or carrier pigeon." As it turned out, two weeks later she got a fax from Schlesinger because he was at a conference in Yalta to which President George W. Bush had given a speech. Schlesinger explained why he thought Bush was deluded in contending that the Yalta conference of 1945 caused or ratified the division of Europe. HuffPo gave Schlesinger a voice.

Huffington says, "I also wanted to have new voices—people who no one had ever heard, but who had potential to be great." Today HuffPo has over three thousand bloggers who have proven themselves trustworthy and can post their blogs directly on the site without scrutiny by editors. The site also receives and reviews blog submissions from thousands more people every month. During the 2008 presidential election the site launched the "Off the Bus" project. "Most journalists are on the bus and are fed the candidates' official lines," she recalls. "With Off the Bus we had over twelve thousand citizen journalists acting as the real eyes and ears of the campaigns." The innovation was popular with readers. During the last election not only was *The Huffington Post* the most-viewed stand-alone political news site, but HuffPo's readership doubled in 2008 compared with 2007 and redoubled in 2009.[5]

The HuffPo also has media monitors keeping watch on TV and radio shows, pointing out moments and outrages worthy of comment. When the so-called Tea Parties were organized across the country in 2009, HuffPo had thousands of citizen journalists sending photographs and stories about what was happening on the ground. During the 2009 auto company bankruptcies, readers created an interactive map of GM dealerships that were closing down. HuffPo also made public the now-famous recording of Morgan Stanley's CEO announcing that they were renaming the word "bonuses" to call them "retention awards."[6]

Huffington denies that the popularity of sites like hers contributes to the decline of newspapers. "The issue is not how we save newspapers," she maintains. "It's how we save journalism." HuffPo employs half a dozen reporters and in March 2009, Huffington contributed to a $1.75 million fund to promote investigative journalism.[7] "I hope we can not only save investigative journalism, but strengthen it," she told ABC News. "Because I think it's amazing that journalists missed the two biggest stories of our time. You know, the lead up to the war in Iraq and the economic meltdown."

So far, HuffPo is a great example of how journalism can be profitable and prosper—at least for the publisher—in a digital world. But it raises as many questions as it answers. Eighty-nine-year-old Helen Thomas, perhaps the most famous White House correspondent of the last fifty years, is worried about the decline of professional journalism. "Everyone with a laptop thinks they're a journalist. Everyone with a cell phone thinks they're a photographer." She calls this trend "frightening because you can ruin lives and reputations willy-nilly without realizing it. No editors. No standards. No ethics. We're at the crossroads. [In my day] when your mother told you she loved you, a good journalist would check it out. Now so many newspapers that are so valuable are going down the drain. It's a crisis."[8]

The HuffPo seems to have a viable business model based on advertising. But what happens to all the journalists who are welcome to write for the publication but do not get paid? Diane Brady, senior editor and content chief at *BusinessWeek*, expresses the concern of many when she says: "It's fine for those writers who simply want a platform. But journalists need to make a living too. What is the business model for them?"

The digital revolution has affected no sector more than media and entertainment. Newspapers throughout the United States and Canada are collapsing. On the surface it looks like bad business decisions might be to blame. What, you might ask, were the managers of *The New York Times* thinking when they borrowed hundreds of millions of dollars to buy lavish real estate and other dubious properties like *The Boston Globe*? Others say that crashing circulation and revenues are caused by the tough economic climate. But no amount of rationalization or denial can hide the looming truth that the collapse of the newspapers is not coincidental, conjunctural, or containable. It is systemic—rooted in the digital revolution. So the leaders of the old media should take a deep breath and get going on the kind of experimentation required to forge some new approaches that are sustainable. Helen Thomas is right to be concerned. In this chapter we explore the vexing challenges the new media raises for news and media companies, including some tough questions for all of us. As the business model for journalism collapses, what happens to our social model for getting news and informing ourselves of what's important in society? What will happen to quality in this democratized world when the traditional filters for accuracy, balance, and journalistic standards are gone? Will investigative journalists who are losing their jobs be replaced in society by public relations flacks? With no one seemingly accountable for the truth, whom can we trust? Will the decentralization of media cause a balkanization of human perspectives and values?

The reaction of many mainstream media (MSM) companies has been good and bad, but mainly ugly. As they scramble to address the new economics of the digital world, many have attempted to create the equivalent of a movie by filming a stage play (minus a few notable exceptions discussed below). The law of new paradigms has kicked into play: leaders of the old paradigms have great difficulty creating the new. In hindsight their record is pretty predictable, and even pathetic. Why didn't Rupert Murdoch create *The Huffington Post*? Why didn't AT&T launch Twitter? Yellow Pages could have built Facebook. Microsoft had the resources to come up with Google's business model. Why didn't NBC invent YouTube? Sony should have preempted Apple with iTunes. Craigslist would have been a perfect venture for *The New York Times* or any regional newspaper. As the media becomes democratized and as those vested in the old paradigm fight change, a historic period of calamity is in the making. At the same time there are fresh, sometimes breathtaking new innovations in every medium—experiments and juggernauts that show us the way forward.

THE RISE OF PARTICIPATORY NEWS AND CULTURE

The basic threat to traditional media is not just that bits cost less than atoms. If plummeting costs were the only change, then media companies could exploit new technologies and become far more profitable using this new model. However, information produced and disseminated online behaves differently from physical books, magazines, and newspapers. With the new Web, the Internet is no longer about idly surfing and passively reading, listening, or watching. It's about peering: sharing, socializing, collaborating, and, most of all, creating within loosely connected communities. Nowhere is this truer than in the media and entertainment business, where prosumers are deeply involved in aggregating, rating, and commenting on the news—and, increasingly, in creating the news themselves. This has now extended to virtually all media, including music, film, television, and radio. Indeed, the capability of humanity to report, analyze, compose, create, act, perform, produce, and share has never been greater. There are over 700 tweets produced a second, and Twitter now boasts over 13 billion entries.[9] 2.5 billion photos are added to Facebook monthly[10] and YouTube serves over a billion video views every day[11]—meaning that the majority of "television shows" and "movies" watched are now less than five minutes long. The shakeup is profound, and the numbers are so huge as to almost be incomprehensible.

Over the same period, this global coffeehouse has become a business for many people. Technorati reports in its 2009 State of the Blogosphere that a whopping 28 percent of bloggers report earning income from their blog.

Half of these "professional bloggers" write full-time either to support their business or as a business itself. Three-quarters of these bloggers have college degrees and 40 percent have graduate degrees. A remarkable 17 percent say blogging is their primary source of income.[12] Though the majority of blogs or other citizen-producer media are typically not of a quality to compete with commercial media, they point to the increasing ease with which anyone can create their own news and entertainment and bypass established sources. The term "user-generated content" has now become anachronistic. It was based on an old paradigm where the world was divided into owners, creators, and producers of content on the one hand and "users" on the other. Today, the new creators of content are not primarily "users." They are becoming the new *producers* of media in an economic sense.

These new forms of media creativity and entrepreneurship have many benefits. More people can develop a voice and build an audience for their message. Cell-phone-wielding citizens can help document important events, as witnessed in the aftermath of Haiti's tragic earthquake and in the youthful rebellion that followed Iran's disputed election. There are also big opportunities to reinvent the whole concept of the newspaper—opportunities that simply didn't exist until recently. Rather than treating a newspaper like a static document, whose inked pages mark a point in time, publishers can turn their traditional "newspapers" into an evolving digital collection of facts, data, articles, images, and videos that can be endlessly repurposed into intriguing combinations. New media services could be custom built for specialized audiences, leading to new partnerships and new business models. And rather than lock up content behind a pay wall, newspapers could become platforms for media innovation, much the way Apple's App Store has become a mecca for mobile applications built by thousands of independent developers. Some newspapers, like *The Guardian*, are already experimenting with open platforms. We'll tell that story later in this chapter.

But the same democratizing forces that have opened up the production and consumption of media also raise the prospect of great peril, not just for the traditional media companies but for societies thrust into a time of disruption, uncertainty, even calamity, as a trusted old order with its benefits and well-understood rules is smashed. Media organizations that, for good reasons, could not react to the writing on the wall are being bypassed by a new generation of media-savvy prosumers. And to make matters worse, the public's trust in journalists and the media is at an all-time low. The percentage of Americans who say news organizations get their facts straight has plummeted from 55 percent in 1985 to 29 percent today.[13] People, especially young people, have greater faith in information they receive via tools such as blogs, tweets, and information circulated on Facebook. But while the smartest and most savvy media consumers may have the authentication skills and BS detectors to distinguish truth from reality, the average Web surfer has to wade through a cauldron bubbling over with inaccurate, misleading, scurrilous media garbage of both professional and amateur origin. Even Google's CEO, Eric Schmidt, says, "it's a sewer out there."[14] The way we navigate these changes will have far-reaching implications—from the way we inform and foster democratic debate, to the way we educate our children.

THE NEWSPAPER IS DEAD, LONG LIVE THE NEWSPAPER

Newspapers can be an emotional topic. For centuries newspapers have been the foundation of free speech and social critique. Great investigative print journalism has exposed the problems in society more than any other medium. Great writing is a joy to read. It informs, enrages, calms, stimulates, and engages us in society. Even today, in vast parts of the world, such as Western Europe, many people love the daily press.

But twenty-five-year-old Rahaf Harfoush doesn't read newspapers. Her view: "Why would you? They come out once a day, they don't have hot links and they're not multimedia. And you end up with some kind of black gunk all over your fingers." Rahaf is well informed and has an elite job with Geneva's World Economic Forum. She's even written a book about social media and she travels the globe, in demand from executive audiences eager to learn about the new world of marketing. She has views on the financial crisis, the Middle East, the problem of climate change, the health care crisis, and just about every other crisis facing humanity. But she is also deeply knowledgeable about film, popular music, sports, and popular culture. How does Rahaf stay informed? Like millions of her generation, she has created her own digital newspaper, featuring a sophisticated, personalized set of information-gathering tools, such as Twitter, Viigo (a BlackBerry app), Google Reader, Reddit, and *The Huffington Post*. They provide her with real-time, on-demand access to dozens of information sources. "The news is no longer a one-stop trip," she says. "I think the changing nature of the story, and the constant updating of the Internet, make it possible to sample a wide variety of opinions and perspectives. I rely on all these different pieces to triangulate the issues I care about and get to the heart of things."

Unfortunately for the traditional print newspaper, the generational shift in how information is aggregated and processed has reached a tipping point. The data is striking. According to the Newspaper Association of America, there were 470 fewer newspapers in 2008 than there were in 1940.[15] But this is more than industry consolidation. Seventy-two newspapers have ceased to exist since 2000, and many premier papers are in trouble.[16] Last year, industry-wide circulation fell 10.6 percent from the year before. This comes on top of losses of 4.6 percent in 2008. The American newspaper industry has lost 25.6 percent in daily circulation since 2000.[17] Those declines, however, pale by comparison to the loss in revenues, which represent a more significant problem.

Compared with the year prior, in 2009 advertising sales dropped 26 percent.[18] Advertising typically accounts for 70 percent of a newspaper's revenue, so when advertising disappears, so do newsroom jobs. Newspaper employment has plunged by more than a third over the last twenty years.[19] Although the population of the United States is over 305 million, the number of people who read daily newspapers has not grown since 1966 when the country's population was 196 million.[20] And the audience migration to the Internet is accelerating.[21]

Because the Web is dropping transaction and collaboration costs, what used to be inside corporate boundaries can now happen outside. Papers used to have the reach to sell classified ads. That was before a San Francisco do-gooder named Craig Newmark set up a community Web site whereby people could buy and sell stuff for free. It wasn't long before Craigslist had pretty much destroyed the classified revenue of every North American newspaper. Newspapers used to be the only organizations that could aggregate reporters together—all under one roof—to generate enough content to attract a mass audience, and therefore attract advertisers. Now aggregators like *The Huffington Post* can do this at a fraction of the cost.

Optimistic defenders of the status quo insist the carnage in the publishing industry is simply due to the 2009–10 recession. To be sure, the collapse of car companies, banks, and others has played a role. Auto dealerships, home builders, and retail stores—all devastated by the recession—were typically the mainstay of local newspapers. Throw in the axis of evil in classified ad revenue—Craigslist, Monster, and eBay—and the carnage is complete. But there is a much more accurate explanation. The recession was not the cause of the crisis in publishing—it was an accelerant. It sped up and amplified the impact of deep historical trends in the unfolding of the digital age.

To the consternation of newspaper and other publishing executives, the chorus of media pundits who argue that the industry has no hope is getting louder. Writes Bob Garfield, author of *The Chaos Scenario*: "Traditional media are in a stage of dire retrenchment as a prelude to complete collapse.

Newspapers, magazines and . . . TV as we currently know them are fundamentally doomed."[22] Clay Shirky, who teaches new media at New York University, famously wrote, "Round and round this goes, with the people committed to saving newspapers demanding to know 'If the old model is broken, what will work in its place?' To which the answer is: Nothing. Nothing will work. There is no general model for newspapers to replace the one the Internet just broke." Shirky argues, "With the old economics destroyed, organizational forms perfected for industrial production have to be replaced with structures optimized for digital data. It makes increasingly less sense even to talk about a publishing industry, because the core problem publishing solves—the incredible difficulty, complexity, and expense of making something available to the public—has stopped being a problem."[23]

For sure, no one has adequately answered the question: "What should the newspapers do to stay alive?" In fact this is the wrong question, because the Web is providing myriad content, in effect replacing the traditional newspaper under a different business model. But that's of little comfort to owners, or for that matter, the readers of *The New York Times* who like things the way they are, or at least, have always been.

Democratization of News Media: Monopolies of Power

It is becoming impossible to generate revenue for traditional newspapers and the reason is simple. The planet is becoming covered with human sensors—hundreds of millions of people who increasingly alert the rest of us to news events. When something important happens it gets blogged, tweeted, Digged, YouTubed, and otherwise reported, scrutinized, authenticated, analyzed, discussed, and then re-reported with a momentum typically relative to its newsworthiness. Note we didn't say relative to some predetermined concept of importance, whatever that is. Markets or interested people determine whether something is important, that is, worthy of absorbing and of passing along to the next person. The business of news, as it's currently constructed, is vanishing. Efforts to copyright news stories, charge for them, or somehow protect them behind a newspaper "pay wall" are futile. As one young college student famously said in a focus group, "If news is important to me it will find me."[24]

With democracy comes power. Marshall McLuhan's teacher and mentor, University of Toronto political economist Harold Innis, was the real pioneer of media theory. He said that, by understanding the characteristics of media technologies, we can understand the nature of societies and unlock the secrets of history.[25] While this sweeping premise may be an overstatement, his analysis of media is rich with insights for the publishing industry today. Innis's core insight was that new information and communications media empower rising social forces with knowledge at the expense of old monopolies. These newly successful forces create even bigger knowledge monopolies, causing a never-ending cycle of social conflict and challenges. For example, Innis argues that in ancient Mesopotamia, the use of clay favored a dominant role for the temples with an emphasis on priesthood and religion, while libraries housing parchment scriptures were built up in Babylon and Nineveh to strengthen the power of monarchy. Printing brought renewed emphasis on the book and the rise of the Reformation. In turn, new methods of communication, like radio and television, weakened the worship of the book and opened the way for new ideologies. Each transition gave rise to cultural disturbances as old monopolies or oligopolies crumbled, while the improved technologies strengthened new forces.[26]

Consider the "cultural disturbance" to the efforts of a corporation in the United Kingdom to keep public proceedings in the British House of Commons secret. Editors of *The Guardian* newspaper in England were stunned in October 2009 when they received a sweeping court-ordered ban on publishing part of the proceedings of the House of Commons in which a member of the British cabinet was going to answer a question from a member of Parliament. It was akin to slapping a court order on a newspaper in the United States that promised its readers it would raise a topic at President Obama's next news conference. As the paper explained to its morning readers, it couldn't identify "who has asked the question, what the question is, which minister might answer it, or where the question is to be found. *The Guardian* is also forbidden from telling its readers why the paper is prevented—for the first time in memory—from reporting parliament. Legal obstacles, which cannot be identified, involve proceedings, which cannot be mentioned, on behalf of a client who must remain secret."[27] Such is the stuff of despotic regimes, but the United Kingdom?

Enter thousands of Twitter users, who took up the cause. Trafigura, a London-based oil trader connected with dumping toxic waste in the Ivory Coast in 2006, was identified as instigating the gag order.[28] Tweets about the company flew back and forth, and Trafigura quickly became the most-used word in Twitter traffic that week. This was picked up by a number of prominent blogs, and by midday the company had such a high profile that it gave up trying to muzzle the media. As *The Guardian* wrote: "It might be a bit too exaggerated to call it a historic moment, but surely the real-time web passed its test today."[29] While the traditional media was shackled by legal constraints, information consumers rallied together via digital tools to reveal the forbidden information and share it widely.

Reinventing *The New York Times*

As newspapers across the country have been hit hard by dropping circulation and falling ad revenue, one of America's biggest papers has fared better than most.[30] For the year ending March 31, 2009, *The New York Times*' weekday circulation fell just 3.6 percent compared with the year before (the industry average is down 7 percent),[31] and its circulation revenue actually went up thanks to increases in the paper's price. Nevertheless, the picture is grim. The company is losing money fast. Advertising revenue cratered about 30 percent in 2009–10. Classified advertising fell off a cliff: help wanted dropped 60 percent, real estate was down 47.6 percent, and automotive fell 43.2 percent. Even Internet ad revenue dropped 14.3 percent in the second quarter. Digital revenue now accounts for 13.4 percent of total company revenue, only up from 12.3 percent a year ago.[32]

Four years ago the *Times* saw the writing on the wall and appointed Jonathan Landman as deputy managing editor, with the mandate to reinvent the paper around the Internet. Today the *Times*' Web site is widely regarded as one of the best and most innovative media sites in the world. Landman says he built on two of the *Times*' greatest strengths: the top-flight journalism it produces, and equally important, the top-flight readers such journalism attracts. "Our quality proposition is central to who we are," says Landman. All reader feedback passes through the hands of an editor. The name-calling and ad hominem drivel typical of most newspaper comment pages isn't tolerated. "We are not shy about moderating things out," says Landman. "There is no constitutional right to have your comments published. And certainly if it's abusive or stupid or something, well then, what's the point, why is that a good thing?" The result is an articulate discussion by readers with the paper, and readers with one another. The high-quality dialogue also improves the editorial content. "Editors look at the comments and they then draw conclusions about the kind of editing it may still require. You can tell, for example, that people may not understand something. It's not clear. So you fix it. Either you may get additional facts or you may get an interpretation challenged and see that you have to adjust it someway." So in a very real sense, *Times* readers are participating in the editing even if they don't know they're doing it.

The challenge, according to Landman, is to create an environment in which the right people participate with high-quality contributions. "Wikipedia has done a miraculous job of preserving standards in a collaborative way," he says, "and to me the great accomplishment of Wikipedia is not so much that it gets a lot of people to participate. That's actually relatively easy. It's that it's able to enforce a clear set of accepted standards and that it's able to get the community to enforce those standards." Figuring out how to marshal readers' insights, according to Landman, is something the *Times* is still learning. "Things are going really fast," he says, "and you only have a short time to get things right because then the news moves on."

As it moves into uncharted waters, the *Times* has become an experimentation machine. It regularly introduces new techniques to produce more variety and engage the reader, all the while meeting the paper's high standards of editorial integrity as its core value proposition. Whereas twenty years ago it published a couple of op-ed pieces each day, today the *Times* has innovations such as "Talking Heads," where smart people who don't work at the *Times* debate topical issues with one another while recording the conversation on a webcam. In fact the *Times* sees itself not just as a content creator but as a community builder. Its slogan "All the News That's Fit to Print" has been replaced with "It's All About the Conversation." It's fair to say the *Times* is dancing as fast as it can.

THE DARK SIDE

It's pretty clear that newspapers, at least in their present form, will not survive. Those that do will have to adopt a much different model. But what are the implications for society when these pillars of pluralism and free speech are gone? Will new mechanisms emerge to replace the traditional filters for accuracy, balance, and journalistic standards that newspapers currently provide? Who will create the quality journalism we have come to expect, particularly the investigative journalism critical to the health of our democratic system? And will society become increasingly balkanized without the social glue that newspapers provide today? For those of us who care about the role an independent media plays in a free society, these are among the most vexing questions we face.

What Happens to Quality Journalism?

For some, "citizen journalism" is an oxymoron, right up there with military music and jumbo shrimp. They pine for the old days when the major papers determined what news was "fit to print" and millions of readers trusted their judgment. Now via the Internet you get "all the news," whether it's fit to print or not. And for skeptics like Andrew Keen this democratization creates a problem, namely a vast heap of mediocrity that crowds out the good stuff and confuses consumers. In his book *The Cult of the Amateur: How Today's Internet Is Killing Our Culture*, Keen argues that user-generated content is destroying journalism and, for that matter, society.[33] To Keen, "The more layers you have between the originator of content and the recipient of content, the better. Because that means more editing, correction, and improvement."

On the other side of the spectrum are the optimists, the ones who subscribe to the "don't worry, it'll all come out in the wash" school. Cody Brown argues that distributed networks of everyday people will do a better job at maintaining quality than the traditional media. "News is important. It's so important that leaving it to a group of people in an office downtown is and has always been irresponsible," he says. To Brown, when a public can talk to itself it "can be counted on to share and disseminate its own news." What's missing is a better mechanism to do this, so he's launched kommons.com, a news company at the intersection of journalism, massive real-time collaboration, and mobile technology. The insight, according to Brown, is simple: "Instead of telling a public what is news, we create a space for them to tell each other."[34]

Recent data actually provides some evidence to support the optimists' view. Among all those amateur journalists are growing ranks of serious and competent people. In fact the dichotomy between amateurs and professionals is blurring. Many bloggers (28 percent) are now professionals in that they make a living from blogging. They are professional in another sense too: 40 percent of these commercial bloggers have worked within traditional media.[35] They have formal journalism experience, training, and credentials. This fact undermines the critique of Andrew Keen. Rather than professionals being displaced by rank amateurs, it turns out the professionals are simply shifting employers—from the mainstream media to new media, self-employment, and new entrepreneurial journalistic ventures.

Moreover, when it comes to the so-called balance of the mainstream media it is all too easy to get misty-eyed and think that the news that was "fit to print" was only determined by some lofty standard. Peter Drucker famously explained that "the purpose of a business is to create a customer" and that "a company's primary responsibility is to serve its customers."[36] Newspapers understand who their primary customers are—the advertisers who account for most of their revenue. Consider your local newspaper. There is a section on cars and another on new homes, and another on movies. The page count, structure, balance, and content of the paper is designed to attract ad revenue—not to fit some preconceived notion of what's important in news. Of course, one of the ways to attract ads is to have a quality product that creates a mass market, and there have been noble publishers. Most managing editors that run publications today care a lot about quality and fairness and good journalism. But as Clay Shirky likes to point out, a newspaper may have had good coverage of events in China, not primarily because they were doing a service to society but because Wal-mart was prepared to advertise and in doing so funded a Beijing bureau. Fortunately, good quality has often been a requirement to get a large enough readership to attract these customers.

Furthermore, quality is in the eyes of the beholder, and a new set of standards is emerging—not necessarily lower or higher standards, just different ways of defining quality. *The Huffington Post* has a lot of content, with its three thousand accredited writers and twelve thousand "Eyes and Ears" correspondents and the millions of citizens who express their views on the Web site. By traditional journalistic standards, some of this content is inferior, but much of it is thoughtful and worthy of consideration. Some is arguably superior. And because of HuffPo's capacity to engage many people—from regular folks to Nobel Prize winners—they can create real-time, extensive, and comprehensive coverage that trumps the mainstream media.

And let's not forget the occasions when the MSM completely missed the obvious, such as the two biggest stories of the past decade. One was the deception practiced by the George W. Bush administration in the run-up to the Iraq war, and the second was the worst financial crisis since the Great Depression. Take the financial crisis for example. Where were the teams of investigative reporters doing due diligence on Wall Street as the housing market and credit bubble grew to dizzying heights? Particularly since this was happening in the wake of the boom-and-bust frenzy of the dot-coms a few short years ago. But the sad truth is that the media that covers Wall Street is embedded and vested in Wall Street. Jim Cramer of *Mad Money*'s deplorable dismissal of everyone's precrash concerns is an almost laughable example.

Conflicted media poses a new danger. If nature abhors a vacuum, so do advertising and public relations agencies. If there is a decline in standards in the new Wild West of journalism, there is a wonderful opportunity for advertisers to provide a new kind of "content leadership" and insert the messages of their clients everywhere. Will we see product placement in articles funded by unknown sources with something to sell? Will advertorials, that are not identified as such, become a mainstream form of journalism? Or if most journalism is produced by a decentralized network of autonomous individuals and groups, wouldn't it be easier to game the system? None other than Martin Sorrell, the CEO of WPP—the world's largest conglomerate of ad agencies—openly says "yes." In a keynote address to the November 2009 ad:tech conference, he predicted that ad agencies would be getting "very much more involved" in the development of content and that the lines between advertising and editorial are going to get "much more blurred" than they already are today.[37]

So how will we know whom to trust? At the very least, we can count on the fact that the importance of reputation in today's media environment provides an incentive for bloggers, Internet newspapers, newsfeeds, muckrakers, and others to act with integrity and protect their brand. When TMZ, the Los Angeles–based celebrity gossip site, broke the news about Michael Jackson's death, some doubted the story's authenticity. But the MSM soon had to acknowledge that TMZ has a very strong reputation for accuracy, and as it turns out, they were right. Companies like TMZ, no matter what you think of their genre, are scrupulous about accuracy and their brand. A single false alarm or bad story can seriously undermine trust and their ability to be influential.

In the past we were used to having our news handed to us by pros and assuming it to be true. We perceived no need to authenticate. But today there is a new imperative to do so. Perhaps we'll only believe information that supports our preexisting point of view. On the other hand, if we're bitten enough times by false information it's likely that we'll become shy about blindly accepting all we read and hear. Enough bad experiences will encourage us to develop our BS detectors, scrutinize information better, and become more sophisticated consumers of Internet-based information. Teaching young people how to scrutinize, validate, and put things in context will be among the toughest tasks for educators. On the other hand, there is reason to believe that young people have an inherent advantage. Having grown up in a culture of a million hoax e-mails, scams, and attempts to deceive, they've learned that either you get savvy or you pay the price.

Who Will Do Investigative Reporting?

The *San Francisco Chronicle* is Northern California's largest newspaper, and it's barely clinging to life. It's hemorrhaging red ink, circulation has nosedived, and the newsroom has lost many staff. "If journalism is failing here, in one of the United States' most engaged and educated constituencies, it is an ominous sign," commented another newspaper, *The Globe and Mail*. "For more than a century, newspapers have functioned as a civic conscience, instrumental in promoting democratic ideals, whether by checking corruption, fostering accountability or simply informing the public."[38] Many editors see their mission as "comforting the afflicted, and afflicting the comfortable."[39]

James Madison, writing more than two hundred years ago, insisted: "The right of freely examining public characters and measures, and of free communication among the people thereon . . . has ever been justly deemed the only effectual guardian of every other right."[40] For a long time the media in general, and newspapers in particular, fulfilled this role. To be sure, many publications are still uncovering corruption and exposing truths that the powerful would prefer stay hidden. But unfortunately, investigative reporting is a badge of honor that is increasingly difficult to sustain. Investigating improper behavior by politicians, government agencies, and corporations is a costly and time-consuming practice. Expenses include the salaries of a team of journalists and researchers, months of research and scores of interviews, and sometimes frequent and far travel. When budgets are tight, it's investigative reporting that is usually cut back. If newspapers can no longer do the job, who will?

Some argue that the way to protect good journalism and investigative reporting is through an NGO model. The U.S. Congress is considering legislation to allow newspapers to become nonprofit organizations, thus enabling tax-deductible donations. In this scenario the mission of a publication could be more securely set, free of the demands of stockholders or commercial return. Or perhaps it's new media to the rescue. As noted earlier, *The Huffington Post* has established an Investigative Journalism Fund. It's an innovative nonprofit/for-profit hybrid that attempts to fill the void created by the closing of so many newspaper investigative units. It also provides new opportunities for seasoned journalists who have been laid off or forced into early retirement. What's more, the stories that come out of the HuffPo's Investigative Fund will be free for anyone to publish.

One organization that may be positioned to benefit is ProPublica, an online, independent, nonprofit newsroom that produces investigative journalism in the public interest. ProPublica's Manhattan newsroom has thirty-two journalists, funded through a major, multiyear commitment by the Sandler Foundation and other philanthropic contributions. Many stories are offered exclusively to a traditional news organization, free of charge, for publication or broadcast. In 2009 the site published 138 such stories with 38 different partners.[41] In April 2010, ProPublica reporter Sheri Fink was awarded a Pulitzer Prize for a piece that told about "the urgent life-and-death decisions made by one hospital's exhausted doctors when they were cut off by the floodwaters of Hurricane Katrina." ProPublica had partnered with *The New York Times* on the story.[42]

Some have even suggested that newspapers should be candidates for government bailouts, right up there with the failing financial institutions.

However, a survey by Sacred Heart University in September 2009 found that eight out of ten Americans would oppose any plan to spend tax dollars to aid failing newspapers, even though 64 percent believed that good journalism ensures a healthy democracy.[43] Although the public doesn't want to help newspapers with their tax dollars, many are willing to actively assist papers to produce a quality product. Consider what a British newspaper, *The Guardian*, did during the recent expenses scandal of UK politicians. *The Guardian*'s archrival, the *Telegraph*, was printing daily front-page jaw-dropping revelations of elected officials claiming outrageously indefensible expenses. The *Telegraph* had a team of reporters working for a month with leaked documents. In response to the public outrage, the government announced it was going to release online more than a million scanned documents and receipts, a digital tsunami that would overwhelm the resources of any news organization.

Still smarting from having been scooped day after day by the *Telegraph*, *The Guardian* asked readers to help it sift through the million documents and find as-yet undisclosed malfeasance. Open-source software running on the paper's Web site let readers go through the documents one-by-one and characterize a receipt four ways: "interesting," "not interesting," "interesting but known," and "investigate this!" More than 20,000 readers pitched in, and 170,000 documents were reviewed in the first eighty hours.[44] A progress bar on the front page gave the researchers a goal to share. Then a list was developed of top-performing volunteers, which helped motivate them further. When *The Guardian* added to each receipt a photo of the associated member of Parliament, reader response soared. They even rented inexpensive server space from Amazon to handle the initial avalanche of interest. The total out-of-pocket expenses for the paper: less than $150. And the initiative helped the newspaper build a much stronger rapport with its readers. The upshot is that investigative journalism needn't become a casualty of the digital age. In fact, news organizations that think creatively about how to deliver it could be in a position to deliver more investigative journalism than ever.

Without the Social Glue of Newspapers, Won't Society Become Balkanized?

Walter Cronkite, anchor of the *CBS Evening News* from 1962 to 1981, was often respectfully referred to as "the most trusted man in America." Cronkite died in July 2009, which caused many journalists to reflect on the state of the media in America today. "We have gone from three networks to being barraged by cable channels offering infotainment, news, comedy and celebrity gossip 24-7-365, Internet sites beyond number, podcasts, talk radio, and plenty of other sources of both news and 'news,'" editorialized the Peoria *Journal Star*. "In that hyper-competitive atmosphere, no one working today has been able to summon credibility across the boundaries of rich and poor, blue state and red state in a balkanized America to be able to say, 'And that's the way it is,' and be almost universally believed, unscathed by parody."[45]

It's a different world, and one in which many media organizations have abandoned any pretense of impartiality. For conservatives, there never was such impartiality to begin with. Many said that Cronkite was part and parcel of the rat's nest of godless liberals who for decades dominated the "mainstream media." The result is that many conservatives today seek out right-wing news organizations such as FOX News or RushLimbaugh.com. Similarly, Democrats increasingly turn to center and left-of-center news organizations such as *The New York Times*, *The Huffington Post*, and MSNBC. More and more news organizations are preaching to the converted. Tomer Strolight, president of media company Torstar Digital, says, "People can now spend their time in a media landscape that gives them comfort. When someone enjoys FOX News, are they being informed or just comforted?"

During the 2008 presidential election, candidate Barack Obama championed the notion of bipartisanship. He promised to "reach across the aisle," and encourage both parties to work cooperatively rather than constantly deriding each other. He urged Americans to view themselves not in terms of living in red states or blue states, but in the United States. Despite his best efforts, so far he has failed. Legislative votes closely follow party lines.

But we should ask, is the increasingly polarized media exacerbating the balkanization of America, or is it simply reflecting what is happening for other reasons? In his book *The Big Sort*, Bill Bishop argues that Americans of like mind began to cluster many years ago. "We have built a country," Bishop writes, "where everyone can choose the neighbors (and church and news shows) most compatible with his or her lifestyle and beliefs. And we are living with the consequences of this segregation by way of life: pockets of like-minded citizens that have become so ideologically inbred that we don't know, can't understand, and can barely conceive of 'those people' who live just a few miles away."[46]

The Internet is a reflection and also amplifier of everything in society. It will be used by every perspective and group to advocate their particular point of view. This has some benefits. After all, Cronkite may have been trusted by mainstream American society, but the old voice of the so-called mainstream paved over dissenting views and poorly represented the experiences and struggles of minorities. If you were an African American in the

1960s and '70s, you wouldn't necessarily have seen your views and your experiences accurately represented by the mainstream press. Many argue that's still the case today. But a more democratized media is a more diverse media, which means the formerly voiceless can develop their voice and build a following among like-minded peers. Of course, this does little to promote political and social cross-pollination and, unfortunately, it's also true that a democratized media gives voice to racists and Holocaust deniers as well.

The hope is that people everywhere will want their points of view heard beyond their own ideological or cultural circles. The challenge of any change agents is not simply to preach to their converted, but to engage the skeptics and doubters in a conversation and perhaps even change their minds. But will anyone listen in today's increasingly fragmented environment, where switching perspectives is as easy as clicking a button? No one knows for sure, but our research on young people's media habits suggests that a growing number understand this challenge and intentionally seek perspectives with which they may not agree.[47] Twenty-five-year-old Rahaf Harfoush says she "syndicates feeds of various points of view, including those I'm unlikely to support. It helps me form an opinion." But this kind of behavior takes smarts and it takes work. Not every young person is as diligent as Rahaf, so we ought to double our efforts to instill these values at a young age. Just as we need to teach greater media literacy, we ought to reinforce the importance of pluralism in our educational institutions.

NEW BUSINESS MODELS FOR NEWSPAPERS

It cannot be said that the newspaper industry didn't see the dramatic changes on the horizon. It was one of the first to leap into the digital world. In the early 1980s a consortium of newspapers launched groundbreaking experiments in the computer-based creation and distribution of papers. An astonishing television report on San Francisco's KRON TV, predating the launch of the PC, began: "Imagine, if you will, sitting down to your morning coffee and turning on your home computer to read the day's newspaper."[48] A dozen years later when the World Wide Web came on the scene, newspapers were among the first to "repurpose" their content on the Internet. Some even encouraged their writers to blog. But most newspapers stopped there.

Now they're in a pickle. And the solution is not clear. The industry likes to discuss the "monetization of content"—a nice piece of jargon referring to "how can we make money from our content?" Another way of putting the question is "how can we make money from selling a product that, in its current form, interests fewer and fewer people?" Better questions would be:

"What new products and services should we create based on our existing expertise?" "What new capability can we orchestrate, to create unique and compelling value for consumers?"

In the hunt for new business models and new audiences, few newspapers have gone as far as *The Guardian* has in embracing a degree of openness that would see most newspaper executives committed to the asylum by their shareholders. Rather than lock up their content behind a pay wall, as Rupert Murdoch has done with papers like *The Wall Street Journal*, *The Guardian* has opened up its vast collection of data sets, articles, video, and images—including over a million stories going back to 1999—and invited the world to start remixing. This isn't just altruism. Emily Bell, until recently *The Guardian*'s director of digital content, thinks sharing their content will help unlock new services and new sources of revenue. The whole idea, she says, is to get *The Guardian*'s content "woven into the fabric of the Internet."[49] Just as YouTube videos have become a ubiquitous feature on sites across the Web, Bell wants *The Guardian*'s articles, images, and data sets to become a pervasive feature in the blogosphere, social networks, and other popular destinations. After all, why should *The Guardian* assume that only its in-house team can envision and profitably implement all of the potentially compelling ways its content could be packaged and delivered to readers today and in the future? Wouldn't its chances of success be greater if hundreds and perhaps even thousands of collaborators had the opportunity and the motive to innovate on top of *The Guardian*'s content?

In contrast with many newspaper groups, *The Guardian* is effectively letting go of its control on content in order to maximize its reach and maximize innovation. The more people who repurpose *The Guardian*'s content across the Web, the greater the number of eyeballs that see its brand and ultimately the ads that will accompany its content through its proprietary ad network. And just as other popular platforms such as Apple's App Store create lucrative opportunities for developers who are clever enough to extend the iPhone's functionality, there is a big prize awaiting the smart minds that turn this valuable new slice of the information commons into useful and engaging services. Indeed, if a developer figures out how to build an audience and make money using *The Guardian*'s open platform, everyone shares in the profits.

How about the journalists who write for *The Guardian*? They are no longer just producers of content; they're becoming curators who establish a context in which readers and other partners help push the boundaries of what a newspaper is and how it's created and consumed. Indeed, savvy journalists could reinvent themselves as media curators, leveraging open platforms like *The Guardian* to present customized views to clients who need help navigating through the increasingly cluttered media environment.

While *The Guardian* sees its open platform as the key to reinvention, others like Jeff Jarvis have been thinking about whether there are business models that could support a streamlined news organization that uses the Web as a platform for conducting high-quality journalism. With a team of business analysts and journalists in the City University of New York Graduate School of Journalism, Jarvis conducted a fascinating experiment. He asked his students to determine what might happen to journalism in a top-25 metro market should a major newspaper go bankrupt. Can journalism be sustained? And how? Jarvis wanted to find out whether there would be audience demand for investigative journalism, and if so, who would pay for it and how much.

Using detailed industry and survey data, they hypothesized different scenarios where the community could continue to be served well, but using models very different from those used today.[50] The most interesting don't involve an organization in the traditional sense but rather ecosystems of bloggers and journalists who could generate enough traffic and advertising revenue to work as full-time journalists—albeit with a much smaller cost structure.[51] Some bloggers today can earn over $200,000 of revenue by themselves.[52] So it's feasible to think they could do even better in a network. Innovative models like these won't preserve the newspapers we know today, but they offer hope for journalists and for the values they represent.

We believe a new, more dynamic news industry looms on the horizon. But how can newspaper executives reinvent their value proposition and their business models? Four key strategies suggest themselves.

First, listen to today's youth, because within their culture is the new culture of news and information. At first this may sound strange. After all, most young people don't read newspapers, at least not in the traditional way. They find their (commodity) news free on the Internet. And they are more likely to discover news through a hyperlink embedded in someone's Twitter feed than they are through browsing a traditional newspaper online. But if this is how most news will be consumed in the future, newspapers are just delaying the inevitable by not getting with the program now. News execs should use the new media themselves, as personal use is the precondition to any comprehension. Give up on the old model—no one will pay for commodity news or some bundle of content that they can get elsewhere. And unless you've got something really unique (see next rule), forget about walled gardens and micropayments—they don't work with the new generation and eventually won't work for anyone.

Second, commodity news won't cut it for any audience, let alone youth, so create a distinct offering. People will pay for unique value, that much is certain. Indeed, lots of content providers sell their content profitably. Thomson Reuters is growing as one of the world's leading sources of specialized information for businesses and professionals. *The Economist* is also bucking the trend. While news magazines are atrophying, *The Economist* insists that you subscribe to their dead-tree publication to get access to their digital content. And they're getting away with it. Subscriptions have been growing and, unthinkably, so have newsstand sales with more than 75,000 copies sold every week in the United States alone at a price of $6.99! Supposedly a British magazine complete with UK spelling and journalistic sensibilities, *The Economist* now has 800,000 subscribers in the United States and is positioned to surpass *Newsweek*.[53] The key to their success? *The Economist* has a unique value proposition—they provide something that no one else does. It's the only reliable place to get international news presented under a clear viewpoint of fiscally conservative, socially liberal thinking. *Newsweek* is an optional read. But for people who care about what's happening in the world, and who share their perspective, *The Economist* is a must-have.

Third, develop rich, multimedia experiences for new digital platforms and devices. In fact, new possibilities for newspapers and magazines are appearing daily as iPad or tablet-style devices come into the marketplace. Chris Anderson, editor-in-chief of *Wired* magazine, argues that the current versions of digital magazines on a laptop or desktop lose the "coherence and majesty" of the printed version. But the additional functionality of tablets, such as 360-degree views, iPhone-like response to hand movements on the screen, collapsing and layering content, including video and audio, all adds up to a compelling reader experience.[54] He's right, and these devices are particularly attractive to the Net Generation. Indeed, after only nine days of sales, *Wired* magazine reported that sales of its iPad edition had surpassed the print edition. Moreover, the enhanced functionality of the tablet will apply not just to the editorial content but to advertising as well. This means advertisers could potentially obtain greater insight into reader behavior by measuring what readers did on a page and what engaged them.

Finally, embrace collaborative innovation. Take a page out of *The Guardian*'s playbook: create an open platform so that others can help you invent new sources of value. Just as IBM's participation in the Linux community was a factor in helping the company transition from commodity operating systems to lucrative consulting services, newspapers can exploit the power of collaborative innovation to move from commodity news to exciting new models and services that give readers and other participants a major role in value creation. Imagine the news ecosystem of the future, with thousands and perhaps millions of contributors plugging in at different levels: citizen journalists who upload photos, videos, and eyewitness accounts; feature bloggers who only get paid based on advertising revenue; professional reporters who focus on higher-value activities like investigative journalism; and a new layer of knowledge curators who repurpose or remix all this content into new offerings. With organizations like *The Huffington Post*, this vision is becoming reality and many more are sure to follow.

None of this will be easy, particularly for the old guard. Indeed, with legacy cultures, infrastructure, and entrenched old-media attitudes, one thing is clear—there will be many more failures along the road. As for journalists, not all are crying the blues. For many there is a lot of pent-up anger about the old media and a real sense of empowerment and even excitement about the road ahead.

12. INSIDE THE FUTURE OF MUSIC: PROSUMERS TAKE CENTER STAGE

Playing for Change is an outstanding Web site that shows the power of digital technologies being used to help create innovative and exceptional music. Through videos, photos, and blog entries, the site tracks the travels of award-winning filmmaker and producer Mark Johnson, who has circled the world to showcase the world's incredibly rich diversity of musical genres. From a blind street singer in New Orleans to Native American drummers and South African choral singers, Johnson uses the power of the Internet to bring hundreds of talented and colorful musicians from dozens of regions to the world's attention. Are you interested in Champeta roots music from the Caribbean coast of Colombia? Or how about Kashmiri folk songs from the remote beauty of the Himalayas? Johnson's got it all on his Web site and he's inspired some pretty unique musical collaborations as a result.

Johnson's first recording was of a street musician in California singing "Stand by Me." He then recorded and filmed thirty-seven other street musicians in different countries singing or playing the same song. Each musician could hear and play along to the music of the other musicians already recorded. When the individual recordings were woven together, the result was a true musical tapestry and an unparalleled international musical collaboration. The song was a hit too—its riveting video was a favorite on YouTube and Facebook and has been viewed more than 30 million times.[1] "Each of the musicians is aware that this is a project about making the song the best it can be, rather than highlighting the individual," says Johnson. "I sit down and show them where the song is up to on a video iPod, and we discuss how can they contribute to making the whole thing better."[2] Perhaps more important, Playing for Change highlights a broader role for music in a global society. "Wherever they are, people use music to persevere through struggle," Johnson observes. "Whether it's in post-Katrina New Orleans, or between Catholic and Protestant kids in Northern Ireland or the Israel-Palestine conflict, the idea is to unite and uplift people all over the world through the power of music."

Andres Lopera, a twenty-six-year-old trombone player from a small town in Colombia, is part of an equally remarkable story. The seventh child of a father who works as a recycler and a stay-at-home mother founded and now leads a thirty-five-musician chamber orchestra at the University of Texas.[3] He's also performed at prestigious venues such as the Vatican and the Hall of the Americas in the Organization of American States. Andres, like many other aspiring musicians, didn't come from a background that would allow him to attend the most prestigious conservatories in the United States or Europe—schools that lead to performing careers in the most renowned orchestras of the world. However, his talent, combined with a revolutionary new kind of orchestra enabled by the Web, has given him a chance to shine.

"Our home is the Internet," says Hilda Ochoa-Brillembourg, founder of the Youth Orchestra of the Americas (YOA). Ochoa-Brillembourg started the organization in 2001 in order to give young people such as Andres a chance to reach their full potential—both as musicians and as leaders in their communities. "We want to incent our musicians to be more than just musicians. We want them to become political leaders, business leaders and community builders." And similar to El Sistema, a widely successful network of youth orchestras in Venezuela, which the YOA was modeled on, the program is achieving some amazing results.

To date, over ten thousand young musicians from twenty countries across the Western Hemisphere have auditioned for the YOA. Each year one hundred of the top players are selected to participate in a five-week program which brings them together to practice and perform.[4] The audition process has been greatly streamlined by the Internet, with the YOA being able to reach young musicians in faraway places like Montevideo, Uruguay, and Quetzaltenango, Guatemala. "We've started with three hundred applications per year and now we're getting over one thousand," says Ochoa-Brillembourg.

In the fall of each year the YOA uploads scores that the musicians use to compete for a spot in the orchestra. Audition tapes, two thirds of which come in via YouTube, are vetted by panels from various conservatories and music schools. The collaboration and community building starts as soon as the orchestra is chosen. "We've been using chat rooms for many years to help the musicians meet and talk to each other. Although more recently these have been overtaken by Facebook which is now much more popular." The YOA's Facebook page has more than five thousand fans and is growing by five hundred per week. The orchestra's seven hundred alumni, which include renowned artists such as Gustavo Dudamel, also have their own portal, which helps members stay connected after their stints with the YOA.

Ochoa-Brillembourg's latest project is to take the YOA Web experience to a new level. If you can't make it to one of the YOA's concerts in person, you will soon be able to listen to live performances in their new virtual concert hall. "We have an orchestra that is truly transforming communities across the Americas but we don't have a piece of real estate, such as Carnegie or the Lincoln Center, which people can associate with us," she says. "That is why we are building a virtual hall that can transcend the usual constraints of space and materials. It will be a truly extraordinary experience." Indeed, ten years ago it would have been hard to understand how an orchestra with an annual turnover of over 50 percent and no physical home could be possible. Now, thanks to the Internet, the YOA is pioneering new forms of musical expression and collaboration while transforming the lives of young artists along the way.

WIKINOMICS AND THE MUSIC INDUSTRY

As the Playing for Change project and the Youth Orchestra of the Americas show, the Internet opens new doors for creative collaboration and new models in the arts. Just about everywhere, the Web and the principles of wikinomics are changing the way music is produced and disseminated, deepening the bond between performers and music lovers. In this chapter we document: a radical new idea for a streaming music service that would give listeners access to all recorded music on any combination of devices they like; a new online community where music fans take over the A&R function of major labels; emerging creative platforms where amateur community members remix one another's artistic works into powerful new combinations; and a rising indie label called Nettwerk that is finding clever new ways to monetize the emotional connection fans make with the music its artists produce.

All of this and much more would be possible on a larger scale today except the foundation of the contemporary music industry, the major record labels, is standing in the way. Instead of seeing the Web as an opportunity, the record companies cling belligerently to their old analog business model, and the industry has become the poster child of failed digital opportunities. The percentage of Americans labeled "active music buyers" (those who purchase more than four CDs per year) has plummeted to just under 20 percent a year for the past three years.[5] Despite the efforts of Apple and Steve Jobs, online music sales, initially opposed by the record labels, make up a tiny portion of the lost revenue. Only a fraction of the billions of songs downloaded in the United States every year are paid for. The International Federation of the Phonographic Industry, a trade body, estimates that

95 percent of music downloads worldwide are illegal.[6] And industry insiders tell us that after sales of CDs and online songs, lawsuits against customers are the third largest source of revenue. The labels' attitude toward digital disruption causes them to engage in one irrational behavior after another. The sad upshot is that the industry that brought us Elvis and the Beatles is now hated by many of its customers and has taken up suing children. The industry's collapse is rich with lessons about building successful enterprises in the wiki world.

The labels' problems predated the Internet. Recorded music is a bloated industry with huge costs from its physical distribution model. It also has many intermediaries, such as distributors and promoters, looking for a cut of the action. To take a band from obscurity to popularity is hugely expensive, but that's what companies have had to do to be given coveted shelf space at the record store. The upshot is that the record companies have been in constant quest of superstars, since less than 10 percent of CDs released actually make a profit.[7] Revenues generated by the best sellers cover the losses incurred by their poorer-selling cousins.

It's as if the game of baseball only counted home runs. Anything less is considered a strikeout, so if you don't look like a home run hitter, you don't get to play. In this context, the Internet should have been a godsend, allowing labels to distribute a digital copy of a song to hundreds of millions of listeners at virtually no cost. By sidestepping the industrial age infrastructure, many more musicians can be profitable. Continuing the baseball analogy, suddenly players capable of hitting only singles or doubles are moneymakers. Even a musician who only bunts could make a living. Record companies, reinvented as digital networks of artists, fans, and entrepreneurs, could nurture many small artists, rather than focusing all their energies on potential superstars. As a society and culture, we would be much better served by such an approach.

EVERYWHERE INTERNET AUDIO

The solution to restoring the music industry to economic health is not to sell songs at a dollar a pop. Instead of clinging to late-twentieth-century distribution technologies, like the digital disk and the downloaded file, the music business should move into the twenty-first century with a revamped business model using innovative technology. The industry needs to think wikinomics. As we've advocated for many years,[8] there is one obvious solution: music should stop being a product that you buy and instead be a service that you subscribe to. Instead of purchasing tunes, listeners would pay a small fee—say $4 per month—for access to all the songs in the world. Recordings would be streamed to them via the Internet to any appliance of their choosing—such as their laptop, mobile device, car, or home stereo.

Call it Everywhere Internet Audio. In effect, every customer would have their own channel and could slice and dice the massive musical database any way they liked—by artist, by genre, by year, by songwriter, by popularity, and so on. The Everywhere Internet Audio service would know what each user likes, based on what they've chosen in the past. You could vote, thumbs up or down, on tunes while you were listening (if you cared to), which would further refine your playlist. You could ask your Everywhere Internet Audio service to suggest new artists that resemble your known favorites or to create a new playlist called "Eric Clapton's current favorites" or "J Dilla's classic beats." As the services became more sophisticated, you could ask the service to retrieve upbeat tunes for you from your high school years that feature great guitar riffs.

Musicians, songwriters, and even their labels would be compensated through systems that track their popularity. But this would require collaboration and even sharing. All the music would be pooled, and using actuarial economics, the total pie would be divided up according to the number of times the songs of a given artist were streamed. Technologies and companies already exist that can do this.[9]

Everywhere Internet Audio would dramatically reduce "illegal downloading" without putting a single teenager in jail because the problem of copyright protection would vanish. Who would ever bother to "steal" music? Why would you take possession of a song when you could listen to any song at any time on any device? But rather than build bold new approaches for digital entertainment, the industry has done everything in its power to defend the status quo. Sadly, obsession with control, piracy, and proprietary standards on the part of large industry players has only served to further alienate and anger music listeners.

There is growing support for streaming, and initial variations of Everywhere Internet Audio are already in business. Services such as Last.fm, We7, Rhapsody, Spotify, and Pandora stream music to their subscribers, along themes of music and also individual tunes, on demand. But these companies aren't offering the wide array of personalized services that we envision, at least not yet. But last year Apple bought digital music provider Lala, and it's likely Apple will pump up Lala's services to offer many of the features we foresee.

Reality Check: Will Streaming Audio Work?

Not everyone agrees with us that streaming audio is the best solution. For a reality check, we went to the most sophisticated critic we could find and engaged him in a conversation. Cory Doctorow is not just a Web pundit on his tremendously popular site BoingBoing.net, but a social critic and novelist. Doctorow says the service would fail because people like to collect things. People want to have music collections and fill their iPods, and the amount of music that can be stored cheaply continues to increase. As he put it: "People download music today and don't show any sign of stopping." But that's not always the case. People used to like collecting CDs, but that's less true today. And people certainly don't want to own YouTube videos. Why bother, since they're always available online? Being able to stream them any time makes ownership irrelevant. Music should be the same.

Doctorow also believes that "wireless companies are consistently rated the worst in the world. They treat their customers very, very badly." We won't quibble with that, because service is often unpredictable and also quite expensive. We understand his concern about the quality of 3G networks, but we take as a given that individuals will soon have a high-quality, always-on connection to the Internet while stationary or mobile. Maybe it will be 4G or Wi-Fi or a combination of different technologies. But it's coming.

Doctorow also worries about a third party knowing all your music preferences, and that some would see this as an invasion of their privacy. Do you want some central source knowing you like Islamic music and could you be profiled as a result? But the way to address this issue is to design privacy safeguards into the system. You could insist your service provider not keep your music requests on record. You may lose some of the benefits of Everywhere Internet Audio, but for some people this would be a small price for privacy comfort.

Users might also worry about the service going bankrupt and losing their playlists. This is a legitimate concern. Some people have lost their photographs because they entrusted them to online photo hosting services. But users could take steps to safeguard their musical preferences. You would own your playlist and this information could be stored on your device. Should your ISP or music provider go bankrupt, your preferences would still be safe. You could pass them on to your next music service provider.

Other approaches could solve the industry's problem, but again, they require wikinomics thinking—experimentation and a spirit of collaboration—traits the labels have failed to demonstrate. Intellectual property scholars William Fisher and Neil Netanel argue that peer-to-peer music sites should be allowed to distribute music for free. But the providers of such services, including Internet service providers (ISPs) and device manufacturers, would be charged a fee. As with Everywhere Internet Audio, artists would be compensated according to the popularity of downloads.[10] Alternatively, the Electronic Frontier Foundation has proposed a "voluntary collective license" that would give the purchaser immunity from prosecution for noncommercial file sharing.[11] Again, the fees from the license would be pooled and divvied out to artists.

Fresh thinking like this is supported by a growing number of musicians. The Songwriters Association of Canada is proposing an end-user license fee—something President Eddie Schwartz says is supported by the vast majority of global music lovers.[12] The fee, estimated at $4 a month for access to the world of recorded music on demand, would be administered by the ISPs. This is not some kind of coercive socialistic tax, as some might suggest. Consumers, creators, and rights owners could opt out. "This proposal would do more than eliminate the problem of so-called 'stealing' of music," says Schwartz. "It would enable musicians, songwriters and their agents to be fairly compensated for their work. And with more artists able to make a living, there would be a wonderful explosion of creativity and everyone would be winners." The record labels should be listening.

MUSICIANS CHOOSE MASS COLLABORATION

While the record labels are stuck in a rut, creative artists are forging ahead. For example, when drummer Josh Freese (of Nine Inch Nails and Devo fame) released his second solo record, *Since 1972*, he offered his fans a sliding scale of "limited edition" offers. For $7 fans got a conventional digital download, including three videos. There were offers for every size pocketbook. But check out the $75,000 package:[13] You could not only get a signed CD and T-shirt, you could go on tour with Josh for a few days. You could have Josh write, record, and release a five-song EP about you and your life story and you could even take home one of his drum sets (only one, but you could choose which one). If you happen to be a musician, Josh would join your band for a month . . . play shows, record, party with groupies, etc. And for the really adventurous, Josh offered to take shrooms and cruise Hollywood in a Lamborghini or take his limo down to Tijuana (exactly what that entails we can't legally get into here). Of course, Freese's over-the-top offers are tongue-in-cheek. But at the same time, his unusual antics are symbolic of larger trends, like the increasing personalization of the relationships between fans and artists and the growing number of collaborations between them.

Indeed, when you look past the entire file-sharing hullabaloo, the real opportunities for innovation start to reveal themselves. Social networks that allow artists to connect directly with fans are poised to take over large parts of the traditional music industry, ceding the advantage to new players. Take the process of discovering new artists, once the exclusive domain of recording industry's A&R (artist and repertoire) representatives. The music industry is being sidelined by sites like OurStage, which let fans choose their favorite artists for themselves. When OurStage was founded in 2006, Ben Campbell had one goal in mind: bring true democracy to entertainment. "The internet was supposed to [. . . help] artists go around the big record labels and around the studios and go straight to their fans," Campbell explains. "But it hasn't worked out that way because tens of millions of people showed up to the Internet party at the same time with no way to sort the quality from the noise."[14]

To deal with the problem, Campbell devised a system that connects millions of fans to striving musicians to help them discover and promote the most promising artists. "Rather than having some guy in a suit in NYC or LA make the decisions, we let the fans decide who is great."[15] Participation is easy. Artists upload sample tracks to the OurStage platform describing the genre and adding a few personal details. Music fans can then listen and compare the tracks to those of other artists. OurStage's algorithms do the rest by creating rankings or charts highlighting the most popular selections.

There are benefits for both fans and musicians. Music lovers get access to the latest bands, and can connect with artists and other fans and win prizes for ranking songs. Musicians get exposure to millions of potential supporters and can win as much as $5,000 in competitions and build their band's brand name. OurStage also has unique partnerships with the likes of MTV and Live Nation, which give musicians a shot at coveted gigs, formal recording contracts, and mentoring sessions with major artists. "Our goal is to provide real, meaningful opportunities for our artists [. . .] including: live performances at festivals like Bonnaroo, Virgin Music & Bumbershoot; entrance into competitions like Starbucks Music Makers and John Lennon Songwriting; spots on CD samplers like Relix, Paste & CMJ . . ." says Campbell.[16] OurStage is also building an online marketplace to connect artists with potential gigs, another shot across the bow of music industry incumbents.

Over the past few years OurStage has seen tremendous growth. The site has over 120,000 participating artists and an average of 4 million visitors a month.[17] OurStage's footprint and influence goes well beyond the site, with top OurStage artists showcased on MTV and playing leading music festivals, as well as on radio stations. The site reaches out to other social networks such as Facebook and MySpace to engage their users and promote OurStage musicians. Most important, OurStage has successfully launched and advanced the careers of countless artists while providing fans with *the* place to go on the Internet for new music.

While OurStage is one of the more sweeping examples of the emerging "new world order" for music companies, other sites are also harnessing mass collaboration to help artists and deliver music to fans. Another site called Slicethepie tells each user to "help yourself to a piece of the music industry" by directly financing promising artists. The artists can use the money to record an album and if it's successful, a portion of the proceeds trickles down to the original backers. The Web site Topspin (topspinmedia.com) helps artists connect directly with their fans by offering a platform to manage music catalogs, promote their work, and sell music without the usual intermediaries. Similarly, Nimbit is a direct-to-fan sales, marketing, distribution, and artist management platform that strives to give control over creative work back to musicians.

While the traditional music industry mammoths are unlikely to fall overnight, and most will try to adapt through partnerships or acquisitions, the writing for the old model is on the wall. Artists, as well as music fans, are increasingly finding new ways of connecting and collaborating using tools like OurStage and Slicethepie. And this doesn't leave much room for those standing in the middle.

Indeed, Radiohead shocked the industry when in October 2007 they ignored every industry convention and released their album *In Rainbows* as a digital download exclusively through their own Web site. Then Radiohead did something else that few artists or labels have dared: they let customers decide what to pay for it! Contrary to conventional wisdom, estimates suggest that millions of fans paid between an average of $6 and $8 for the album. Since Radiohead is independent, they don't share that revenue with a label. It sure beats the average of less than 10 percent that the typical major label artist earns from a CD sale. EMI's chairman, Guy Hands, called Radiohead's actions "a wake-up call which we should all welcome and respond to with creativity and energy."[18] Sadly, we're still waiting to see some evidence that the broader industry shares his sense of urgency.

MUSICAL PROSUMERS

While fans can get more engaged in promoting the artists they love, music lovers can increasingly become artists themselves. The Web is changing our experience of music. Instead of just passively listening, people everywhere are becoming prosumers, engaged in expressing and creating musical works. Aspiring musicians can now have the equivalent of a million-dollar recording studio on their Mac or PC at home. Consumers of music can now remix and even produce it themselves—and they do, in increasingly large numbers, using the Web as a stage for their performances. Some people do this for profit, but most people just do it for fun.

If you love to play guitars like we do, you'll probably love RiffWorld, an online community where guitar players around the world collaborate to crank out old hits and even cowrite original music. When we spoke to founder Doug Wright, he talked about how RiffWorld is breaking down musical barriers and giving rise to global, self-organized groups of guitar players who come together spontaneously in the online version of the Friday night jam session.

Normally, if you are a guitar player and you want to play music with people, you put a classified ad out or you look for someone to jam with in your immediate environment. It takes a lot of time and effort to set up schedules to play with other people and get together with all your gear, so you limit your search to only those people who are the best match for your age, taste, style, or genre. "But with online collaboration," says Wright, "finding other players is much easier, and so people are more flexible in their boundaries. We have sixteen-year-old metal fanatics playing with sixty-year-old blues guys and both are having a good time at it. So we are not only crossing geographical boundaries, we are crossing generation gaps, genre rifts, and style canons."

For another view on how collaboration is changing the experience of music for consumers, ask Amanda Kistindey, a single mom and an avid music lover who is also a hobby composer and record producer. Not in the traditional sense, though, as she doesn't do it for commercial gain. Rather, Amanda is part of an extraordinary community of 6,500 people[19] who collaborate using a Web platform called hitRECord to compose, remix, record, and produce music and other artistic content, simply because they love doing it. She got deeply engaged in the fall of 2009 when she was able to remix an instrumental arrangement from a user in the United Kingdom with the lyrics of a poem written by a user in the United States to create an entirely new work of art.

HitRECord.org was established in May of 2005 when professional actor Joseph Gordon-Levitt needed a place to store his musical ideas and experiments. Having originally set it up as a creative storage locker on his brother's office server, he began to share it with others, and it wasn't long before it grew in popularity and evolved into something more substantial. In 2007, a forum was established and, from that, a new community for artists was born. "The purpose of hitRECord," as described by Gordon-Levitt, "is to facilitate a collaborative creative process resulting in collectively RECorded and remixed works of multimedia."[20] Records can be any form of documented expressionism—including photography, written word, video, and music—users are encouraged to take this media and remix, rework, and release their new creative mash-up back into the community. Since then, thousands of RECorders have joined the site and a total of 30,868 completely original articles of media have been released. Seeing the potential, Gordon-Levitt turned hitRECord into a professional production company in January 2010.[21] Now, when he sees something really special, hitRECord will invest the resources to turn it into a money-making production. Fifty percent of revenues goes to the creators once costs are covered.[22] "The media used to be a monologue, now it's becoming a dialogue," says Gordon-Levitt. "And there's a really big difference," he says, "between a relatively small industry broadcasting its signal all over the world and a global interconnected one where everyone has a chance to play."[23]

Collaboration is not only welcomed at hitRECord but is the foundation of their community. For Amanda the pool of 31,000 various recorded resources (photographs, videos, music, and written word) serves not only as a means for creating collaborative results but also as "a springboard for inspiration." One person's doodle is uploaded, another user animates it, another writes a song about it, and before you know it, three individuals have created a music video from different locations around the world. She describes it as "a true coming together of like-minded individuals to celebrate a common purpose: the rebirth of artistic creation in a digital world."

Indeed, through communities like RiffWorld and hitRECord, we are beginning to see a shift in the way that amateur artists and creators share and develop ideas. Rather than toiling in isolation, they are flocking to the Web to form communities with their peers. "There's 20 million guitar players in America but there's only about 500,000 bands," says Wright. "That means there's 19 million guitar players who have day jobs and school and families. They have an hour a day, maybe an hour a week, to play the guitar. But with the Internet they can experience what it's like to play music and write music with other people who are as passionate about guitar as they are." And that, in the end, is the ultimate expression of musical prosumerism: not just listening to music, but making it with a community on the Web.

NEW BUSINESS MODELS FOR MUSIC: NETTWERK RECORDS SHOWS THE WAY FORWARD

If there is one thing that stands in the way of exciting new musical collaborations, it's the industry's very rigid views on intellectual property rights. Our research on the Net Generation over the years has really reinforced that view. The world first got a taste of the Net Gen's changing intellectual property norms as millions of technology-literate kids and teenagers flocked to Napster (and later Kazaa, BitTorrent, and LimeWire) and revolutionized the distribution of music, television shows, software, and movies as a result. File sharing now accounts for half of the world's Internet traffic—much to the chagrin of Hollywood—signaling that the Net Gen is renegotiating the definitions of copyright and intellectual property. The most promising aspect of this development, however, is that, like the guitar players and musical innovators described above, a growing number of Net Geners are applying their new skills in the online media world to engender a new creative renaissance on the Web.

The oft-cited Pew Internet Project's survey of U.S. teenagers confirms that more than half (some 64 percent in December 2007, up from 57 percent in 2004) of online teens are what the project calls "content creators."[24] These content creators report having engaged in myriad activities, including creating blogs or personal Web sites, sharing original content such as artwork, photos, stories, or videos, and remixing content found online into a new creation. New online services like Jumpcut make it ludicrously easy for this generation to make and remix movies, while thousands of bedroom DJs are taking advantage of new technologies that put a recording studio on their personal computers.

So, are changed attitudes toward intellectual property a threat or an opportunity? Intellectual property hawks (like the Business Software Alliance) brand file sharers as a bunch of pirates and worry that they are increasingly indifferent to copyright law. But is it really piracy when DJ Copycat spends hours mashing up the Sex Pistols' "Pretty Vacant" with tracks from The Charlatans and Visage into a new creation that he then shares with his followers online? Or is this exactly the kind of serendipitous creativity we should be encouraging as we look to invent new usercentric business models in the media industries? While we disagree with copyright fundamentalists such as the BSA and the RIAA, it's clear we need to find a way whereby creators and inventors can benefit monetarily from their work. But shouldn't this include people who spend hundreds of hours making or remixing music, photos, and video for sites such as Flickr, MySpace, and YouTube?

At least some labels are beginning to think differently about intellectual property and how to make money in today's industry. Take Nettwerk Records, home to artists such as Avril Lavigne, Sarah McLachlan, Ron Sexsmith, the Barenaked Ladies, Sum 41, and Dido. The label's cofounder, Terry McBride, thinks the key to success in an industry reshaped by consumer-controlled distribution is to monetize the one thing you can't commoditize—the emotive value of the music experience. And although McBride supports copyrights, he does not toe the usual industry line on intellectual property rights. He says the solution is not to stem the flow of technology, but to understand which technologies consumers are using and then find creative ways to make music part of the experience.

If kids are listening to music while playing video games, McBride wants Nettwerk involved. If people are downloading songs while text messaging, McBride wants Nettwerk music to be available. He's not particularly fussy about the medium—the content is his message. McBride's recipe for success is short and simple: "We're going to support every platform and format that comes, and we're going to allow the fan to consume our music however they want to consume it. We're not going to tell them how to consume it." In other words, "music should be like water: let it flow in every way possible."[25]

It sounds reminiscent of the cyber-libertarian refrain: "music wants to be free." But McBride has established a serious track record in the business, making him a force to contend with. He claims to have had the same knee-jerk reaction the rest of the business did when file-sharing technologies like Napster emerged. "But once we started understanding it, getting educated on it and talking to the fans, we realized this is a whole different ball of wax and there is a way to make this work." McBride seems to concede that copies of his artists' music will inevitably circulate on the Internet for free. But he's bullish about the ability of Nettwerk to find creative ways of monetizing the fans' emotional connection to the music his artists produce.

Fans of the Barenaked Ladies, for example, can buy the recording of the concert they just saw. "Copies of the live show are available within five minutes of the concert ending," says McBride. During a tour of the United Kingdom in 2007, USB live sales were about 70 percent of the Barenaked Ladies' media sales. McBride estimates that 5 to 10 percent of the people who showed up to the show bought a USB stick before they went home. "That's a huge conversion rate," he says; "we're usually lucky if we get a 1 percent or 2 percent conversion rate on CD sales."

Nettwerk has also experimented with increasingly open models of music production. When it came time to produce a Sarah McLachlan Christmas remix single, for example, the traditional approach would have been to spend

$30,000 to $40,000 hiring a producer to do the remix. McBride figured that was too much to pay for a single that would sit in the marketplace for six weeks, so he organized a remix contest on a major DJ site. Hundreds of DJs submitted entries and thousands of fans voted for their favorites. McBride purchased the top three remixes and started to sell them. "It cost us nothing besides the file purchase price, which was only a couple of thousand dollars," says McBride, "and we made more than what we spent on just the digital sales within a two-week window."

Opening up the process to external DJs also led to copious free promotion. "I know for a fact that every DJ that spent a lot of time and effort to make their mix would have played it in their own sets at the clubs that they were playing at," says McBride, who figures that the same prosumption approach could be applied to everything from T-shirts to album cover designs. "There's many different ways to harness the emotional connection with your audience and involve them in the creative process."

McBride's open approach appears to be working. Online sales account for 40 percent of Nettwerk's sales, which are in the tens of millions, compared with the 10 percent to 15 percent his competitors average. If more labels followed Nettwerk's lead, they would look at young people's proclivity to hack and remix digital products as a business opportunity. For example, why not create or encourage the growth of online communities like RiffWorld or hitRECord, where participants could pay a fee to get access to powerful new editing tools along with the raw materials (including the latest music and movies) to fashion their own media creations?

TIPPING THE BALANCE OF COPYRIGHT LAW

Unfortunately, if most record execs had their way, it would be all too easy to extinguish potentially lucrative sources of user-driven innovation and creativity on the flawed assumption that any form of creative "remixing" of intellectual property is tantamount to piracy. Some execs are not even shy about admitting it. "The recorded music industry has for too long been dependent on how many CDs can be sold," says EMI's Guy Hands. "The industry, rather than embracing digitalization and the opportunities it brings for promotion of product and distribution through multiple channels, has stuck its head in the sand."[26] In fact, the real story is worse. The recording industry increasingly uses existing patent and copyright protections as a shield against innovation and change in their markets. And, for as long as new business models remain elusive, media behemoths seem intent on exploiting the heavy hand of regulation to enforce consumer behaviors that are in fact contrary to consumer interests.

As we write these words, the EU is considering legislation that would enable companies to cut off consumers' Internet access if they are found to be "illegally" downloading files. The lawmakers minimally decided that consumers should be given due process and not arbitrarily unplugged. This was in response to French president Nicolas Sarkozy, who had advocated, and whose government passed, a "three strikes and you're out" law, under which Internet use would be tracked and users caught downloading would be warned twice before their Internet access would be cut off for a year.[27]

At least one government has abandoned any pretense that citizens have the right to make fair use of copyrighted material, and is no longer interested in trying to find the proper balance between the rights of consumers and the rights of the music labels and movie studios. In April, the UK government passed the draconian Digital Economy Act.[28] Cory Doctorow, the writer we mentioned earlier, wrote an online column entitled "Digital Economy Act: This means war." As a working artist, Doctorow says he wants just copyright rules that provide a sound framework for his negotiations with big publishers, film studios, and similar institutions. "My whole life revolves around the digital economy: running entrepreneurial businesses that thrive on copying and that exploit the net's powerful efficiencies to realize a better return on investment."[29] But instead, the new law stifles his creativity and establishes an unprecedented realm of Web censorship in Britain. The law means that a family can be disconnected from the net on the say-so of an entertainment giant, access to certain Web sites can be blocked, and free Wi-Fi hotspots are effectively prohibited.

But there is a growing chorus of forward-thinking people who are calling for lawmakers to roll back, or at least counterbalance, the ever-tightening regime of copyright law. For years *Guardian* columnist Victor Keegan has lamented the efforts of entrenched corporate interests to further extend the lifespan of copyright protection. He calls for lawmakers to tip the balance of copyright law in favor of users and amateur creators. Keegan writes:

> We are now living in a digital age of instant and cheap availability, meshing and remixing and of mass creativity, with increasing numbers of creators prepared to give their services free (as in much of the open source movement). We need fresh regulations for a new age before we cave into the demands of the producers as they try to get draconian rules put in place before the shutters come down on the old

> world. . . . The creative economy is vitally important, but the way to nurture it is to follow the winds of the information revolution and not the desire of existing corporations to preserve a business model that has been turned upside down by the revolution taking place in virtually every creative industry.[30]

Keegan's right—we need a level playing field on which new forms of co-creation and innovation can flourish. If Hollywood titans succeed in locking up their vast archives of media content it would certainly be a setback for open-content initiatives, but it will most certainly not be a death knell. The regulatory measures deployed by incumbents amount to temporary stopgaps, not sources of enduring competitive advantage. In any case, a more balanced approach to the way we recognize, encourage, and reward knowledge production will be driven by viable new business models that can fuel value creation, not by putting up legal barriers to inhibit innovation.

As we explained in *Wikinomics*, reticence displayed by the media industries is not surprising, and in some cases it's entirely rational. Publishers of media, music, software, and other digitized goods have a very legitimate business problem: they can't reasonably adopt open approaches that would cannibalize existing revenues without a viable means to shore up their ailing income streams. Industry pundit Jim Griffin calls it "Tarzan economics." "We cling to the vine that holds us off the jungle floor," he says, "and we can't let go of the one we've got until we've got the next vine firmly in our hand."

The problem is that media incumbents are moving too slowly. They're getting mired in the thick underbrush of thorny contractual agreements and outdated and costly infrastructures. What's worse is that the economic foundation of the industry is based on a business model suited for the era of analog publishing, not for a world of user-driven digital creation and distribution. These institutions are powerful and deeply ingrained in the industry's social and economic contract. It's hard for senior executives to imagine a world where their companies could lose control of the very resources that they have monopolized for so long.

The dilemma media executives face is that they can't replicate the phenomenal successes of Linux, Wikipedia, YouTube, and other collaborative communities without also accepting the radical openness with which these communities share their intellectual property. The result is that new business models for open content will come not from traditional media establishments, but from companies such as Apple and Google and smart labels like Nettwerk Records. This new generation of companies is not burdened by the legacies that inhibit the music publishing incumbents, so they can be much more agile in responding to customer demands.

In the end, the saga unfolding in the media industries serves to illustrate a fundamental principle: customer value, not control, is the answer in the digital economy. The media industry—and all industries for that matter—must resist the temptation to impose its will on consumers as a matter of convenience, or worse, as a result of a lack of ingenuity and agility. Rather, music labels and other media entities should develop Internet business models and offerings with the right combination of "free" goods, consumer control, versioning, and ancillary products and services. This includes new platforms for fan remixes, mash-ups, and other forms of customer participation in media creation and distribution. Indeed, what a growing number of artists seem to understand is that you don't need to control the quantity and destiny of bits if you can provide a superior value proposition for customers. Free content is a reality that is here to stay. So artists will have to give customers a product that is better than free—a product that is more convenient, more compelling, and that provides more value than customers can get from illegitimate sources. For successful companies and artists, the ability to deliver exceptional value is their hallmark.

Joseph Schumpeter, the great economic theorist, maintained that businesses must either embrace new technologies by giving up old methods and products or cede the market share to those who will.[31] Of course, Schumpeter's theories of creative destruction assume free competition: that firms will bring new technologies and products to market in a competitive environment and that customers, not government, will be the ultimate arbiters of who wins. The bottom line is that intellectual property laws were never intended to put the brakes on technological change—quite the opposite. But today, the content industry is using these laws perversely—not to promote innovation, but to quash anything that threatens the status quo. This change in intellectual property law threatens the chain of creativity and innovation on which we (and future generations) depend.

Companies in every industry should listen up. Take a page from IBM, which gave away $400 million of its software to Linux and changed the dynamics of the IT industry. Or Goldcorp, which placed its geological data in the public domain and increased its market value by two orders of magnitude. Don't let your business or industry be the next canary in the mineshaft.

13. THE FUTURE OF TELEVISION AND FILM: JUST ANOTHER COOL INTERNET APPLICATION?

Leo Laporte is an influential technology author and Emmy Award–winning broadcaster whose audience grows weekly. But his production facility, called TWiT Cottage (short for This Week in Tech), is a lot different from a typical radio or television studio. The Cottage is on a quiet street in the idyllic small city of Petaluma, sixty miles north of San Francisco. A friendly twenty-something greets your car on the street and escorts you into the lobby (the front hall), offering a cup of tea or cappuccino. In the main studio (the living room) author and broadcaster Laporte sits headphoned on an exercise ball, talking energetically into a microphone surrounded by a wall of technology. There is no one else in the room except his guest panelists and a second staffer who manages eight cameras from a control center (the dining room). Add in company administrators and there is a global human resource of seven people.[1]

Laporte hosts and produces some of the most popular podcasts in the world, including This Week in Tech, Security Now, net@night, The Daily Giz Wiz, Windows Weekly, MacBreak Weekly, Jumping Monkeys, and Munchcast under the TWiT banner. In May 2008, Laporte launched a live streaming video version of TWiT called TWiT Live with twenty-five hours of original programming each week; 2.6 million people watched TWiT Live in its first month.[2]

Laporte interviewed Don in studio with a panel of other experts (and 200,000 of his friends) chatting about how the world is changing. When we followed up a few months later for a book interview, Laporte interrupted our conversation, saying: "By the way, we're live right now—do you mind if we keep the cameras going?" It was the first time either of us had interviewed someone for a book, on TV!

It is a long and winding road that led Laporte to where he is today. A former broadcaster for the mainstream media, he has a solid background in technology television shows, including CNBC's *The Personal Computing*

Show, MSNBC's *The Site*, ZDTV's *The Screen Savers* and *Call for Help*, and *The Lab with Leo Laporte* on Canada's G4techTV. He has also written extensively for technology magazines, and has published a number of best-selling technology books. How did Laporte realize he wanted to run his own "TV show" on the Internet? He says the epiphany happened in 2004 when he made a presentation to the management at the cable network TechTV. They asked for his thoughts on how to boost ratings. "So I came in with a presentation that said, 'If you're going to do a tech channel what you should do is super-serve the audience.' They wanted a larger audience, but they thought the way to do this was to dumb down the content. And I wanted to give them the contrarian point of view that the best way to reach this audience was to be more geeky, more techie, to super-serve their interest."

His boss at the time was Joe Gillespie, and Laporte recalls: "Joe said to me, 'Brand is the refuge of the ignorant.' He said that only stupid people fall for branding messages; that intelligent people will make their decision based on features and benefits." Joe was explaining the philosophy of his advertisers: we can't build a network for intelligent people because advertisers won't be interested. "They only want dummies," says Laporte, "and when that sunk in, I thought I am in the wrong business. You know, here I am trying to explain to them why we should go for the intelligent audience and he's telling me, 'It will never work!'" Laporte did the math and concluded that creating for smart people was a feasible business proposition. He figured that 14 million Americans work as computer programmers. To be profitable, he would only need to reach about 5 or 10 percent of them. So he decided to super-serve a geek audience and to do it at a fraction of the cost by using the Internet to distribute it and by using off-the-shelf consumer technology, like computers, to create it. "The digital revolution makes content creation cheap and the Internet makes the distribution universal."

This is very disruptive to traditional models. Laporte says, "I'm creating a business out of nothing, at a much lower capital cost than mainstream media because bandwidth is cheap and plentiful. If it weren't, none of this would work. And you know, it took a few years but within three years, we built a larger audience than TechTV ever had, by not pandering." It's taken a little longer for advertisers to get the message. "Advertisers now are finally realizing that they would prefer to have an engaged audience of intelligent people than a loosely engaged audience that's not really paying attention. Today my advertisers want me to produce more shows so they can buy more advertising time." In fact advertisers are paying more for ads on TWiT than they did on TechTV, Laporte's revenue has doubled in each of the last two years, and he expects it to double again next year.

Laporte understands how he's part of making the media more of a meritocracy. "As you wrest control of mass media away from the moguls and you democratize, what happens is everybody gets a voice and I think the best voice is being heard." And he uses the wisdom of the crowd to his own benefit. "I think Twitter is the beginnings of an Internet nervous system where, if I follow the right people on Twitter, and I've carefully chosen 1,500 people, I won't miss anything. If anything excellent comes along, it will surface through this group. I use them as 1,500 editors, who are constantly surfacing great content for me." In that sense he has 7 production staff and an additional 1,500 in research, making his total human resources 1,507.

TWiT Cottage is emblematic of how wikinomics revolutionizes more than just the creation, production, and distribution of media entertainment. It is changing the nature of the media themselves, transforming the audience into producers, in turn challenging the economics and even existence of these industries.

Today's version of network television is doomed. The networks suffered their first body blow when cable and satellite splintered the audience across 500 channels. But that was just the beginning. Shows such as Laporte's, delivered via the Internet, give us a 500,000-channel universe. From one perspective, the future of television has never been brighter. TV sales are at an all-time high. The TV is being turned on in the home more hours a day than ever before.[3] And some special event programs are drawing record audiences. But don't be deceived. Television is on track to become just one of the Web's many attractive applications.

The movie industry has fared a lot better, but storm warnings loom. As increased bandwidth becomes available, most films will be accessed through the Internet, and the same challenges of peer-to-peer sharing that upended the record business will challenge Hollywood, Bollywood, and Berlin. Legions of gleeful hackers and frustrated teenagers are dying to immobilize the next Digital Rights Management copyright protection technologies. And as with music, customers want to get engaged with films, remixing them or even creating them. YouTube filmmakers are perfecting the three-minute short and shifting audience time away from traditional models.

At the same time, the video game industry continues to grow, not just as a challenge to traditional entertainment spending, but to the model of media entertainment itself. As gaming becomes increasingly multiuser and networked, it underlines the message that people, especially young people who represent the future, want to be engaged, rather than passively consume media.

Although disruption and calamity abound, in all of these industries there are radical, sometimes inspiring initiatives for their reinvention. Just as publishing is becoming democratized, people everywhere are getting engaged with media, becoming prosumers and, in turn, upending the traditional models. As Jonathan Miller, chief digital officer of Rupert Murdoch's company News Corp., told us: "The Internet has become the primary medium today, not the television, not the newspaper, not the telephone." He describes TV news as being the "highlight reel" of news on the Web.

FILM 2.0

Just as MP3s and the Internet have turned the music industry upside down, emerging technologies in the film industry promise equally dramatic changes coming soon to a theater near you. Or to your computer screen. Or big-screen TV. Or mobile phone. Or minivan. Or sunglasses. The current model of the film—two hours in a dark theater with popcorn—won't disappear, but it will become just a small piece of the bigger market that is best described as Film 2.0. Movies, video games, digital effects, and networking will all mesh to change where and how we watch films. And how films are created, distributed, and monetized. Wikinomics is beginning to transform the movie, the movie business, and how people interact with video content.

Today Hollywood makes money mainly from box-office revenues and DVD sales. Box-office revenues totaled a record $10.6 billion in 2009 for the United States and Canada. Worldwide revenue reached $29.9 billion.[4] It's not just higher ticket prices that account for the revenue increase; actual attendance in the United States and Canada jumped after declining in 2008. During the economic crunch of 2009, going to the movies was seen as an affordable form of escapism.

Unfortunately for the studios, DVD sales have been dropping.[5] The reason: fewer and fewer people want to take possession of a film. They'd like to borrow bits rather than purchase atoms. Renting films is much less lucrative for the studios. With options such as Netflix, dollar-a-day redbox rentals, video-on-demand, and more important, movies becoming downloadable and streamable in real time, consumers can increasingly watch a film wherever and whenever they want. In Europe, where the cellular phone system is more advanced and residential broadband is faster, consumers routinely stream movies to their cell phones, computer screens, or home theaters, all for the same price.

Of course, traditional movie theaters are alarmed and are looking for ways to lure couch potatoes away from their sophisticated home theaters. But unlike other media companies, the film industry has been more open to embracing new technologies and a new generation of digital natives. It realizes that consumer tastes are changing, and many are itching to do more than simply sit back and watch a movie. They like producing content. Witness the twenty hours of video being uploaded *every minute* to YouTube,[6] the equivalent of Hollywood releasing more than 110,000 new full-length movies into theaters each week.

In fact, the whole notion of what *is* a film is changing. The narrative of the full-length feature film, lasting over an hour and running on multiple reels, dates back a hundred years or more to movies like *The Story of the Kelly Gang* (1906), *The Battleship Potemkin* (1925), and *The Jazz Singer* (1927). Since then, movies have improved in video quality but the notion that a single film should last one and a half to two and a half hours has remained fixed. This length justified the price of admission and attracted the audience to make the effort to go to the theater. Yes, a small number of short films were also made, but the audience for such films was limited, largely because theaters didn't want to show them. But by looking at Net Gener interaction with video content online, we see a much more open mind as to what constitutes acceptable entertainment. Tomorrow's films will be more varied in length. Short films will be increasingly popular, watched not only on laptops but also by young people on tiny screens. The three-minute movie already dominates YouTube, which, as we discuss in a moment, will soon be part of consumers' typical viewing diets on their living room televisions.

Prosuming Movies

More important than being open-minded toward films of widely varying length, young consumers want to interact with their film entertainment the same way they do with video games and other content. Previously, this was the preserve of theme parks, such as the Finding Nemo Adventure at Disneyland, where Crush the turtle talks with the audience. Because his performance is live and interactive, Crush gives a different show at every showing.

George Lucas is pioneering a new model of relationship between filmmakers and their fans. Rather than trying to prevent fans from creating and sharing videos related to his popular movie series, Lucas has made material available online that fans can use to create their own Star Wars tributes. Lucas also gave his blessing to a project run by New Yorker Casey Pugh, who chopped up Lucas's 1977 blockbuster, *Star Wars: Episode IV*, into 472 fifteen-second segments and put them online.[7, 8] Star Wars fans from around the world were invited to submit their version of those scenes. Pugh then sifted through the submissions, picked the best, and then wove them together to make an entirely new film: *Star Wars Uncut.* It premiered in Copenhagen, Denmark, in April 2010 and will be made available for all to view at StarWarsUncut.com.

Some may see this extensive collaboration with fans as revolutionary, and in relation to what most other companies are doing it really is. At the same time, it was really a commonsense baby step. When one considers all the positives that come from it in terms of marketing, community engagement, innovation, and fan loyalty, the question shouldn't be why did Lucasfilm approve this—but rather why have most other companies and content creators been slow to follow suit?

The bigger innovation is coming as video games and Hollywood special effects movies morph into entertaining new combinations. Each generation of games looks more realistic, and many are structured around a plot. Video game sales are far more lucrative than films, and the big film studios want in on the action. Jerry Bruckheimer, the famous Hollywood producer of such action films as *Top Gun* and *National Treasure*, reflects the new approach: "We're in the entertainment business. We will entertain you in the theaters, on TV and on your game platforms."[9]

It's not hard to imagine Film 2.0—an interactive experience. It has an underlying sense of destiny and plot along with rich characters played by talented actors. But being digital and interactive, the viewer becomes an integral part of the film, perhaps a protagonist, love interest, family member, or bad guy. And because it is networked, the user collaboratively generates the storyline and character development. Collaboration changes the narrative as well. Take World of Warcraft, an MMOG (Massively Multiplayer Online Game) that engages over 11.5 million people who collaborate across more than 58,000 team guilds.[10, 11] Like the old-style movie, there is a plot and character development. Only the viewer becomes a character in this digital fantasy universe.

While video games are becoming more like movies in their own right, some innovative filmmakers are using real-time, three-dimensional virtual environments like Second Life as a "stage" for their productions. The first so-called machinima film, entitled *Diary of a Camper*, was originally created in 1996 by the Rangers team clan (United Ranger Films) using the Quake video game engine and was inspired by 1993's Doom, which gave players the ability to record game play.[12] In 2007, machinima reached new heights when HBO decided to purchase the rights to a Second Life machinima series entitled *My Second Life: The Video Diaries of Molotov Alva*.[13] Originally appearing on YouTube on March 2, 2007, the Second Life avatar of Douglas Gayeton has created quite the media buzz with over 475,000 hits since then.[14] The "documentary" features a mysterious resident who leaves his real life behind to search for existential answers in the virtual world of Second Life. So far, this deal is the highest profile example of an SL-to-RL (Second Life to Real Life) rights deal to date. This also leverages Linden Lab's policy in which residents may retain the underlying intellectual property rights for content that they create in the virtual world.

Innovative filmmaking experiments of all kinds are growing in numbers as digitization further democratizes the making of movies by reducing cost. The big studios may be releasing fewer films each year, but independents are producing more. Independents are also getting a boost from Web 2.0 sites such as IndieGoGo.com. The site promises "filmocracy" by providing filmmakers with an open platform to pitch their projects to the world. Fans can contribute money, suggestions to improve the film, or both. Rather than the filmmaker looking for a small number of wealthy investors, IndieGoGo promotes the idea of many investors providing a small amount of money each—much like the crowd funding of the Obama campaign. And much like Obama supporters, once fans donate money, they are much more committed to seeing the project succeed. They can use social Web technologies to build online buzz for the film they've just invested in.

SnagFilms.com makes it much easier for independent documentary films to find an audience. The site currently offers for free viewing 850 full-length documentaries, from established heavyweights to first-timers.[15] By streaming films worldwide, on-demand, 24-7, and with no software installation or downloading required, SnagFilms' potential audience is huge. The site's objective: "We make it easy for you to find a film that shines a light on a cause you care about. You can then open a virtual movie theater on any Web site." Widgets let viewers open a free virtual movie theater in their Web page, blog, or Facebook or MySpace pages. "At their best, documentaries don't merely entertain us, they engage and inspire us to action," says SnagFilms. "We provide a link for you to a charity related to the topic of each film (many of them selected by the filmmaker) so you can get involved, immediately. And just by embedding our widgets, you've donated your pixels and helped support independent film."[16]

If the creation process can become collaborative, why can't the theater experience become interactive and engaging too? After all, the growing percentage of movies with digital projectors will give movie directors unprecedented flexibility. At home, viewers of DVDs are accustomed to bonus features such as extended versions of the movie scenes, outtakes, the "director's cut," different soundtracks, and in some cases, different plot twists. With digital distribution in movie theaters, directors will be able to play with interactive features, such as letting the audience vote on key plot elements. Multiple versions of a film will be introduced to theaters simultaneously, offering different features for different audiences, such as a film with subtitles for those hard of hearing, or a version for adults only and a tamer version for teenagers.

As we look forward, it's clear that many kinds of filmmaking and film viewing will be possible. Each will occupy its own time and place, and its own audiences, in a broader constellation of interactive media. In some cases, viewers will be content to sit back and become engaged by great cinematic storytelling. Traditional films will continue to do that. But at the same time, there is a whole new world of power and influence opening up for both filmmakers and people who love great entertainment. Indeed, as most aspects of filmmaking become collaborative—including everything from funding and creation to marketing and distribution—the whole world of visual entertainment will only get more and more interesting.

TELEVISION IN THE DIGITAL AGE

Has TV Really Been "Resilient" to the Internet?

As noted earlier, some aspects of the television industry look healthy. Large-screen televisions are flying off Best Buy's shelves at a record pace. Television ownership in the United States is at an all-time high, at 2.93 TV sets per household. The number of U.S. homes with three or more TV sets is now at 55 percent.[17] Those TVs are being "viewed" by more people, for more hours a day, than ever before.[18] Viewers are still gathering in front of the big-screen monitor to watch their favorite TV shows, particularly live sporting events. And the quality of some of the television content has never been better. This has led some observers to say television has been enormously "resilient" to the effects of the Internet.[19]

Indeed, everywhere one looks, television has been dancing as fast as it can to stay competitive in an interactive world. Shows increasingly solicit audience feedback, as when reality shows such as *American Idol* invite the audience to vote for their favorite contestants. The technology has also evolved to keep pace with increasingly sophisticated tastes, going from the bulky black-and-white box in the fifties to offering color in the sixties to flat screens in recent years. And if manufacturers have their way, we will all soon be upgrading our big screens for 3-D television. The current buzz on 3-D is that sports will be the killer application.

But behind the scenes, the very nature of television—how its shows are being produced, who pays for them, and the technology used to get the content to that big screen—is being completely transformed. The whole concept of television broadcasting is soon to be gone, swallowed by competition from content delivered by the Internet to computer screens, iPads, smartphone screens, and, of course, the large LCD or plasma screen in the living room.

In terms of viewership, the numbers aren't as rosy as they look. Yes, the television is being turned on, but increasingly viewers use television as background entertainment while they focus on more engaging matters, such as their Facebook wall on their laptop, or text messages on their smartphone. Such media multitasking is particularly true of the Net Generation. In the early years of the television industry, the three networks, ABC, CBS, and NBC, towered over all other media producers. Americans were enthralled by television, and the incredibly lucrative and large viewership pie only had to be cut three ways. In 1953 one episode of *I Love Lucy* played on 71 percent of all television sets in the United States. In today's 500-channel universe, a show can be renewed with a paltry share of 5 percent or less.[20] In the mid-1980s, the top-rated show *M*A*S*H* attracted as many as 105 million viewers an episode.[21] By contrast, TV's top attraction for the past three years was Fox's *American Idol*, and even its viewership is in a downward spiral. This year's ninth-season premiere attracted 29.9 million viewers,[22] down from last year's 30.1 million viewers, which was down from 33.4 million during the 2008 premiere, which in turn was down from 37.4 million in 2007.[23]

Few programs illustrate better the television networks' weakening ability to command large audiences for flagship programming than the major evening newscasts. The news programs on ABC, CBS, and NBC have watched their audiences evaporate in the past three decades. In 1980, total viewership for the three broadcasts was close to 50 million. Today the three programs together attract a meager 22 million viewers, and the average viewer is in his or her sixties.[24]

Television: A Cool New Internet Application

In 2001, the denouement of the dot-com frenzy, Americans were asked which they would give up first, the Internet or television. It was a slam-dunk: 72 percent of respondents said they could do without Internet and 26 percent said they'd give up TV. No more. Earlier this year the survey was repeated, and 49 percent said they would drop television, and 48 percent said they'd give up their Internet.[25] And ten years from now, the question will be moot. Television will no longer exist. The Internet will swallow television, and today's programming will become just another application on the Web. Just as today's consumers turn to the Web to book a hotel or flight, communicate with their friends or download a song, soon they'll look to the Web for all their video content. The notion of gathering with family or friends to watch a certain program on a certain TV channel will apply only to live sports events, and shows like the Academy Awards that have real-time news value. Prime time will finally be anytime. After all, viewers don't gather at a certain hour to watch videos on YouTube. Movie fans seem to cope with the idea that not everyone can attend a movie premiere. The 500,000 viewers of Lady Gaga's most recent online video in the first day of its release didn't synchronize watches to make sure they were all sharing the same experience simultaneously (incidentally, in March 2010 Lady Gaga became the first artist to reach 1 billion views online).[26]

It's not hard to see why the Internet will absorb TV. Television only offers television. The Web offers not only television content but almost everything else in the world that has been digitized. And it's interactive. When eighteen-year-old Laura Gillies moved into the students' residence at the beginning of her freshman year at McGill University, she and her roommate quickly agreed to put the television in the closet. So did all her friends in residence. "Why would I want a TV?" says Laura. "I would have to pay for cable, which would make no sense, because I can watch the few television shows I follow, such as *House* or *Gossip Girl*, online whenever I want. And the Internet gives me access to a lot more stuff to watch that is a lot more interesting."

Hulu.com and similar Web sites set up by existing television channels offer current programming as well as old episodes. Hulu is financed by advertising, but has fewer advertisements per program than traditional television. Hulu worries that mouse-clicking viewers will abandon the site if it inserts too many ads in the course of a show. Other streaming content sites are experimenting with charging a fee. Online is clearly the way of the future, and currently Hulu is by far the most successful. In December 2009, it passed the milestone of having TV fans watch more than 1 billion videos on the service.[27] When Hulu began, it just had shows from NBC, FOX, and a few independents, but today it has more than 220 content partners. Because it's online, Hulu can offer interactive features traditional television can only dream of. Hulu recently added closed-captioning for shows. Hulu surfers can search the captions to find specific places in a program. So you can search *The Tonight Show* to find Conan O'Brien's digs at his former employer. Or find out how many times in an episode Homer Simpson mentions "doughnuts" and call up the specific sound bites.[28]

The fact that TV's most successful producer is migrating to the Web reinforces evidence of traditional TV's demise. After years of impressive viewer numbers on network television, *American Idol* creator Simon Fuller moved his most recent program completely to the Web. *If I Can Dream* aims to integrate the Internet and social networking. "This is a weekly, network-quality documentary on Hulu.com and a show streaming on the Web in real time," says executive producer Michael Herwick. "Kids are not watching TV these days—they're on their computers, Facebook, MySpace and Twitter," he says. "This is a new entertainment property for the younger generation that the older generation can appreciate."[29] The show revolves around five people—a musician, two actresses, an actor and a model—living in a Hollywood mansion and tracked by cameras 24/7. Viewers will follow their lives as they try to make it in their respective industries and viewers can also interact with the stars on social media sites.[30] John West, founder of the new Kids Sports Entertainment Network called The Whistle, is now following in Fuller's footsteps. When West first got going he assumed he would be launching a television network with a good Web site. "Now we're a Web property, a set of mobile applications, and a supporting TV presence," he says. When he investigated how young people would want to watch sports, he concluded that "watching" was a small part of the value. "The norms of this generation have made us focus on interactivity, personalization, and collaboration on the Web and mobile platforms," he said. It wasn't just his corporate market research that led him to this conclusion either. "I noticed how my own four- and seven-year-old kids spend more time on my wife's iPhone and iPad than on television."

Even though live broadcasts of big events such as the Super Bowl or the Academy Awards are still able to draw large audiences, we'd argue that their continued success is linked to the fact that interactive components are making these traditional broadcasts more interesting. For example, last February's Super Bowl XLIV, the most-watched program in the history of U.S. television, drew an average audience of 106.5 million viewers.[31] With an audience like this, the NFL is in a league of its own.[32] But a growing proportion of this audience is not just watching passively. Sports fans increasingly share their views through Twitter, blogs, and Facebook while live events are being broadcast. The same is true of people watching the Oscars, Emmys, and Grammys, which after years of decline, are now on the rebound.[33] Why? Because viewers feel more engaged with the event itself and the opportunity to contribute additional content makes watching the show more rewarding. "It enhances the experience of talking about, and talking at, the TV," says James Poniewozik, television columnist for *Time* magazine. "If you follow enough interesting people in your Twitter feed, it's kind of like having a real-life, real-time DVD commentary."[34]

Until recently, for many people the drawback to watching television programs on the Internet was that you were confined to the small computer screen, and often just a viewing window much smaller than the screen itself. The computer monitor couldn't offer a bigger video of sufficient viewing quality because the Internet connection wasn't fast enough to carry larger, data-intensive video. Accordingly, viewers tended to prefer watching sports, drama, and comedy on the traditional television. They chose TV because it had better picture quality, better sound, and gave them an overall better experience. Why do viewers turn to the Internet? Because they missed an episode on TV, it's convenient, and it has fewer ads.[35]

But this is changing. Broadband connections to the home are becoming much faster and able to carry a signal that can fill a large-screen television with a high-definition image. This desire for a fast connection is relentless. Earlier this year, Google announced it would build trial ultra-high-speed broadband networks in a small number of locations across the United States. The project promised fiber-to-the-home Internet speeds of 1 gigabit per second, more than one hundred times faster than what most Americans have today. At this speed, a high-definition movie can be downloaded in less than thirty seconds, which leaves cable and satellite TV twisting in the wind. Google said it planned to offer the service at a competitive price to at least 50,000 and potentially up to 500,000 people.[36] More than 1,100 communities and almost 200,000 individuals across the country formally responded to say they wanted to be part of the project.[37]

Google is not alone in trying to put the Internet on steroids. Verizon gambled $23 billion on building an ultra-high-speed broadband television and phone network called FiOS.[38] Last year Cisco estimated consumer Internet traffic will grow at a compound annual growth rate of 40 percent a year between 2008 and 2013.[39] Internet video will grow even faster, to 60 percent of consumer traffic by 2013, up from about one third in mid-2009.[40] To help this all happen, this spring Cisco unveiled a new routing system that will allow Web users to download videos and files far faster than ever before. How fast? Cisco claims that the router could deliver the entire printed collection of the U.S. Library of Congress—the world's biggest library—in just over a second.[41]

With the Internet soon to be the principal information pipeline into the home, most new television-related products today, including LCD and plasma monitors, Blu-ray players, and TiVo-style devices, have been upgraded to connect directly to the Net. An estimated 9 million Americans already connect their laptops to big-screen TVs to watch Hulu.com programs and YouTube videos.[42] But don't think all YouTube videos are shaky footage of skateboarders doing their stuff. YouTube now offers professionally produced content, which creators put online in exchange for a share of the advertising that YouTube (Google) sells.

On October 9, 2009, YouTube's cofounder and CEO, Chad Hurley, announced in his blog that YouTube was streaming more than 1 billion video clips a day.[43] Hurley wrote that one of YouTube's basic principles was that "Clip culture is here to stay: Short clips are voraciously consumed and perfect for watching a wide variety of content." Hurley also argued that more and more people were using YouTube to make money, turning their hobby of creating YouTube content into paying jobs. Case in point: Michael Buckley, the host/writer/producer of *What the Buck*, one of the most popular entertainment shows on YouTube. On his Web site, Buckley says his show covers "important topics such as Britney's Beaver and Tyra's Latest Weave." In 2007, Buckley won the YouTube award for Best Commentary. In 2008, he was one of the NewTeeVee Breakout Stars of the Year. He made the front page of *The New York Times* and has appeared on all the big networks.

Prior to becoming an online celebrity, Buckley was an administrative assistant for a music promotion company, and part-time host of a weekly show on a Connecticut public access television channel. Buckley's cousin began posting on YouTube some clips of Buckley's humorous rants about celebrities. At the time Buckley didn't know about YouTube and didn't even have a broadband Internet connection. But he loved the positive feedback he was getting from his YouTube viewers, so he decided to build a studio at home to record rants specifically for YouTube. His investment: A $6 backdrop from Walmart, $40 work lights from Home Depot, and a $2,000 Canon video camera.[44] His popularity continued to rise, so Buckley became a YouTube Partner, which lets YouTube sell advertising within and around his videos, and Buckley and YouTube split the proceeds (there are now thousands of YouTube Partners). Buckley quit his old job in September 2008, and now posts two videos a week, earning more than $100,000 a year.[45]

Even traditionally conservative broadcasters like the BBC are getting their feet wet in the new world of collaborative TV. Although it's now standard practice for television networks to offer online forums where viewers give feedback on programs they've just seen, the British broadcaster recently went a step further with an experiment that involves its viewers in the actual scripting of a four-part documentary on the impact of the Internet on society. "For a documentary about how the Web is changing the way society works, it seemed to be a great opportunity to explore how the audience might collaborate," says Dan Gluckman, the program's producer. The program aired its episodes between January and March 2010 and on April 13 was awarded a Digital Emmy for its innovative production process.[46] But in a radical departure from traditional documentary making, the show is posting ten-minute video excerpts that viewers are free to reuse for personal, educational, or charitable causes.

While the audience for traditional network television continues to decline, experiments like these can help freshen up the genre. At the same time, the number of people like Michael Buckley or Leo Laporte who make a living from producing video content is set to increase further. The ever-escalating membership in the YouTube Partner program and the proliferation of advertising-supported Web sites such as TheOnion.com or CollegeHumor.com means that the appetite for video content continues to grow. Instead of being viewed solely on computer monitors, online devices in the living room will bring the Internet to big-screen TV. Just as tens of thousands of people now make a living from selling products on eBay, years after it started out as an online flea market, similarly the skateboard YouTube videos are making room for professional content produced by videographers who have never stepped inside a television station.

So what's a TV network to do when the whole modus operandi of television is being turned upside down? Get with the program and embrace wikinomics! Turn viewers into producers. Create platforms where partners can co-innovate value. Share some of your intellectual property. Don't be just a content provider, be a curator. Create context for the world to self-organize on your platform. Create vibrant communities. When it comes to media entertainment, consumers want to collaborate, share, be engaged, and prosume.

VI

REBOOTING THE PUBLIC SQUARE

14. CREATING PUBLIC VALUE: GOVERNMENT AS A PLATFORM FOR SOCIAL ACHIEVEMENT

It is autumn 2009 in the District of Columbia and federal chief information officer (CIO) Vivek Kundra's day is awash with data. At his meeting with the federal council of CIOs, he is working through the entire $76 billion portfolio of federal technology projects under his management, which are being monitored in real time. They can see how much was budgeted, how much was spent, and how much is left over. Kundra calls it the IT Dashboard and says it works a bit like a stock market: "We make the decisions on which ones to sell, which ones to buy, which ones to . . . sink more investment into."

Back in his office, there are plasma screens displaying information about tasks happening in the schools, streets, and administrative offices of the country. New tasks on the big spreadsheets come up in yellow, past due tasks in red, and completed tasks in green. Federal administrators are watching it too. They'll be talking over the performance data at a management accountability meeting later in the day with Jeffrey Zients, the U.S. government's new chief performance officer. Across town, the head of a D.C.-based government watchdog is preparing for her prime time media appearance by downloading exactly the same information. In the meantime she's plotting trends on Google Earth and releasing new insights on her Twitter feed.

Many employees find this new openness striking, even unnerving. And it's true; the innovations Kundra is pursuing are genuinely remarkable at a time when most people associate government with waste, inefficiency, and graft. Where most governments build mainframes and buy expensive software, Kundra is encouraging federal agencies to use free Google services and open-source wikis for everything from word processing to performance measurement to service improvement. He calls it the government cloud, but think "app store for government"—a place where employees can access a vast ecosystem of secure applications and data sets for doing their jobs. It may sound like a no-brainer, but it's an enormous improvement over the stubborn industrial age models that still prevail throughout much of government.

THE RISE AND FALL OF INDUSTRIAL AGE GOVERNMENT

Despite being a decade into the twenty-first century, the unfortunate reality is that most governments still reflect industrial age organizational thinking, based on the same command-and-control model as industrial age enterprises. After all, bureaucracy and the industrial economy arose hand in hand. The economy needed roads, sewers, electrification, railways, and a sophisticated military. As government got bigger, and the revenue of government increased, it became necessary to build more elaborate procedures, structures, and controls, all run by new layers of professional managers. These bureaucracies operated like individual "stovepipes"—with information only flowing vertically and rarely between departments.

During the last forty years, governments, like corporations, applied computers to their work as each agency acquired and built data processing systems to meet their automation needs. You'd think computers might have made things better. But, in reality, old procedures, processes, and organizational forms were just encoded in software. Huge, unwieldy mainframe beasts not only cemented old ways of working, they required still greater levels of bureaucracy to plan, implement, operate, and control. But even the most surgical IT experts have utterly failed to resolve the chaos of inconsistent databases, dueling spreadsheets, and other data anomalies that plague most government agencies.

Despite numerous attempts, very little has changed in the past half century. Even with the rise of the Internet, government's internal machinery may as well be carrying a machete and donning bulletproof armor. It seems even the most irresistible force of our times is struggling to dislodge the immovable mass of government bureaucracy. The result? Too many government organizations are still largely locked into old structures and outmoded ways of working.

The good news is that just as new waves of innovation are washing over the private sector, opportunities to harness new models of collaboration and innovation are arriving at the doorstep of governments everywhere. Indeed, if mass collaboration is changing how enterprises innovate, orchestrate capability, and engage their stakeholders, why can't the public sector seize networked business models to cut across departmental silos, improve policy outcomes, reduce costs, and increase public value? Market forces alone cannot meet the needs of today's citizens. But neither can the old industrial age models of government. Aging demographics, rising deficits, and a hatful of planetary challenges demand new models of government and democracy based on the five principles of wikinomics—openness, collaboration, sharing, integrity, and interdependence. If politicians and public servants are to ensure their relevance and authority going forward, they must move quickly to meet rising expectations for openness, accountability, effectiveness, and efficiency. This chapter explains how.

GOVERNMENT BY AND FOR THE PEOPLE

Transforming the deep structures and processes of government is a daunting challenge, but there is hope with people like Kundra in charge. If he gets his way, within five years there will be huge changes to the institutions and modus operandi of government based on a very different view of information technology. There will be no department-by-department mainframes or data centers anywhere in the U.S. government. Most of the information and applications that underpin public office will be run in a virtual cloud of computing capability that stretches across the whole of government. Employees and citizens will search for information from their desktop, just like they do when they're at home.

It's ambitious, but Kundra has already done it once in a similar role for the District of Columbia. One of his first initiatives in D.C. was to migrate from expensive enterprise platforms to Web-based solutions, in particular, Google Apps. From e-mail provision to Web-based software for word processing and spreadsheets to YouTube for video hosting—the costs dropped by 90 percent! A big believer in the power of transparency to make government better, Kundra led efforts to increase accountability by improving the citizen's access to public data. Among other innovations, he created a city-wide data warehouse that enables all government employees and stakeholders to see and help analyze what isn't working in the whole community. "I wanted people to hold us accountable," he said, "whether they're a student or an expert. Releasing data is integral to analyzing our operations and seeing where we can improve, where we have improved and where we have failed. Eighty percent of government data can be shared and thus it will be shared."

Kundra did more than just release the data; he put up $20,000 in prize money and launched an innovation contest called "Apps for Democracy." Contestants were encouraged to tap into the city-wide data warehouse to invent new kinds of Web-based public services. The experiment yielded forty-seven Web, iPhone, and Facebook apps in thirty days—delivering $2,300,000 in value to the city at a total cost of $50,000.[1] Now Kundra is bringing the same ethos of openness, participation, and collaboration to the entire U.S. government.

Kundra's track record with the District of Columbia may have earned him a senior role in the Obama administration, directing technology projects for the entire federal government; but why should citizens and policy makers around the world take note? Kundra's experience and continued leadership demonstrate what can happen by pursuing a strategy of transparency and collaboration in government—a strategy that has happened without the huge investments in technology foretold just a few short years ago, or, indeed, without the investments that continue to be made in many quarters of government.

The experience is all the more relevant because today's citizens can self-organize to do many of the things that governments do today, only they often do them better. In education, a few hundred students with Asperger's syndrome around the world can form a self-help group using Facebook. In The Student Room, a UK-based network where high school and university students share academic and social knowledge, users have posted 20 million forum posts about issues such as homework and university applications.[2] Parenting Web sites like Netmums operate as online communities, with over 750,000 users providing advice to prospective and current parents.[3] In the consumer field, MoneySavingExpert now has 2.5 million newsletter recipients and 6 million unique users per month with many sharing information on the latest money-saving tips and tricks.[4]

This is not your granddaddy's paternalistic government; it's literally government by and for the people. Call it governing by network, or Government 2.0, it all adds up to the same thing. In the wiki world, government shouldn't even try to create all of this capability internally when it can leverage the social networks that are already flourishing. Instead of creating new departments and new layers of management, governments should be creating platforms for social achievement.

Open Sourcing Government

We both began researching how the Internet was going to transform government and democracy during a landmark global study launched in 1999 and funded by twenty government agencies in sixteen countries. The study raised all sorts of new questions about government and its changing roles and form in the future. Could societies "open-source" government much the way thousands of dispersed Linux programmers converged on the Internet to develop one of the world's leading computer operating systems? Would large-scale, Web-enabled consultations improve political decision making or channel greater ingenuity and urgency into efforts to solve global challenges like terrorism and climate change? What about the provision of public services; could public agencies use ongoing collaboration with citizens, civil society, and the private sector to achieve better results at a lower cost? These were bold ideas back then, and there was much support among the political leadership in many countries. But as is often the case, the bureaucracy wasn't ready for the change and governments spent most of the last decade paving the cow paths as they replicated the same old structures on the Web.

Not all was lost, however. To embellish Victor Hugo's aphorism, there is nothing so powerful as an idea whose time has come, *again*. Thanks to the new Web, and a new generation of social innovators, open-source models of government are not just possible; they're often the best way to get things done.

Just ask the citizens of Estonia. When Estonians regained independence from the former Soviet Union in 1991, they not only acquired new political freedoms, they inherited a mass of rubbish—thousands and thousands of tons of it scattered across illegal dumping sites around the country. When concerned citizens decided that the time had come to clean it up, they turned not to the government, but to tens of thousands of their peers. Using a combination of global positioning systems and Google Maps, two entrepreneurs (Skype guru Ahti Heinla and Microlink and Delfi founder Rainer Nõlvak) enlisted volunteers to plot the locations of over ten thousand illegal dump sites, including detailed descriptions and photos. That, in itself, was ambitious. Phase II of the cleanup initiative was, by their own admission, rather outrageous: clean up all of the illegal sites in one day, using mass collaboration.

So, on May 3, 2008, over fifty thousand people scoured fields, streets, forests, and riverbanks across the country, picking up everything from tractor batteries to paint cans.[5] Much of this junk was ferried to central dumps, often in the vehicles of volunteers. If fifty thousand Estonians can clean up their country in one day (albeit a relatively small one), what else could they do? As Tiina Urm, a spokesperson for the initiative, put it: "It is not really about the rubbish. It is about changing people's mindsets. Next year it might be something else."[6]

The Five Principles in Government

As we look into the future, the innovations happening in places like Washington and Estonia demonstrate how knowledge, information, talent, and energy are being moved, shaped, and channeled in brand new ways, inside, across, and outside of the boundaries of government. While industrial age government was based on monopoly power, and structured around rigid hierarchies, today's governments need to distribute power broadly and leverage innovation, knowledge, and value from the private sector and civil society.

Franklin Roosevelt and Winston Churchill wanted stronger government. Ronald Reagan and Margaret Thatcher wanted less. Thanks to the Internet we can now have it both ways. In the United States and many other jurisdictions, government is becoming a stronger part of the social ecosystem that binds individuals, communities, and businesses—not by absorbing new responsibilities or building additional layers of bureaucracy, but through its willingness to open up formerly closed processes to broader input and innovation. In other words, government becomes a platform for social innovation. It provides resources, sets rules, and mediates disputes, but allows citizens, nonprofits, and the private sector to share in the heavy lifting.

As this new philosophy of open government picks up steam, Kundra tells us that federal agencies are doing things very differently. "A great deal of the innovation is now happening on the front lines of our organization," says Kundra. "Previously this was the last place you'd expect it to come from, but now we've opened up channels of communication with front-line, customer-facing employees, and they've responded with anything from the smallest innovations to the most complex ideas we've contemplated. Our job is to take these ideas and figure out which ones we want to scale rapidly." Indeed, when done right, this open approach delivers tangible benefits.

When President Obama called on the United States Citizenship and Immigration Services (USCIS) to find ways to improve the immigrant application experience in 2009, Kundra saw an opportunity to use transparency to drive change. He wondered why someone could track a FedEx package around the world, but when it came to applying for citizenship, one practically had to hire an attorney or contact a member of Congress to determine his or her application's status. So Kundra convinced the USCIS to publish its turnaround times on a public dashboard, which immediately had two effects. First, immigrants could now check on processing times at specific USCIS field offices around the country, or send a text message with their receipt number and receive an update on their application status. Second, placing that data online put pressure on the USCIS as a whole to make its operations more efficient. The dashboard itself was built in ninety days, and though it involved shifting some money around, it didn't end up requiring any additional budget expenditure.

Thanks to Kundra, and other leaders we profile in this chapter, there is a new kind of public sector organization emerging—one that opens its doors to the world; co-innovates with everyone, especially citizens; shares resources that were previously closely guarded; harnesses the power of mass collaboration; and behaves not as an isolated department or jurisdiction, but as something new: a truly integrated organization. Today, it's a radical notion, but perhaps it's only as fantastic as the current version of government would seem to a feudal prince from the Middle Ages visiting us now. Or perhaps it's just as improbable as a European Union would have sounded to an early twentieth-century European.

WIKI-STYLE COLLABORATION FOR EMERGENCY RESPONDERS

Still don't believe that governments can embrace the wiki model yet? We can't blame you—only 26 percent of Americans say they are satisfied with the way their nation is being governed.[7] To be fair, most citizens don't even know about it yet. But it turns out that the same brand of wiki-style innovation practiced by U.S. CIO Vivek Kundra is catching on like a virus in other parts of the United States. In the state of Alabama, Hurricane Katrina—the worst in U.S. history—helped spur the creation of a revolutionary platform for data sharing that has turned a small, quiet state into a leader in government innovation.

As Hurricane Katrina washed over the southeast shores of the United States in late August 2005, it quickly became evident that neither local, state, nor federal officials possessed enough information about the evolving conditions or the actions of their counterparts to respond in a timely and effective manner. They lacked basic tools for communicating in real time across teams and departments. There was no infrastructure for quickly assessing the damage or tracking the deployment of people and assets as first responders raced to help people in need. This inability to share information hampered relief efforts, and ultimately led to much public criticism of the officials in charge.

In the immediate aftermath of the hurricane, Alabama governor Bob Riley asked state homeland security director James Walker to dig up aerial photographs of the coastline before the storm so that they could assess the damage. Walker knew such photographs existed. The state government had given millions of dollars to county governments to collect aerial imagery. But when he needed it most, the photographs proved hard to locate. So Riley tasked Walker with finding a digital solution that would give every state and county agency ready access to all of this imagery. What emerged at the end of the process was a remarkable platform for shared decision making called Virtual Alabama.

Mass Collaboration in Three Dimensions

To picture Virtual Alabama, first picture Google Earth, except rather than user-generated photos and restaurant locations, imagine being able to browse loads of government information from across the state with a few clicks. Local and state government agencies can not only access each other's aerial imagery, they can now see property valuations, 3D models of buildings, real-time locations of emergency vehicles, live security camera feeds, and a growing number of additional data sets. Within two hours of a storm, every first responder in Alabama—from the state's director of homeland security to local police officers and firefighters—can now access aerial photography of the disaster area, see full damage assessments, and track resource deployments across every level of government in real time.

If a fire breaks out at an engineering building at the University of Alabama in Huntsville, for example, firefighters can use Virtual Alabama to access floor plans for the building and find out which classrooms are occupied at that hour. Or suppose another tornado were to hit Enterprise, Alabama. Officials could use Virtual Alabama to access aerial photographs from before and after the tornado to assess the damage, determine the property tax valuation of each damaged structure, and quickly put together a disaster assistance request for the federal government.

Like other interactive geospatial mapping efforts, the applications are broad and potentially transformative. Virtual Alabama has not only revolutionized crisis ops; it's changing the way state and local governments cooperate in other areas such as education, law enforcement, environmental management, and economic development. For example, Walker and his team are in the process of overlaying tax maps, forestry maps, utility line maps, and other maps with tax records, up-to-the-minute weather data, and live cameras on public buildings and highways. Now when the state of Alabama competes with other jurisdictions to attract investment opportunities, it uses Virtual Alabama to collect and present data more effectively. As part of its bid for a plant built by German steelmaker ThyssenKrupp, the state used Virtual Alabama to construct a 3D model of the proposed factory on top of aerial imagery of the area. It then added roads and other infrastructure the state was willing to build, and mapped out population centers, transportation routes, schools, and other data that was of interest to the company. This helped Alabama win the plant project, the largest nongovernmental economic development project in U.S. history.

A New Breed of Public Sector Innovators

Virtual Alabama represents an important innovation for government and highlights several lessons on mass collaboration and openness in the public sector. The serious deficiencies that the response to Hurricane Katrina exposed have, in Alabama at least, been overcome. And a wave of innovation has been unleashed that promises to facilitate more efficient and more effective decision making across all levels of government in the state. "Every day we hear new ideas on how the system is being used, and that's very exciting for us because that is truly the power of mass collaboration," says Chris Johnson, the project manager for Virtual Alabama. "If we were to have sat down and scripted all this out, we wouldn't have come anywhere close to what has happened."

Walker and Johnson represent a new breed of public sector innovators. They built Virtual Alabama for only $150,000 on top of Google Earth compared to an estimated $40 million it would have cost the government had it tried to build its own platform from scratch. And rather than try to control everything, Walker and Johnson empowered front-line employees to use and contribute data themselves, flattening the traditional pyramid-shaped hierarchy in the process. "People in state and federal governments tend to underestimate the ingenuity, the patriotism, and the 'wanting to do the right thing' of local government workers and first responders," says Walker. "Washington doesn't have to do everything. Locals will solve problems if we empower them."

Walker still thinks senior officials should take responsibility for creating the conditions in which innovations can flourish. "Technology is what's going to drive the future," says Walker, "but you have to be able to research and develop, test and evaluate new ideas." To do that, employees need support and trust from the top. Walker claims the trust placed in his team by the governor provided latitude to pursue new innovations and freed him from thinking, "I can't fail. I can't waste . . ." Without this level of support, Virtual Alabama would never have gotten off the ground.

PUBLIC SERVICES MEETS THE PROSUMER

"Traditionally, public services have been designed by governments and rolled out to citizens who are expected to comply with the terms and conditions of a program," says Maryantonett Flumian, a former senior official in the Canadian government who's now president of the Ottawa-based Institute On Governance. "Typically, the service is designed to be the same for everyone. It is always linear. The model is judged by outputs—how many checks got in the mail, how many people got back to work, how many calls got answered. Compliance with the rules and regulations that form the design of the service is paramount especially in transactional services," she says.

Today, there is an opportunity to move from a transactional approach to a "holistic," citizen-centered perspective where citizens themselves play a more active and ongoing role in defining and even assembling the basket of services they need. Citizencentricity has often been talked about in government, but very little of substance has been done about it. Flumian argues that true citizencentricity would mean redefining what it means to provide a service (e.g., delivering a benefits check) and a shift in focus from the process (say, the rules governing the dispersion of public benefits) to the outcomes, such as reducing poverty. It would also mean treating the citizen as an active participant rather than an inert recipient with little to contribute in return.

Under Flumian's leadership the Canadian government launched Service Canada in 2005, a citizen-centered service provision model that created one single agency focusing on development, management, and delivery of all social services for all citizens. Over the years, the organization shifted away from the unbending administrative, compliance-based culture of government toward a professional service culture where the public receives integrated and seamless services that are delivered in partnership with other federal departments and levels of government.

The Citizen Prosumer

Government getting its act together behind the scenes is only the beginning of what's possible. In the new model of public service delivery, the "citizen-collaborator" becomes a prosumer of services—identifying needs and helping to shape their fulfillment. The technology and tools become a means of finding better ways to integrate service—taking into account a person's preferences, his or her community's needs, and the places and spaces where services are needed most. The result is a dramatic improvement in the responsiveness of public systems, and an increased ability to focus the energy of all those involved—from officials to stakeholders to citizens themselves—in setting and achieving goals together.

To be sure, Service Canada is not your average public sector organization. Traditionally, innovation's home is at headquarters, where the "strategic" view directs the work of a department. The front lines are tasked with delivering the programs directed by the policy set at the center. There is a sharp division of labor: policy steers, services row. Citizens are participants at election time. Otherwise they are passive consumers of services.

Like Walker in Alabama, Flumian insisted on a new kind of public sector culture that isn't dependent on organizational charts. An attitude of openness, transparency, accessible information, and performance data combines with collaborative tools to make space for collective evaluation, strategizing, and action. Analysts meets counter staff meets Web administrators, and their different perspectives help create new insights that lead to new ideas. Citizens also participate, sharing the same information that is guiding decisions inside the administrative offices of government.

Other governments are taking action too, and leadership is coming from all sectors. Initiatives such as FixMyStreet.com in the United Kingdom enable residents to submit concerns about safety, vandalism, or other local issues directly to their municipal council. Set up by a nonprofit called mySo ciety.org, the aim, according to its founders, is to "give people simple, tangible benefits in the civic and community aspects of their lives."[8] Using the site, British residents play a more active role in increasing public welfare, while helping local government officials to be more attentive to issues that require resolution within their jurisdiction. Similar initiatives like SeeClick Fix.com are flourishing in North America. Paul Bass, *The Independent*'s editor and publisher, called the community-powered site "journalism, in its purest and rawest form."

Collecting feedback "out in the open" is a good first step. Asking citizens for ideas on how to allocate spending or improve service quality can be even more powerful. In fact, citizens have already proven themselves able to make measured, well-reasoned decisions about budgetary issues in cities around the world. The Brazilian city of Belo Horizonte, for example, has been doing participatory budgeting since 1993 and today allocates some $43 million for public works projects that are selected by citizens in nine individual districts.[9] In Zeguo Township, China, citizens have been convened through statistically random sampling to establish spending priorities for road building and construction projects.[10] The German city of Hamburg is perhaps the most technically sophisticated: Its participatory budgeting exercises conducted in 2006 and 2009 featured an online budget app with sliders that citizens could manipulate up or down to increase or decrease the level of funding for twenty-two budget items. The site attracted 50,000 visitors who generated over 2,100 draft citizen budgets, a selection of which were published in local newspapers and used as a basis for discussion in the local parliament.[11]

Not every attempt to engage the public has been successful. Having arrived at the seat of power on a wave of popular participation, the incoming

Obama administration created the Citizen's Briefing Book to field ideas from the public that could inform their policy platform. These digitally enabled conversations were designed to look and feel a lot like Digg.com, the popular technology news aggregator. Citizens posted suggestions and the community of users voted until the most popular ideas rose to the top. The intention was right, but this ill-conceived attempt to engage the public had some disappointing results. In the middle of two wars and an economic meltdown, the highest-ranking idea was to legalize marijuana, an idea nearly twice as popular as repealing the Bush tax cuts on the wealthy!

This experience doesn't disprove the model. It demonstrates that you can't simply cut and paste Web 2.0 phenomena into government. You need to introduce ground rules, work with trusted intermediaries, make sure you tap into knowledgeable communities, and build in processes to cope with issues of representation and accountability. Rather than throw out an open and indiscriminate call for ideas, the most successful attempts to involve the public engage specific segments (e.g., youth, seniors, families, people with disabilities) in a conversation about their needs and how they can best be met.

Dismantling the Cult of the Policy Expert

Getting the process for engaging citizens right is just the beginning. Mass collaboration often encounters resistance internally because of something we call the "cult of the policy expert." Policy makers and senior administrators tend to think of themselves as an elite group who are in a unique position to make dispassionate decisions in the public interest. As experts, they assume that they have access to the best information—or at least better information than the public. While that may have been broadly true in the past, it is not necessarily true today. Ubiquitous information networks can enable organizations to tap the insights of large numbers of people to arrive at decisions and outcomes that are superior to those presided over by individuals and small groups of decision makers.

Some leading agencies have already developed innovative models where external experts augment the work of the bureaucracy. The Peer-to-Patent: Community Patent Review initiative, for example, is the brainchild of Beth Noveck, another public sector innovator whose efforts won her a leadership role in crafting the Obama administration's directive on open government. The initiative enables patent examiners at the United States Patent and Trademark Office (USPTO) to draw on the expertise of external field experts using online collaboration tools to increase the speed and quality of their decisions.

This network of external experts advises the Patent Office on prior art. That's industry jargon for preexisting applications of ideas submitted for review that would disqualify a patent application. Since patents are only supposed to be issued for truly novel and original inventions, finding prior art is really job number one for the examiner. The problem is that assessing the state of the art has gotten a lot harder since the days of Thomas Edison, when the pace of innovation was glacial by today's standards. Fields ranging from software to nanotechnology are advancing so quickly today that only people at the leading edge have a realistic impression of what's truly innovative. Peer-to-Patent was set up to help patent examiners wade through the complexity using a combination of wikis, social reputation technologies, collaborative filtering, and information visualization tools that enable them to tap into crowd wisdom.

While skeptics worry that external participation could compromise impartiality, most observers agree that Peer-to-Patent is a welcome innovation for the USPTO. A dramatic increase in the number of patent applications since the late 1980s has put enormous strain on the agency. Critics argue that the quality of patents has decreased as underpaid and overwhelmed patent examiners struggle with a backlog of nearly 1,200,000 applications and growing.[12] Mounting pressure to plow through this backlog means that the average patent application receives fewer than twelve hours of review by a single examiner—a fact that perturbs many in government and industry alike.

The potential economic and administrative benefits of Peer-to-Patent are immense. It currently takes an average of forty-four months before a patent application even reaches the desk of an examiner—far too long in an era when technology advances at breakneck speed. With Peer-to-Patent, the USPTO expects the same application will be processed in twenty-three months or less.[13] But administrative efficiency is not the only goal. "The idea is to create a living, breathing ecosystem of scientific expertise," says Noveck, "where the community of inventors decides on its own criteria for assessing patent applications." Greater participation from experts familiar with the technological domains under consideration not only speeds things up, it increases patent quality. "This will make it easier to protect the inventor's investment while safeguarding the marketplace of ideas," she says.

Remixing Government: Open Data Drives Social Innovation

Though one can glean the positive potential from examples like Peer-to-Patent, it seems substantive changes in the way government delivers public services may not arrive until government's monopoly on things like driver's licenses, tax collection, and social security is loosened. There are good reasons to have government monopolies on the use of armed force. But in other domains, the case for government control is much less compelling than it once was.

While controversial in some quarters, the reality is that self-organized citizens and entrepreneurial firms can invent extensions to existing public services with fewer inhibitions and greater speed than government. The Power of Information Task Force, which was established to advise the UK government on how to take advantage of new developments in digital media, recommended that all public agencies in the UK create online innovation spaces where the general public and staff can co-create information-based public services, much like the way companies such as Amazon, Flickr, and Apple enable third-party developers to build extensions to their software platforms. The task force also recommended the UK government create a public service R&D function with a "modest fund for leading-edge R&D to continue to test ideas and incubate new capabilities."[14]

Tom Steinberg, a member of the task force and the founder of mySociety.org, sees this as the beginning of a sea of change in the way governments create value for citizens. "When enough people can collect, re-use and distribute public sector information, people organise around it in new ways, creating new enterprises and new communities."[15] In the past, only large companies, government, or universities were able to reuse and recombine information, notes Steinberg. "Now, virtually anyone with an Internet connection can mix and 'mash' data to design new ways of solving old problems."

With public sector organizations stockpiling huge quantities of data, the challenge is increasingly to extract insights that can inform action. Some changes, like the relationship between health and GDP, are so gradual that they're nearly imperceptible to us. Animating nearly two hundred years' worth of data depicting the relationship between infant mortality rates and GDP per capita in a forty-five-second clip is much more revealing than a static chart, says Professor Hans Rosling, the creator of Trendalyzer, a tool that allows users to turn spreadsheet data into rich, interactive visualizations on the Web.[16] If you present the same data set without animation, just using "before" and "after" graphs, people somehow disbelieve it, says Rosling. These animations don't tell a story: they're "story busters," because they correctly convey the richness and diversity of the data without oversimplifying.

In reality, both governmental and nongovernmental entities stand to benefit from technologies that make governmental data increasingly available in attractive, bold graphic forms that anyone can understand, dissect, and debate. What surprises most public officials is the degree to which participants use the data in completely unexpected ways. In the city of Sault Ste. Marie in Ontario, mapping the location of residential pools helped public health officials to pinpoint mosquito breeding grounds to target in their efforts to combat West Nile virus. When Congress empowered the EPA to collect data on 328 deadly chemicals in use in commerce and report it to the public via computers, it provided a major boost to environmental groups who soon recognized its enormous power and began building Web-based interfaces that made the data easy for the public to understand.

Today, sites like MapEcos.org and Scorecard.org provide visual, Web-based interfaces that enable citizens to see toxic emissions data and more, all in one place. Visitors to MapEcos can investigate up to twenty thousand U.S. industrial facilities and plot their emissions data on Google Maps with color-coded markers that make it easier to pick out the worst offenders.[17] Click on a location, and it will tell you what the place is called, what it emits, and how it compares with other businesses in the county, state, and country. You can search the site by pollutant, industry, and hazard level. Or simply type in a zip code and pull up a list of polluters in their neighborhood.

Data sharing also drives economic opportunity and, on occasion, has been known to spark new growth industries. When the National Institutes of Health released data from the Human Genome Project, it spurred massive innovation around a new era of personalized medicine. President Reagan's directive to provide free and open access to the Defense Department's GPS signals gave rise to a plethora of commercial uses ranging from mapmaking, land surveying, scientific analysis, and surveillance to hobbies such as geocaching and waymarking.

The Obama administration hopes Data.gov—an open hub for federal data—will amplify and accelerate this tradition of innovation. "In a global age where information can move in milliseconds anywhere in the world," says Kundra, "you can't necessarily limit innovation across national boundaries. You have the ability to tap into innovation all over the world."

Building the Ultimate MyGovernment: Opening Up Free Trade in Public Services

Given examples like the Human Genome Project and the global positioning system, it's not a stretch to imagine that successful business models could be formed around the exploitation of open public service platforms to create entirely new services or even replace some of those currently delivered by government today. The idea is to emphasize variety and choice in service venues, providers, and options, rather than the old industrial age model of one-size-fits-all government.

Like the Personal Health Page described in chapter 10, imagine every citizen was granted their own MyGovernment page from birth—an interactive space through which they channeled all of their interactions with government, whether it be renewing a driver's license, filing taxes, finding a new doctor, or registering a business. Using a combination of widgets, RSS feeds, and other social technologies, service providers from all sectors—public, private, and nonprofit—could offer up services and citizens could assemble their own customizable packages using their tax dollars. There would be true competition among service providers and genuine choice for consumers. Government could continue to develop its own apps to address constituent needs that may be underserved by the market. But other areas—licensing and permits, small business loans and grants, consumer protection, location-based services, for example—would be ripe for competition.

Opening up a domestic market for public services makes sense, but why stop there? Why not take the ultimate MyGovernment a step further and allow citizens to seek out their own solutions to issues globally rather than expecting a packaged response from a single entity: health care from the Netherlands, business incorporation in Malaysia, marriage licensing from a municipality in the United States, education from a worldwide virtual network. In other words, why not have free trade in government services? It's no longer a question of whether such a world is technically possible. A truly personalized service would provide a window to all these possibilities, beamed right to the citizen's platform of choice, including their desktops, mobile phones, or their favorite social media sites.

Private-sector investment into offerings such as these could catalyze much-needed innovation and potentially lessen the burden on taxpayers. But to manage this change effectively, government leaders must think carefully about when and where to retain control and how to leverage government's core competencies to create a sustainable and appropriate service strategy. Among other things, governments will need to guard against threats to consumer privacy, data security, and the potential misrepresentation of government content and services by the private sector. If companies and nonprofits build new interfaces to popular services, who's accountable if something goes wrong? If a nonprofit or business folds, who will ensure service continuity or take custody of personal data? Will governments guarantee that privatized services remain accessible to citizens irrespective of their income bracket?

These are all tough questions. But consider them implementation challenges, not fundamental reasons to delay action. New collaborative models of service delivery are not just possible, but imperative. As Flumian puts it, "The big question is, what roles and responsibilities will government, citizens, not-for-profits, and business assume in a society where knowledge is everywhere, where hierarchies are anachronisms, and where 'the state' is no longer king of the jungle, but part of an ecosystem energized by mass collaboration?"

BOOTSTRAPPING THE BUREAUCRACY

If exceptional public service innovations are hard to come by today, it's largely because governments were built for stability, not change. Jessica McDonald, former head of British Columbia's public service, observes that the cyclical nature of political change conditions the public service to focus on their ability to deliver services in a predictable way. "It stops us from thinking creatively about what it is we're really trying to achieve and all the different ways we could get there." Enormous institutional momentum also buffers government from a sense of urgency. Dan Mintz, former CIO of the U.S. Department of Transportation, put it bluntly: "We aren't going to go out of business anytime soon."[18]

Perhaps he's right. But it's no excuse for being complacent. Over the next twenty years, an entire generation of baby boomers will retire from government, creating an exodus of knowledge and skills. In the United States, this demographic shift will see more than sixty thousand civil service employees exit annually between now and 2015. Large departments like Defense will lose 20 percent of their workforce. Many of these people hold executive, managerial, or key administrative positions—replacing them will be nearly impossible.

Recruiting and retaining a younger generation of public servants won't be easy either. Just when government most needs an infusion of fresh-thinking talent, there's a growing lack of interest in public administration as a profession among young people. Although managers typically fret at the prospect, this attrition may not be such a bad thing. Rather than worry about head counts, governments should look for new ways to ignite innovation.

The good news—and yes, there is *some* good news—is that many of the young people who have landed jobs in the public service sector aren't just passively allowing the weight of bureaucracy to stifle their creativity. Drawing on their experiences on sites such as Facebook and YouTube, young government professionals are using blogs, wikis, and social-networking tools to create ad hoc communities that are bringing a new ethos of innovation and collaboration to life. This new generation of public servants didn't wait for anyone's permission; they just did it. It's part of their new culture of work. And the more this culture permeates the public sector at large, the more dividends this bottom-up approach to bootstrapping the bureaucracy will pay.

Facebook for Government

Exhibit one in this new generation is Steve Ressler, the founder of the world's fastest growing network of public servants and probably one of the most impressive young government leaders we've met in a decade. But when Ressler landed a job in the newly formed Department of Homeland Security, he wasn't overly impressed with what he found. The first thing he noticed was that there were few other young people around. And everywhere he looked there were silos. Departments weren't talking to one another and there was no online home for connecting with colleagues or sharing ideas. When the department banned Facebook, Ressler called it the single most demoralizing thing management had ever done. "It said to us we don't get collaboration, your tools, your generation, and we don't trust you." In college, he majored in social network analysis and he was an early adopter of networks like Friendster, MySpace, and Facebook. He'd hoped that by joining DHS he might get an opportunity to apply his expertise to the business of tracking down terrorist cells. But as a rookie, he quickly got bogged down in bureaucratic minutiae.

So Ressler applied his skills in the next best way he could think of—he started organizing social events for other pent-up innovators in government. Two years and a few wine and cheese parties later, Ressler and some friends had spawned a new informal network called Young Government Leaders with almost two thousand members and seven local chapters across the nation.[19] The achievement won him much kudos and many invitations to sit on prestigious panels, but he had grander designs in mind. Ressler wanted a Facebook for government: a place where like-minded peers could connect and collaborate. And since nobody in government seemed interested, he decided to build one himself. With virtually no money to spend, he built the backbone of his community on Ning, an open platform that allows individuals and organizations to develop their own hubs for social networking. He crowdsourced the site design, and three months later on Memorial Day in 2008 he officially unveiled GovLoop to an enthusiastic crowd of early adopters. With expenses having recently peaked at only $100 a month, Ressler says, "It goes to show that you don't need to spend a million dollars to launch a network."

Even on this shoestring budget, GovLoop now serves over thirty thousand members.[20] Though most members are from the United States, public servants have joined from Brazil, Canada, Israel, and the United Kingdom. "These people are passionate about improving government," says Ressler. "They're in the prime of their career, they get social media and they want to do something positive." In many cases, discussions started on the site have led to new initiatives. Ressler recently quit his job at the DHS to free up more time to work on GovLoop. Challenge number one is growing membership. Thirty thousand "is just the tip of the iceberg," he says. "People really want to learn from what others are doing. If we can strengthen engagement, foster more dialogue, and connect more people, we can turn conversation into knowledge." And that knowledge will provide the fuel for reinventing government from the bottom up.

Reinventing Government from the Bottom Up

The growing popularity and influence of networks like GovLoop are a direct response to some of the well-known deficiencies in government: rigid HR policies, insufficient training, sclerotic decision making, hierarchical management structures, and a lack of interagency collaboration. Young people, in particular, turn to these informal networks to navigate their way through an exceedingly cumbersome institutional environment.

Even the networks inside government are flourishing. NASA launched Spacebook—a social network for space scientists that has proven especially popular with new recruits. The U.S. Department of Homeland Security doesn't have an official network for employees, but an unofficial group on Facebook hosts lively discussions with thousands of members working in all facets of the agency. Back in British Columbia, enthusiasm for this type of open collaboration resulted in a government-wide brainstorming application called Spark. Ideas posted to date range from new options for expanding online citizen self-service opportunities to operational ideas like establishing an internal talent bank to reduce the need for external contractors.

But can we really reinvent government from the bottom up? Nascent networks like Spark, GovLoop, and Spacebook are proving that it's not only possible; it's the way of the future. These are early days, to be sure, but the vast potential is already evident. Informal networks could do 80 percent of what traditional departments do, but arguably they could do it better. Think about it—what's GovLoop at the end of the day: a repository for organizational knowledge; a source of new ideas and innovation; a forum for ad hoc cross-departmental collaboration; a provider of training and mentorship; or all of the above?

"Executives need to support this kind of shift," says BC's McDonald. "But ultimately innovation is not executive-led, it is actually led by the organization."[21] Leaders who overlook social networks or fail to connect with them will miss out on a vital source of engagement and innovation. Leaders who reach out to informal networks stand to make considerable advances in virtually every aspect of public management—from the capacity to share knowledge and connect leaders across government to the ability to recruit top performers and keep them engaged in the important work of creating value for citizens.

A CHALLENGE TO PUBLIC SECTOR LEADERS

As we step into the future, societies are facing incredible challenges of complexity on a global scale. Sustaining societies and economies in the face of climate change, energy shortages, poverty, demographic shifts, and security threats will test the ingenuity of those who wish to see, do, and participate in the public good. In each of these issue areas governments face a reality in which they are increasingly dependent for authority on a network of powers and counterinfluences of which they are just a part. Whether streamlining government service delivery or resolving complex global issues, governments are either actively seeking—or can no longer resist—broader participation from citizens and a diverse array of other stakeholders. Just as the modern multinational corporation sources ideas, parts, and materials from a vast external network of customers, researchers, and suppliers, governments must hone their capacity to integrate skills and knowledge from multiple participants to meet expectations for a more responsive, resourceful, efficient, and accountable form of governance. It's a journey that has only just begun.

The first wave of digitally enabled "e-government" strategies delivered some important benefits. It made government information and services more accessible to citizens while creating administrative and operational efficiencies. But too many of these initiatives simply paved the cow paths—that is, they focused on automating existing processes and moving existing government services online.

This next wave of innovation presents a historic occasion to fundamentally redesign how government operates, how and what the public sector provides, and ultimately, how governments interact and engage with their citizens. Governments can and must rise to these challenges. It is truly a time when government either plays an active and positive role in its own transformation, or change will happen to it. The transformation process is at the same time exhilarating and painful, but the price of inaction is a lost opportunity for government to redefine its role in society and help launch a new era of participatory government.

15. THE RISE OF THE CITIZEN REGULATOR

It seemed like a typical morning commute for most New Yorkers on January 8, 2007, until a strong, gaslike odor permeated parts of New York City and nearby areas of New Jersey, forcing several schools and companies to evacuate, and interrupting traffic along some subway and train lines. Emergency crews were unable to pinpoint any gas leaks or other causes, despite sending out a fleet of fire trucks and a hazardous materials crew. After 140 industrial facilities had been searched, New York mayor Michael Bloomberg declared that the crews had given up hope of finding the source of the mysterious odor.[1] While there were no reported casualties, the uncertainty caused anxiety and fear in a city where pungent odors can raise vague worries about a potential terrorist attack.

Fast-forward into the near future, however, and the series of events that bewildered local officials might have unfolded quite differently according to Eric Paulos, a computer scientist at Carnegie Mellon University. The odor leaves traces of nitrogen dioxide that, in his alternate reality, are automatically detected by tiny wireless air quality sensors embedded in the mobile phones of millions of New Yorkers. A simple mash-up of the data overlaid on a Google map identifies the culprit—a dangerous incinerator that is not supposed to be in use. City officials shut down the plant immediately and issue a text alert to calm nervous residents who fear that some form of chemical attack has been unleashed.

This account may be fictional, but it is not far-fetched. Like Paulos, technologists and science fiction writers have long envisioned a world where a seamless global network of Internet-connected sensors could capture every event, action, and change on earth. With the proliferation of radio-frequency identification (RFID), satellite imagery, cheap personal video recorders, powerful mobile computing devices, and an array of Internet-connected sensors, that vision of millions of New Yorkers participating (perhaps unwittingly) in an act of civic regulation is increasingly plausible. Indeed, the question raised in this chapter is whether a combination of new technologies and citizen participation could unleash an era of participatory regulation, where citizens and other stakeholder groups play an active role in designing and enforcing regulations. The idea is that just about every area of regulation today—from air and water quality to food safety and financial services—could benefit as a result of having a larger crowd of informed and empowered individuals helping to protect the public interest. While we do not envision a scenario where all citizens will suddenly participate en masse, even a small, well-organized group of highly motivated citizens with access to information can use social networks like Facebook and Twitter to make a big impact. Regulatory agencies would need to open up, rethink old processes, and supply the tools and data that citizens would need to contribute meaningfully. But given the depleted state of regulatory agencies around the world, we think greater citizen participation in regulation would be a very good thing indeed.

THE NEED FOR A NEW REGULATORY PARADIGM

A string of events over the past couple of years has underscored just how strained and ineffectual our current systems of regulation have become. In 2008, a salmonella outbreak swept across the United States—claiming at least eight lives, sickening over five hundred people in forty-three states, and spurring a worldwide recall, including TV dinners, snack bars labeled organic, and large pails of peanut butter shipped to schools, military bases, and nursing homes.[2] Some three thousand diverse products would soon be hauled off store shelves—the largest recall in U.S. history.[3]

A *New York Times* exposé in the wake of the unfortunate episode revealed a food safety regulation process that is ill equipped to cope with the increasing complexity of modern food production and distribution.[4] Government records acquired by the *Times* show that state and federal inspectors do not require the peanut industry to inform the public—or even the government—of salmonella contamination in its plants.[5] And industry giants like Kellogg use processed peanuts in a variety of products but rely on their suppliers to perform safety testing and divulge any problems. These days, processed peanuts find their way into more and more foods as a low-cost, tasty additive, making tainted products harder to track.

The FDA's own Science Board concluded in 2007 that the agency did not have the capacity to ensure a safe food supply, with domestic businesses under its purview having risen to 65,500 from 51,000 in 2001.[6] The byzantine nature of global food production and distribution only heightens the agency's capacity problems. The same report noted that the FDA currently inspects just 1 to 2 percent of imported food shipments—not much of a deterrent to the world's less scrupulous food producers.

Though worrisome in itself, food safety is merely the tip of the iceberg. The failure of financial regulators to detect the Madoff scandal or sound the alarm on ludicrous lending practices and overleveraged financial institutions is perhaps the most glaring example. But regulators were also incapable of preventing tainted milk products and toxic toys produced in China from finding their way onto retail shelves in the United States and other countries. And the problems hardly end there. Issues as diverse as climate change, water scarcity, emerging technologies, and infectious diseases demand innovative approaches and each issue comes with an impending sense of urgency.

Some of the issues that challenge today's regulators include the sclerotic pace of rule making, growing economic complexity, increasing international interdependency, the corrosive influence of "junk science" and industry lobbying, and a broadly insufficient capacity for effective oversight. Arguably some of today's troubles are self-inflicted. After dismantling or circumscribing centralized regulatory agencies in the 1980s and 1990s, many governments handed industry the power to police itself in areas ranging from toxic emissions to financial services. The thinking was that government regulation was too burdensome and costly, and the mechanics of updating it were clunky. Delegating rule making to industry bodies would make regulation more responsive to the needs of industries that were evolving quickly and becoming increasingly global in scope. Governments were to be the "regulators of last resort"—stepping in only after self-regulation was deemed to have failed. The problem, in practice, is that most instances of industry self-regulation have deficiencies (like lax rules or inadequate enforcement) and governments (for the most part) have proven unable or unwilling to take swift action when market failures become evident. Indeed, after years of chronic underfunding, it should be no surprise that many regulatory agencies are ill equipped to pick up the slack, let alone confront novel challenges for which they have neither the resources nor the expertise.

This underscores a major dilemma. Without transparency, oversight, and accountability, self-regulation is clearly inadequate. At the same time, the speed, interdependency, and complexity of today's world make a return to centralized rule making and enforcement increasingly implausible. As Robert Cooter, professor of law and economics at the University of California at Berkeley, put it, "An advanced economy involves the production of too many commodities for anyone to manage or regulate. As the economy develops, the information and incentive constraints tighten upon public policy.

These facts suggest that as economies become more complex, efficiency demands more decentralized lawmaking, not less."[7] While these forces don't rule out effective government intervention, they are undermining the traditional approach to regulation whereby a small group of trustworthy and disinterested expert regulators intervene in the economy to further public objectives.

LETTING THE ONLINE CROWD PLAY WATCHDOG

So if the old model of command-and-control regulation is broken, what could replace it and how would it work? We believe effective regulation is more likely to stem from efforts that increase transparency and public participation in a broad swath of areas that affect the health of our children, families, and communities. This isn't the same as allowing a small group of powerful companies to police their own activities. Our proposal is the reverse. We say open up the regulatory process: make everything transparent on the Web and let citizens and other parties contribute their own data and observations. Where possible, let citizens help enforce regulations too, perhaps by changing their buying behavior or by organizing public campaigns that name and shame offenders. This is a bit like the idea in chapter 3 where we described how the Open Models Company is convening a community of financial quants to evaluate the risks associated with complex financial instruments. Except in this chapter we're talking about broadening responsibility for designing and enforcing the rules that govern all economic activity, not just finance.

There is some precedent for this idea. For several decades, transparency has been used to spur behavior change in areas ranging from corporate accounting to nuclear disarmament. In her edited volume *The Right to Know*, for example, Brookings Institution scholar Ann Florini argues that citizens in all parts of the world have shown themselves to be unwilling to tolerate secretive decision making. "As a result," she says, "India, South Africa, the UK, Japan, Mexico and a host of other countries have all adopted major freedom of information laws; intergovernmental organizations such as the World Bank and the IMF have adopted sweeping new disclosure policies; and hundreds of major multinational corporations have adopted voluntary codes that require them to disclose a wide range of information about their environmental, labor, and other practices."[8] Even NGOs and charities are routinely providing more information about their policies and activities to members and stakeholders.

The idea behind this transparency trend is fairly simple: revealing information about the activities of powerful institutions is a potent deterrent to misbehavior. The more people can find out, inform others, and organize, the less politicians and corporate leaders can pursue self-serving behavior or act against the public interest. As a tool of regulation, transparency helps provide reassurance that others are doing what they are supposed to. Investors can be reassured of the quality of their investments, the international community can be certain that no individual nation is polluting beyond its CO_2 emission quotas under Kyoto, while NGOs can be satisfied that the companies they scrutinize are meeting their commitments to ethical conduct. More rigorous and accessible financial disclosure will also allow society to assert some control over a global financial system that has become so large, complex, and opaque that it remains beyond the capacity of even the largest actor in the markets to understand. Even George Soros, the legendary master investor, is said to have been wary of derivatives because he didn't "really understand how they work."[9] If enough individual and institutional investors followed the Soros rule, it would have quickly disciplined the markets. In other words, offer more transparency or you don't get my money.

Skeptics may doubt the capacity of citizens and advocacy groups to help regulatory bodies develop more effective systems of monitoring and enforcement. But a growing number of regulatory agencies are already convinced, as evidenced by the U.S. Environmental Protection Agency's efforts to open up its rule-making processes and the SEC's recent announcement that it is developing new systems for collecting anonymous tips in the investment community.[10] Even when inertia prevails in government, other organizations are taking the lead. The FDA may not require manufacturers of processed foods to label where a product came from, whether it contains genetically modified organisms, or was produced using synthetic hormones, antibiotics, and pesticides. But retailers like Tesco and a legion of online product guides are making this information available anyway. Why? Because customers are demanding transparency!

Now imagine the FDA extending a similar level of openness to real-time product recalls. Indeed, why not set up an open-source platform where a combination of RFID technologies and historical sales data would enable retailers to alert anyone who purchased a recalled product. We bet hundreds of programmers would vie for the opportunity if the retailers agreed to open up a series of data feeds. The FDA should also consider open sourcing risk assessment (they call it Hazard Analysis and Critical Control Points) for the same reasons we argued that it's time for opening up risk assessment in the financial sector. Both are instances where the open-source maxim applies:

focus enough eyes on one problem, and suddenly all bugs are shallow. In this case, government employees could collaborate with the public to identify and prevent potential food hazards in a manner akin to the way "the crowd" evaluates the quality of Wikipedia entries.

We're not saying transparency is a *substitute* for better regulation by national governments and international institutions, but we are convinced that more disclosure and increased civic participation in regulatory systems could be a formidable complement to traditional command-and-control systems. "The shorthand," says Charles Sabel, a legal scholar at Columbia University, "is that you can no longer regulate anything with fixed rules made at the center by a bureaucracy of any kind." Sabel argues that both regulatory authorities and the regulated industries are increasingly less concerned with the need to secure definitive legislation and instead press for the creation of framework agreements within which some kind of learning and continual improvement is possible. Typically, these frameworks include baseline standards, targets, and timelines for improvement, and transparent systems for monitoring and evaluating compliance. In many cases, individuals (whether in the role of citizen, consumer, employee, or investor), civil society organizations, and business enterprises are likely to have as much, if not more, influence over the outcomes as government bureaucrats and policy makers.

Companies may not like this more fluid and complex regulatory reality. Some argue that they could do with less scrutiny, not more. Others, while not inherently opposed to the idea of regulation, fear that greater openness could cause the regulatory agenda to spin out of control. After all, it's a lot harder to lobby a diffuse crowd online than it is to descend upon Washington with promises of campaign funding in exchange for favorable rules. But, like it or not, the forces of transparency can scarcely be avoided and new models of regulation are emerging. As Susanne Stormer, who heads up Stakeholder Relations at Novo Nordisk, put it: "In the long term, not being transparent is really not an option. You can say truth, like pregnancy, cannot be hidden for long. And even now, your stakeholders will find out one way or another, and you can no longer control the communication that's coming in and out of your organization."

PARTICIPATORY URBANISM

Still, you may be asking yourself whether the average citizen has the knowledge or the tools required to do things like regulate food safety or help environmental agencies collect air quality data. In fact, they can, with little more than a cell phone, according to Eric Paulos, the computer scientist at Carnegie Mellon we mentioned earlier. Indeed, with a few simple modifications, Paulos can transform an ordinary mobile device into a powerful personal measurement instrument capable of sensing our natural environment and empowering collective action through everyday grassroots citizen action. Outfitting your average phone with a few simple sensors and some software, says Paulos, will allow users to get answers to a wide range of questions that would otherwise elude the average citizen—questions such as: How healthy is the air I'm breathing? Were pesticides used on these fruits? Are my children's toys free of lead and other toxins? Or, is my new indoor carpeting emitting volatile organic compounds (VOCs)?

Sounds futuristic, but the day when most mobile computing devices come preloaded with sensing capabilities is not that far away. Smartphones already pack a lot of processing power and that power is increasing over time. The network capacity required to carry the data streaming from a plethora of connected sensors is also growing rapidly. Meanwhile companies like Nokia are betting that embedded sensing capabilities could unleash a wide range of new consumer applications. The handset maker has already prototyped an "eco sensor" phone with novel sensing technologies for atmospheric gas levels (including carbon monoxide, particulate matter, and ground-level ozone detectors), ultraviolet radiation, and noise pollution, among other things.[11]

As a first step toward demonstrating the practical potential of urban sensing, Paulos and his team equipped street sweepers in San Francisco with sensors to measure pollution levels as they navigate around the city. The sensors beam data back to a central database that researchers use to create a real-time map of the city's environmental landscape. Once sensors become ubiquitous, the plan is to collect air quality data directly from consenting smartphone users. Interested citizens can already use a text-messaging service to receive daily air quality data for their zip code. But the team is in the process of launching a more full-featured Web site that will provide access to live and historical data, as well as online community features that allow citizens to discuss neighborhood issues and formulate strategies for addressing them.[12]

In Manila, similar projects have been piloted where ordinary cell phone–using citizens join forces with environmental groups in their fight against urban air pollution. While not quite as cutting edge as their San Francisco counterparts, individuals using cell phones and mobile text messaging can report vehicles they see emitting excessive clouds of pollution to a central database called Smokebelchers Watchdog. Amelia Judones, a college student, is among the volunteers. "I do it in the car, on my way to class," she says. "It costs me nothing, and I feel I can contribute something to the fight against pollution."[13]

Most complaints are against trucking and commercial vehicle companies. At the end of each week, the environmental group leading the effort compiles a list of vehicles with five or more complaints against them and sends it to the Land Transportation Office (LTO)—the arm of the Department of Transportation and Communications that issues licenses to such companies. The LTO then summons offending vehicle owners to their offices for an exhaust test. In the first two weeks of the campaign, which launched on June 6, 2002, 123 vehicle owners were called in. "The volume was so great that we now receive the complaints not weekly, but every day," says Roberto Lastimoso, chief assistant secretary of the LTO.[14]

Urban sensing projects like these illustrate how greater civic participation could bolster the monitoring and enforcement capabilities of overstretched agencies, and not just those agencies concerned with air quality. The movements for corporate accountability described next suggest that the role for citizens, industry, and nongovernment organizations is greater still. Not only can citizens and other stakeholders help monitor and enforce the rules set by governments, they can promulgate entirely bottom-up regulatory systems to address issues ranging from climate change to corruption to labor conditions in the overseas factories that manufacture garments for Western consumers.

SELF-ORGANIZED CITIZENS SHINE THE SPOTLIGHT ON CORPORATIONS

For most businesspeople and many leaders in government, the economic benefits of globalization are patently obvious. But a large and increasingly well-organized network of citizens is not so convinced. Rather than net job creation, they see jobs and investment dollars fleeing their jurisdictions. Rather than a global lifting of living standards, they see companies exploiting lax overseas labor laws and exporting pollution to developing countries. And rather than a neutral, rules-based trading regime, they see corporate lawyers writing the rules of international trade to privilege the needs and rights of corporations over those of ordinary citizens and their elected representatives. This strong undercurrent of discontent has been amplified by the Internet and now constitutes a major force in "regulating" corporate behavior around the world. Indeed, whereas corporations once regarded their primary duty as serving shareholders (and perhaps their customers and employees), it is now common for corporate leaders everywhere to regularly reaffirm their commitments to serve society and protect the environment. How did this happen? Not through any traditional form of regulation, but rather through the power of networked citizens to closely scrutinize business behavior and bring the discipline of public opinion to bear on companies in a way in which governments rarely can.

This shift toward citizen oversight got started in the early 1990s, as public concern about the social and environmental performance standards of major multinational corporations gathered momentum. Stories of alleged abuses overseas started to appear in the headlines of major newspapers almost weekly: the sweatshop scandals, Shell's debacle in Nigeria (where questionable links to the repressive dictatorship of Sani Abacha made headlines in the Western news media), and Monsanto's "terminator" seed technology. One dramatic event after another highlighted what some groups in society perceived to be unsavory abuses of corporate power. Soon a global citizens' movement was rocking cities like Seattle, Washington, D.C., and Genoa as a motley collection of trade unionists, college students, environmentalists, and social justice advocates came out in force to protest meetings of the World Bank and the World Trade Organization, which activists saw as the apostles for a corporate agenda.

In reality, multinational companies were just doing what the market demanded: taking advantage of open borders and a new pool of human capital to produce an ever-growing quantity of goods and services at ever-lower prices. In most cases, overseas investments brought jobs, prosperity, and higher standards to communities that had little access to the global economy, even if some companies made missteps along the way. Indeed, many of the grave consequences forecast by activists—for example, that the economic might of global corporations would soon dwarf that of individual nations—proved to be highly alarmist in hindsight.

Still, the feelings of suspicion linger and events like the global financial crisis have not helped endear corporations to the public. In fact, public opinion polls reveal that more than eight out of ten people in a cross-section of twenty-five countries think corporations should take on greater responsibility for solving social and environmental issues.[15] When companies are not responsive to these demands, corporate critics use the Internet to pepper management with detailed inquiries, monitor private-sector behavior around the world, and swap insights and intelligence with one another. Many companies are uncomfortable with such scrutiny, a feeling that only intensifies as social media continues to accelerate the speed at which critical messages can "go viral."

Indeed, as technology advances, so too do the capabilities of organized civil society groups. Peter Gabriel's Witness, a human rights organization, travels deep into the jungles of eastern Burma to document the military's persistent attacks on ethnic minorities and expose corporations like Chevron that still maintain links with the oppressive regime. So far, iron-fisted government control over telecommunications has not prevented its members and supporters from posting numerous grainy, and sometimes gory, videos on YouTube. Witness even has its own platform called the Hub, an interactive community for human rights activists who can upload photos and videos to document abuses. The site's arsenal includes a video advocacy toolkit, an online petitions app, interactive maps that plot the geographic location of user-contributed media, a mobile version, and direct cell-phone-to-Hub uploads.

CorpWatch.org, a San Francisco–based nonprofit, has gone even further. Its sophisticated array of research tools empowers amateur corporate investigators to operate out of the comfort of their living rooms. Crocodyl.org—a corporate malfeasance wiki—covers fifteen issues in thirty-five industries and has detailed profiles on hundreds of companies that are kept up-to-date by volunteers around the world.[16] A recent extension to the site called CrocTail integrates with CorpWatch and provides an interface for browsing SEC filings from several hundred thousand U.S. publicly traded corporations and their many foreign and domestic subsidiaries. Tonya Hennessey, project director at CorpWatch, says, "The CrocTail application has particular relevance at this moment, with the public eye focused on the structural nature of corporate abuses, including multinational tax-avoidance and the use of off-shore subsidiaries to evade responsibility for human rights violations."[17] Launched in partnership with the Sunlight Foundation in June 2009, the app features a world map pinpointing subsidiary locations and an expandable subsidiary tree for navigating corporate hierarchies. Registered researchers can tag subsidiaries with issue notes that are automatically linked to the parent company profiles. And in a bid to spawn more powerful research tools in the future, CorpWatch's open API gives other organizations access to the underlying tools and data.[18]

CorpWatch and Witness are not just manifestations of an isolated or temporary phenomenon. A multitude of citizen-driven initiatives indicate that something more profound is occurring as globalization and interdependence intensify, and as the national governments seem less able or less willing to cope with the consequences. Rather than wait for government, individuals and nongovernmental organizations are developing their own transparency-based regulatory solutions.

Bottom-up Regulatory Solutions for a Global Economy

In response to the incessant pressure from activists, the public relations ploys that were characteristic of corporate responses in the 1990s quickly evolved into codified commitments to voluntary performance standards. In a handful of high-profile industries, firm-level codes of conduct are evolving into industry-wide regimes. And industry-wide regimes in early adopters such as the chemicals industry are spreading to a growing number of industrial sectors. In fact, few industries have been left untouched by these developments as self-regulatory regimes continue to emerge in a growing number of North American, European, and Asian industries.[19]

Major players in the apparel and retail industry, for example, now subscribe to codes of conduct such as the Fair Labor Association and the Worker Rights Consortium that include requirements for third-party plant inspections and regular public reporting on workplace health, safety, and environment issues in overseas suppliers. The forestry industry is being muscled into adopting more sustainable forestry practices as their downstream retail partners, such as Home Depot and Lowe's, work with a third-party certification body, the Forest Stewardship Council, to certify that their retail goods are "old-growth-lumber-free." Unilever, the world's largest producer of packaged fish, partnered with the World Wildlife Foundation to launch the Marine Stewardship Council to certify that a growing percentage of their supply is harvested from sustainable fisheries. Meanwhile, new industry-led initiatives have also emerged to address diverse issues such as climate change (the Carbon Disclosure Project), conflict diamonds (the Kimberley Process), corruption (the Extractive Industries Transparency Initiative), and slavery in the cocoa supply chain (the Cocoa Initiative and the Harkin-Engel Protocol).

Beyond industry-level codes of conduct, supply-chain audits, and product certification schemes lies an even more extensive set of firm-level practices that were virtually unheard of only two decades ago. Oil and gas companies such as Shell, for example, now hold extensive stakeholder consultations prior to engaging in new exploration projects and release detailed annual sustainability reports that track their social and environmental performance around the world. Even the highly secretive biotechnology industry has bowed to demands for greater transparency and dialogue around new products and technologies by reaching out to environmental groups in a bid to reduce public opposition to genetically modified foods in Europe.

The picture emerging from these disparate sources is that of a sea of change in corporate practices and the nature of business regulation. Behind that sea of change is a dense and expanding web of voluntary rules systems that constitute a semiprivate system of corporate governance. These instances of participatory regulation, in turn, herald an extensive shift in the way societies regulate everything from the social performance of economic enterprises to the exploitation of natural resources. Take natural resource management as a further example, as we discuss next.

Opening Up Natural Resource Management

In the past, natural resource management came down to the capacity of an authoritative, centralized body in a geographic territory to monitor and control the exploitation of a given resource, whether it be forests, minerals, or fisheries. Today, three forces are opening up the regulation of natural resources to a much broader global audience, making new models of participatory management a genuine possibility.

First, satellite imagery, coupled with a plethora of intelligent microsensors, allows us to view staggering quantities of data from our natural and built environments.[20] And thanks to tools like Google Earth, much of this information is now freely available on the Internet, which empowers policy makers and practitioners while giving considerable ammunition to conservationists and local communities. Second, there is growing recognition that at least some resource stocks should be considered global public goods, due to the ecological services they provide to the global biosphere. Citizens around the globe have taken an increasing interest in the protection of these public goods, and in doing so they call into question traditional notions of national sovereignty when it comes to planetary stewardship. Third, there are well-organized and increasingly agile networks of conservations and environmental groups that reach across national borders and wield considerable influence in key policy debates.

One of the pioneers in this space is the Washington, D.C.–based World Resources Institute (WRI), which maintains Global Forest Watch (GFW), an interactive Web site that improves transparency and accountability in forest management decisions by increasing the public's access to information on forestry developments around the world. First launched in 1998, GFW's underlying principle is that increasingly powerful information technologies make transparency one of our most potent mechanisms for strengthening the incentives for responsible industry practices and building the capacity for sustainable forest management. The site provides access to a wealth of information about threats to forests and the entities behind those threats. Within minutes, an interested researcher can see the location and duration of a company's logging concessions, look up local forestry laws and regulations, and check whether the logging companies have paid their taxes. Most information can be easily navigated using a visual map interface that taps into a combination of satellite imagery, national forest data sets, and "on-the-ground" reports. More advanced users can download geographical data from their warehouse and manipulate it for their own analyses using third-party apps like Google Earth.

Global Forest Watch is not perfect. Some aspects of the site are clunky to navigate and the most recent data for some countries dates back to 2000.[21] Real-time data would be ideal, but the prohibitive costs of maintaining up-to-date data across all of the countries covered means that capability may be out of reach for the foreseeable future.[22] For many communities that rely heavily on forest ecosystems, however, Global Forest Watch still fills a gaping hole in their capacity to move toward sustainable management.

Take Gabon, a small West African country with an extensive system of lush tropical forests that cover 80 percent of its territory. During past decades, forest data and maps were guarded from the public—sowing confusion on the ground and creating a significant obstacle to sustainable forest management. Citizens living within Gabon's forests were frequently confronting logging operations that had crossed into their communities or customary lands unannounced, posing a serious threat to their livelihoods. Companies operating in logging concessions faced a similar predicament, unable to secure their concession borders and prevent neighboring companies from poaching trees. In the absence of clearly defined, publicly available logging boundaries, forest communities and companies alike had no platform from which to defend their rights. So in 2006, the WRI signed an agreement with the Gabonese Ministry of Forest Economy, Water, Fishing, and Aquaculture (MEFEPA) to collect data and create interactive tools to support sustainable management.[23] These efforts culminated in the publication of a collection of maps and data sets in May 2009. For the first time, policy makers, companies, and citizens have access to an accurate presentation of activity occurring within Gabon's forest sector.[24]

Gabon is just one example. Since 2001, the WRI has built a network of seventy-five environmental groups and universities that monitor forest development activities in nine countries, which encompass over 60 percent of the world's remaining large tracts of intact forest.[25] The network's broader footprint includes hundreds of additional forestry groups that rely on its data to mobilize global concern and build local capacity in their countries. A logical next step could be for the WRI to engage industry directly, perhaps by encouraging companies to use Global Forest Watch to report on their operations directly to the public. Other features could include a utility to allow users to post video clips, pictures, and audio testimonies, thus bolstering their ability to participate more directly in monitoring nearby logging activities. It would also make sense for GFW to integrate more advanced map servers and social media technologies into its platform. The WRI claims to be pursuing funding for these kinds of improvements.

Forest management is only one of many domains where nascent models of participatory regulation have taken hold. Other domains, including workplace health and safety, anticorruption, environmental protection, and climate change, provide useful examples of how participatory regulation can help augment existing regulations or provide a flexible foundation on which binding rules might eventually be built.

PLATFORMS FOR PARTICIPATORY REGULATION

The instances of participatory regulation described in this chapter reveal a potential future where many government policy objectives are achieved through institutionalized forms of social cooperation that pool the resources of a diverse group of stakeholders and leverage ever more powerful information technologies to scale up their impact. The promise of increased stakeholder participation is that more transparent and participatory forms of regulation will help deliver concrete social outcomes without imposing disproportionate costs on either industry or taxpayers. Systems of regulation will become more fluid and timely, responding both to the evolving needs of societies and to the capacity for improvement in industry. Citizens will be more informed to make smart choices and become more empowered to protect their family, friends, and communities from harm.

Of course, there are risks too. Governments could cede control of the policy agenda to unelected interest groups or fail to adequately scrutinize the effectiveness of these alternative regulatory frameworks, leaving them vulnerable to gaming or insufficient enforcement. But the greatest risk is that insufficient innovation in regulatory strategies will undermine the legitimacy and effectiveness of policy and undermine economic performance. Worse, systemic regulatory and market failures (comparable in impact to the financial crisis) could unleash detrimental changes in social, economic, and political order that will further erode global stability. Harnessing expertise and resources from emerging networks in the private sector and civil society will be an essential part of developing effective and forward-looking policy responses.

Five Priorities for Innovation

The closed, hierarchical, and static regulatory structures of today must give way to new processes that embody values of openness, empowerment, inclusiveness, and knowledge sharing. Making the transition to participatory regulation will require regulators and their partners in industry and civil society to master five key areas of innovation:

1. Create a culture of openness

Regulatory agencies often fail to recognize and tackle tough issues because they lack the right experience and perspectives to guide them through new challenges. Fault partly lies in organizational cultures that rarely encourage or reward employees for pursuing innovation. There is also an unfortunate tendency for organizations to only look inward for new ideas and approaches, and thus ignore a much richer tapestry of capability in society that could be leveraged to achieve the government's policy objectives. The net result is that most instances of regulatory innovation in the last decade have been driven by entities acting outside the purview of traditional regulatory frameworks and bodies. So the first objective for regulators, then, is to create an environment that promotes learning and innovation. This means instilling the desire to seek out new information, consider other perspectives, engage other stakeholders, and be more transparent about policies and activities.

When the U.S. Environmental Protection Agency (EPA) set out to produce an action plan for the Puget Sound estuary system in Washington state, it didn't take the usual public policy route—gather a bunch of insiders together to hash out a policy behind closed doors. Instead they threw up a wiki and launched an Information Challenge that invited the broader community to assemble relevant data sources and begin to articulate solutions. More than 600 residents, businesses, environmental groups, and researchers participated and contributed 175 good ideas, according to former EPA CIO Molly O'Neill.[26] The results included, among many other things, a tree ring database from 2006 that provides an excellent baseline from which to monitor the impact of climate change on local tree species, wildlife toxicology maps for the Puget Sound area, and real-time water quality monitoring tools, including water measurements taken from local ferries that could complement existing buoy measurement systems. O'Neill said afterward, "We can actually use these kinds of mass collaboration tools to transform government, not just add layers to government." The kinds of "emergent behavior" you see in cases like the Puget Sound Information Challenge can be applied in nearly all aspects of the regulatory systems, leading to new insights, innovations, and strategies that even the smartest individuals couldn't produce in isolation.

2. Build platforms for participation

Web-based citizen initiatives such as CrocTail and Global Forest Watch show that one of the best and most expeditious ways to enable regulatory innovation is for government to embrace the kind of platform openness that has driven the success of entities like Wikipedia, Apple, and Amazon. In other words, governments should open up their data to broader scrutiny for the same reason organizations such as CorpWatch and the World Resources Institute do: arming interested parties with information so that they can hold both industry and regulators accountable for better outcomes. After all, government can't always anticipate how society's needs may change or all of the creative ways in which regulatory objectives could be achieved in the future. Nor can government necessarily afford to supply an ever-growing field force of inspectors and investigators with the capacity to stay current with the latest technical, scientific, and industry trends. By open-sourcing their approach, and particularly their data, regulatory agencies can stay more attuned to emerging issues and social expectations and also leverage the complementary resources and capabilities needed to address them. To be sure, participatory regulation can succeed only if active and well-resourced citizen movements exist to energize the system.

3. Foster dialogue and continual improvement

The point of participatory regulation is not necessarily to secure prescriptive regulations, but to define a process through which firms can learn and continually improve their performance. The challenge in aligning public and private value, as Charles Sabel of Columbia University suggests, "is to find a way of using experience to create a framework that suggests what's allowable from the point of view of the world's consensus on moral values, and what's feasible by way of improvement [in industry]." Over time, this dialogue between firms and society is updated continually as experience helps inform realistic expectations. Firms reinterpret society's values, while society respecifies what's required by way of improvement. Both judgments become more nuanced as firms learn more about what society wants, and society learns more about what firms can do in different contexts.

The Cocoa Initiative, a multistakeholder effort designed to eradicate slavery from the cocoa supply chain, is a good example of how this kind of iterative dialogue can help establish mutually agreeable expectations about acceptable corporate conduct. The issue first surfaced after investigations by human rights organizations revealed that child slaves were being forced to harvest cocoa in small West African farms where 75 percent of the world's cocoa is grown—cocoa that eventually finds its way into chocolate treats manufactured by companies such as Hershey's, Nestlé, and Cadbury.[27] Chocolate manufacturers initially claimed that they bore no responsibility for slavery since they purchased their cocoa on the commodity markets and had no direct relationships with the cocoa suppliers. NGOs painted the industry as villains and, in some cases, oversimplified the complex and sensitive social issues that underlie child and forced labor in West Africa. The two parties came closer together after months of dialogue, brokered by two members of the U.S. Congress, Senator Tom Harkin and Representative Eliot Engel. The talks resulted in a unique regulatory partnership between NGOs, labor unions, cocoa processors, and the major chocolate brands that is developing a transparent certification process for cocoa suppliers that don't rely on forced labor. The initiative also fosters community-level actions that change the way cocoa is grown.

4. Protect the public interest

As the representative of the electorate in the emerging systems of regulation, it's government's responsibility to protect the public good. Indeed, governments will need to be careful to balance inputs and protect the broader public interest when corporations and nongovernmental organizations take on broader governance roles. Senator Harkin and Representative Engel provide a model for how elected officials can help maintain accountability and mediate between competing interests. But as numerous examples of participatory regulation show, sorting out conflicting notions of the public good can be difficult.

In some of the earliest attempts to impose codes of conduct on the apparel industry, for example, alliances of North American trade unions and NGOs made unrealistic demands for the apparel companies to establish an international minimum wage and create uniform working conditions around the world. When Gap, one of the first companies to experiment with supply chain monitoring, initiated a program to improve working conditions, they found that there were many adverse effects. Not only did Gap's initial efforts dramatically distort the local labor market (in some cases causing doctors to leave their jobs to work in higher-paying factory jobs) and generate resistance from local politicians, they also had the perverse effect of forcing Gap to terminate relationships with suppliers that failed to meet their conditions, thus punishing the employees that Gap, and the alliance of trade unions and NGOs, were ostensibly trying to help.[28]

The Gap example demonstrates that the NGOs and citizen coalitions that have been forceful in monitoring corporate behavior must also accept accountability. Although many civic activists may feel they speak for the public good, the public interest is a highly contested domain. Single-issue NGOs are often myopically focused on their own agendas—they are not always interested in balancing different visions of the public interest, or acknowledging the central role the private sector plays in creating wealth and fueling innovation in modern societies. Having a role in setting a broader regulatory agenda will carry with it a requirement to think and act beyond narrow interests.[29]

5. Organize collective action in industry

The success of participatory regulation will often depend on the ability of influential leaders to rally other firms in the industry around shared goals and challenges. Environmental leaders in industry, for example, often have strong incentives to encourage like-minded companies to join them in adopting higher standards and, where possible, to organize widespread industry cooperation. The success of collective action in industry will depend on many factors, including the size of the group, the degree of stakeholder interdependence, the ability of lead firms to influence laggards, and expectations about cooperation and reciprocity. While not impossible, it takes a combination of dexterity and diplomatic leadership. Roger Martin, dean of the Rotman School of Management at the University of Toronto, suggests that "the most significant impediment to the growth of corporate virtue is a dearth of vision among business leaders."[30] Perhaps, on the positive side of the equation, this lack of vision leaves the playing field open for a new generation of creative and courageous corporate leaders. Their task will be to promote the notion that by working together as industries, and by enrolling the support of civil society and government, they can extend and enhance the benefits of economic activity for society.

MAKING IT HAPPEN

Participatory regulation is based on the notion that concern for the public good, fidelity to norms, and the capacity to operationalize those norms build faster when not induced by threat or coercion. Threats tend to cause individuals and organizations to redefine their interests in opposition to the threat. Coercion can achieve short-term compliance, but often at the expense of undermining long-term commitment.

Today, it is more likely that shifts in corporate mind-sets and behavior will come about not because regulations require them, but because corporations themselves are intrinsically motivated to embrace change. Starting with a voluntary program is often the best way to encourage dialogue, which in turn is the best way to convince firms to consider taking action on a problem. Once you have established concern, you can move toward setting new norms, then to building capacity toward satisfying those norms, then from norms to rules, then to actual enforcement of the rules.

There is political necessity about creating a forum for iterated decision making in which this type of dynamic can take place. As business regulation scholars John Braithwaite and Peter Drahos put it, "There can be no enforcement until there are rules, no rules until global consensus-building has generated norms, no norms until there is concern."[31] The problem, as discussed earlier, is that unlike government regulations, not all firms, given the choice, will participate in institutionalized systems of dialogue and transparency. But if industry leaders can be convinced to adopt higher standards of corporate conduct, then the laggards will be drawn in as rules become encoded in national regulatory systems and corporate management systems such as ISO 14000.

Whether transparency and citizen participation are seen as integral components of effective regulation will depend on our expectations. As Ann Florini points out, "If you think of transparency as a replacement for government regulation, where you're going to get the same rapid transformation of corporate behavior, you're going to be disappointed." Recognizing that transparency has limitations, it's most appropriate to view the phenomenon as part of a broader regulatory toolkit for driving more socially, economically, and environmentally responsible behavior.

In the end, the case for participatory regulation boils down to a realization that both the market and government fail to provide optimal regulatory solutions in times of rapid technological and economic change. Fortunately, we need not "choose" between these alternative social arrangements exclusively; some of the most viable solutions to failures of the market and government are realized through the adaptations of citizens, businesses, and agencies to the limitations of both systems.

16. SOLVING GLOBAL PROBLEMS: BEYOND THE NATION-STATE

When Matt Flannery and his wife, Jessica Jackley, packed their bags for East Africa in 2004, they were in no way prepared for the kind of journey they would end up undertaking. Jessica was going to conduct impact evaluation studies for Village Enterprise Fund (VEF), a nonprofit that combines training, seed capital grants, and mentoring to help rural entrepreneurs across Africa launch small businesses. Matt, whose passion is filmmaking, was planning to document the whole experience as the couple traveled across Uganda, Kenya, and Tanzania.

Like many Westerners who set off to Africa, Jessica and Matt were expecting to encounter hopelessness and desperation. But far from having their worst fears confirmed, what the couple saw imbued them with optimism. Everywhere they looked there was untapped potential. They saw firsthand a number of entrepreneurs who were being helped by VEF, and hundreds more looking for an opportunity to start their own enterprises. "Listening to their stories I was like, 'I want to invest in that, I want to invest in that guy, I want to buy a share in her business!'" said Matt. The couple didn't have much money. But they quickly learned that $100 would make a really big impact on someone's ability to support themselves and their family. "The more we talked about it," says Matt, "the more we realized that if we invested a hundred dollars in their business, they would succeed, it was really exciting, and in turn it was rather addictive!"

Matt started keeping track of his investments in Ugandan businesses on his Web site and pretty soon his Web site morphed into a tool to share the 'underdog stories' of entrepreneurs trying to build businesses, read by his family and an extended network of friends. News spread quickly and that gave the couple an idea. What if they developed an Internet-based marketplace where socially minded individuals could loan directly to these deserving entrepreneurs? They knew there had to be other people like them who would relish the opportunity to invest directly in business opportunities that could transform not only the lives of the entrepreneurs running them, but also the fortunes of a broader community. It was more engaging than donating to charity. And it was an exciting way for people to contribute to alleviating poverty on a continent with no shortage of ideas, but a chronic lack of capital to fund them.

Jessica and Matt called their marketplace Kiva (which means unity in Swahili). And they spent the next year researching and developing a business plan. As a first step, they hired a Ugandan entrepreneur to help track down other worthy entrepreneurs and post their business plans and capital needs to Kiva.org. It took a few months to get traction, but in March 2005, $3,500 in Kiva loans were issued to a goat herder, a fish monger, a cattle farmer, a restaurateur, and three others, all looking for some capital to launch or expand their businesses.[1]

Making small loans to individuals and small businesses in the developing world—a concept called microfinance—was not a new idea when the couple launched Kiva, but it was hardly mainstream either. For most of the past three decades, microfinance was a little-known banking concept being trumpeted by a Bangladeshi scholar turned banker to the poor, Muhammad Yunus. In 1974, in the midst of a horrible famine in Bangladesh, he made a $27 loan to a group of forty-two families in order to allow them to pay off predatory loan sharks and attempt to get back on their feet.[2] Such predatory loans aren't just crippling for the debtor; they can be deadly, even today. In April 2009, incredibly, over 1,500 Indian farmers from a single state committed suicide when their crops failed and they were unable to repay loans due to local moneylenders.[3] With the interest due often amounting to more than the original capital borrowed, such informal lending schemes have often made the poor poorer.

The success of Yunus's first small loan, however, set the path for what became the Grameen Bank in 1976—the world's first microfinance institution and the first viable route to capital for impoverished Bangladeshis. Today Yunus's creation lends more than $1 billion a year to over 8 million borrowers, with a 97 percent repayment rate. That makes it one of the best-known and most established brands in an industry that lends over $20 billion annually to over 100 million households in more than fifty countries.[4]

Grameen Bank may have paved the way, but Kiva was in uncharted territory. Before Matt and Jessica launched their experiment there had never been a way for ordinary individuals to get involved in microfinance and there was certainly no one proposing to do it over the Internet. Kiva was hardly a bank and there were no assurances that investors would ever see their principal back. And yet, six months later each of the loans had been fully repaid, confirming that you don't need to be a banker to succeed in microlending.

Shortly thereafter Kiva.org went live with a much larger portfolio of entrepreneurs seeking loans. And thanks to some timely media coverage it quickly got a big following. According to Jessica, lenders on Kiva see a poor, uneducated farmer in sub-Saharan Africa not as a charity case, but as a strong, hardworking, intelligent individual who has an enormous amount of potential to change his or her own life for the better. Indeed, unlike a donation to a charity, which begins and ends with the writing of a check, the people who connect through Kiva are partners, and that deeper level of engagement seems to be at the core of Kiva's attraction. "Kiva is about more than just peer-to-peer lending," says Matt. "It's a platform for collaboration and a chance to develop genuine business relationships between lenders and the loan recipients." He says most lenders take the opportunity to dig into the business plan, provide advice, and monitor the progress of the recipients. Once the loan is repaid, lenders can re-lend to another entrepreneur, donate their money to Kiva, or withdraw it. Over 90 percent of participants choose to reinvest it.[5] And like the impact of thousands of small loans on Obama's presidential campaign, the aggregation of tens, hundreds, or even thousands of small loans is having a dramatic impact on the lives of others.

Today, Kiva is cited as the leader of a new model of philanthropy and banking. Just five years after being founded, 455,000 individuals, aged one to 101, have used Kiva to channel over $137 million to more than 352,000 entrepreneurs across not just Africa, but the world.[6] These days Flannery himself invests mostly in Eastern European businesses. When we caught up with him in 2009 he'd just invested in a food market in Azerbaijan and a clothing store in Ukraine. "It's the democratization of international philanthropy," he says. "Rather than having a few powerful people make all the decisions about where donor funding should go, microfinance allows little fragments of money from hundreds of thousands of people to be put to work for the people these donors believe should receive it. It creates an extremely empowering and transparent experience."

Of course, in seeking to alleviate poverty and create economic opportunity in the developing world, Jessica and Matt might have tried to reform international development institutions from within. But they choose to work outside of them instead. Today, microfinance has created a parallel banking system that has displaced much of the traditional banking and lending structures in the developing world. The aggregate results, notably 100 million customers with a repayment rate in the high ninetieth percentile, have proven that a networked, and largely self-organized, system of peer-to-peer lending not only can work, it provides a sustainable way to lift millions of people out of poverty. Indeed, the ultimate compliment and confirmation of the success of small microfinance outfits like Kiva is that big banks such as HSBC and Citigroup are now rushing into the field, offering microfinance products that compete with the altruistic visions that motivated the couple.

And this brings us to the central point of this chapter. Kiva is much more than a philanthropic initiative or even a new collaborative approach to banking. It points to something much more profound—a new model of global problem solving. After all, poverty is a global problem. But Kiva was not launched by an international organization like the United Nations or the G20. There was no global summit of national leaders to ordain its existence. Nor were there agency staffers to implement elaborate international reporting systems or preside over stringent management controls. Indeed, Kiva is pretty much the antithesis of the old top-down model of international development. And it is just one of countless examples of how ordinary people and organizations can join forces to do something meaningful about global problems. "This human-to-human connection and collaboration is the most powerful force for change on the planet," says Jessica. To her, Kiva blurs the boundaries that keep people separated—boundaries between rich and poor, donors and beneficiaries, developed and developing world, us and them. "These boundaries prevent us from solving the big problems of our time," she says. But through communities like Kiva all this could change. "It would represent a pretty profound shift in how we see and believe in each other," says Jessica, "but it's the kind of transformation that can change the world."

GLOBAL GOVERNANCE FROM THE GROUND UP

So now imagine a world where new global institutions were created to match the scope of the new economic, environmental, and security challenges. But rather than model them on a bloated and inefficient UN-type model, we modeled them on Kiva—with vast networks of people and ideas united with the full complement of skills and resources needed to translate good ideas into action. It's hardly a trivial proposition. A diverse and growing community of deep thinkers and doers—many of whom you'll meet in this chapter—believe that our current institutions for global problem solving are fundamentally and irreparably broken (more on that soon). And like us, they believe there is a short time window in which to build new networks for global cooperation that embody wikinomics principles like openness, interdependence, and collaboration. We're not talking about making some token efforts to widen the scope and scale of citizen participation in international forums. Nor are we proposing some grandiose vision of a representative global government or a new global bureaucracy. We're talking about ordinary people getting together to create inclusive and participative forums for the generation of ideas and solutions to the most pressing problems facing the world. And ultimately, this means doing away with traditional notions of control and ownership over issues, and going beyond the international silos to create networks of the willing and engaged.

Impossible, you say? Consider the fact that connected digital citizens, awash in information and choices, are already taking action. The NGO sector is exploding in size and influence on the international scene and increasingly setting the agenda in areas such as human rights and the environment. Meanwhile virtual communities linking cultural and ethnic diasporas around the globe are breaking down the boundaries of geography and creating bridges based on values. These worldwide virtual communities not only provide a sense of belonging, they can become a conduit for problem solving by bringing together people who share a heritage or a worldview, but not a physical location.

All of this raises the most fundamental question: Will a completely different, more appropriate form of governance succeed the nation-state, just as the nation-state itself was built on the foundations of early and more limited forms of government? The answer to that question remains anybody's guess, but one fact seems beyond doubt. As we go forward, governance will be increasingly co-owned by a variety of stakeholders, including nongovernmental organizations, multinational corporations, and emerging countries such as Brazil, China, India, and Russia. Even individual citizens have an unprecedented ability to participate and engage in global activities. As former UN secretary-general Kofi Annan once put it, "We [now] live in a world where human problems do not come permanently attached to national passports."[7] In other words, global governance should not be owned by any one nation or governing body. It is, and should be, a challenge owned by all us.

Is this all far-fetched—a mere utopian dream that rubs against the grain of current international norms and conventions? Sure, it's a little out there. But the plan we outline here has three critical elements, or pieces, that are achievable if we put our minds to it. First we need to redesign our systems for global decision making, including the modus operandi for international institutions like the UN and the WTO. Though such an ambitious undertaking is bound to be difficult, an innovative global redesign project led by the World Economic Forum is showing a way forward. Second, we need to identify new sources of capability for solving global problems, especially among the citizen networks that are already organizing around every major issue and challenge on the international agenda. So we check in with some social entrepreneurs who are building a new engine for innovation that is helping citizens and nonprofits contribute more forcefully to solving global challenges. Finally, we need to reduce the democratic deficit and increase citizen input as more decision making shifts to the international level. So we look at a unique policy dialogue where national governments and international organizations engaged with thousands of citizens worldwide to brainstorm new ideas for addressing urban sustainability issues.

By addressing these three issues, we can effect global change in a new way. Rather than stale "talk shops" for an international cadre of elite decision makers, we can engage real people to design and deliver real solutions in their communities. With the Internet, we can surface novel solutions more readily, share knowledge across communities, and quickly amass the people and resources to give good ideas reach and scale. Like Kiva, these networks for global cooperation would be wide open, totally transparent, and largely oblivious to geographic boundaries. We have already seen other instances like this in our chapters on climate change and finance. In this chapter we show how networked problem solving can address a broader array of global issues ranging from water scarcity to human security. But first let's review what's wrong with the way we attempt to solve issues like these today.

THE PROJECT TO REDESIGN THE WORLD

Let's face it. The world is broken. For decades, large international institutions like the United Nations and the World Bank have wrestled with some of the world's most intractable problems—the kinds of problems that don't fit neatly into national or departmental pigeonholes. And while progress has been made in some areas, the overall record is not encouraging. The rapid reconstruction of Europe after World War II and the equally rapid development of India and East Asia via the Green Revolution, for example, are regarded as some of the major successes of international cooperation. But international cooperation and international institutions have utterly failed to extend this rate of economic and social development to the least developed regions of the world. Indeed, while East Asian countries were enjoying an unparalleled level of growth through the 1980s, the IMF and World Bank were presiding over the continued poverty and starvation of millions in sub-Saharan Africa, despite having lent billions and spent billions more trying to solve these challenges. At the same time, the international community has proven equally unable to stem the ambitions of aspiring nuclear nations like North Korea, Pakistan, India, Israel, and Iran. The UN and other bodies couldn't halt the collapse of fish stocks or do very much about the Rwandan genocide in 1994, the collapse of Somalia, the downward spiral in Zimbabwe, or the unresolved civil war in Sudan. Nor could the international community develop a meaningful successor to the Kyoto Protocol during 2009's international fiasco in Copenhagen.

Human rights abuses represent another growing challenge. In principle, the international community agrees that national sovereignty should take a backseat when it comes to gross human rights atrocities. But in practice, routine violations go unnoticed or unpunished by those empowered to stop them. Sadly, the unfortunate events in Guantánamo reveal that when the stakes are high, even the supposed champions of freedom and liberty can become the abusers.

Are global problems simply too hard to solve? Or are the institutions and mechanisms deployed at the international level simply not equipped for today's realities? Arguably it's the latter. Indeed, the international institutions set up after World War II are not only antiquated—they're fundamentally incongruent with the scope and severity of the challenges the world now faces. A 2010 concept paper of the World Economic Forum (WEF) describes the problem well: "The world is facing a paradox. It has become increasingly interdependent, but governments and international organizations are becoming less capable of addressing global issues as large parts of national populations are not yet ready to accept the consequences of policies which must be pursued in the global public interest."[8] Global challenges have become more complex and issues have become more interconnected. Yet, our national and international institutions are siloed, making an integrated approach for complex problem solving extremely difficult. More often than not, national self-interests take priority when challenges like climate change or fisheries management demand solutions that transcend national priorities and borders. Further, as Kiva shows, there are many new nongovernmental players, including civil society organizations, businesses, and individual citizens who can be and should be integrated into the process of addressing global issues. International institutions often claim to be open to "stakeholder input." But in reality, this means extending a few seats at the table to the representatives of major business organizations and prominent NGOs.

There is yet another problem, in that the international institutions created over fifty years ago still largely reflect the power dynamics of the world as it was then, not the world we live in today. While only 12 percent of the world's population lives in North America and Western Europe,[9] these two regions possess overwhelming influence thanks to the size of their economic markets and their grandfathered status as the world's powerbrokers. Not surprisingly, countries like Brazil, China, and India want a decision-making role that is commensurate with their relative economic weight today. But as Klaus Schwab, chairman and founder of WEF, explains, "Countries with a vested interest in the current structures have often been reluctant to agree to changes that would dilute their influence."

When you put the more complex governance agenda together with the inertia and dysfunction in today's institutions, you get the equivalent of global gridlock. "All of this makes for an intractable set of global governance organizations that are unable to satisfy the demands of today's global challenges because they are driven by individualized national priorities," says Schwab. And with his observation comes a warning. "History has shown us that while the diversity of national interests provides breadth of perspective, it too often leads us to the lowest possible common denominator on issues of global importance."

The Great Global Redesign, Again

It's not exactly the first time the world has faced an existential crisis like this. Nation-states are relatively new constructs in human history. The idea of national sovereignty dates back hundreds of years to the 1648 Treaty of Westphalia, which ended nearly a century of brutal wars between European empires. Over the ensuing period, smaller city-states became aggregated into countries and national economies, each with a border, the rule of law, and institutions of government. Italy, for example, was not a unified nation-state until 1861. By 1919, as the dust settled on the remnants of the Ottoman empire in the wake of World War I, U.S. president Woodrow Wilson announced that the world had emerged from an age of empire and autocracy and was quickly moving toward a new international system consisting of modern democratic states.

Not all was rosy, however. Wilson warned that modern nations, like their imperial predecessors, must increasingly put aside their self-interests and forge new institutions that would promote peace and increase the level of cooperation among sovereign states. Indeed, it was increasingly apparent that the anarchy that reigned in the international system was hardly a recipe for the peace and prosperity the world desired. "I can predict with absolute certainty," Wilson proclaimed, "that within another generation there will be another world war if the nations of the world do not concert the method by which to prevent it."[10] That concert never happened—another world war did.

International leaders eventually came to their senses when, in July 1944, a group of world leaders gathered in the small community of Bretton Woods, New Hampshire, where they spent twenty-two days hashing out a new international architecture that would preside over the next sixty-five years of global growth and development. In the years and decades that followed, the world saw the birth of the United Nations, the General Agreement on Tariffs and Trade (GATT), the Geneva Conventions, the International Monetary Fund, the World Bank, and the Universal Declaration of Human Rights, among other institutions.

Fast-forward to the early stages of the twenty-first century and the whole narrative sounds eerily similar. The age of Western preeminence is crumbling and a new world order is emerging in its place. The United States may continue to be the world's primary military power, but the recent rise of countries in Asia, Latin America, and the European Union is shaking up the balance of economic power. By 2050, China's economy will be almost double that of the United States, whose GDP India will have nearly matched on its way to becoming the world's second largest economy.[11] Russia, Brazil, South Korea, and others are emerging as economic powerhouses and the bipolar paradigm of the twentieth century has been replaced by a multipolar world. This change alone demands that we rebuild the international system and all of its institutions. But how can it be done and where will the leadership come from?

The Global Agenda Partnership Shows a Way Forward

The World Economic Forum has an interesting history, and today it is in transformation. Forty years ago, Klaus Schwab was a Swiss academic on a mission to find new ways of "improving the state of the world" and launched the Forum as a meeting place. Since then it has grown in stature and is renowned for its annual summit of world leaders held in the idyllic Swiss ski village of Davos. Over the last few years Professor Schwab and his team noted an alarming disconnect between the scale of the problems the world faced and the apparent capacity of our existing institutions to solve them. Others in the international community had noticed too. And yet, nobody was mobilizing to do anything about it. As the Forum's managing director, Robert Greenhill, put it, "It had become quite evident that there wasn't going to be a Copernican revolution sponsored by the leaders of the old paradigm."

Given its relatively unique vantage point, Schwab and his team felt the Forum was in a good position to broker a fresh debate on global governance. When the financial crisis hit, the moment seemed right for his team to plant the seeds for a little known, but extraordinary project involving thousands of leaders from academia, government, civil society, and corporations around the globe to rethink global cooperation and problem solving. It was audacious, but it worked. A Global Agenda Initiative was officially launched in 2009 and to date, the process involves about 1,200 leaders from various parts of society who have been recruited to 90 Global Agenda Councils.[12] The councils meet not just in Davos, but also in face-to-face sessions around the world and through a global Web platform that provides an informal space for ideation and debate.

The objective of the initiative isn't to present one single road map on global governance, but rather to develop "at least 100 fresh and sometimes alternative proposals on how the global community could fundamentally reorganize itself to deal more successfully with the challenges, opportunities and risks of a more interdependent, complex and time-constrained world," according to Greenhill. In other words, this is a brainstorm about the future by those who have proven to be the most skilled and knowledgeable in relevant fields of study. And in an age when political decisions made at the national level are too often driven by short-term political expediency, the Forum hopes to provide a platform for long-term, future-first thinking about the challenges and opportunities available to us.

One of the participants is Lord Mark Malloch-Brown, former deputy secretary-general of the United Nations and former minister of state for foreign affairs in the UK government. Since leaving government in 2009, he's been advocating for a new model of global governance and working closely with the Forum to achieve this: "Our systems for global cooperation are rooted in national responses. However, we have a global economy, and many issues like finances, health, migration and security defy the old national solution model. And while there's little doubt that for our lifetime the nation-state will be the biggest decision maker in any emerging system, the key is that the nation-states, and the global organizations that are comprised of them, need to learn to share power." For Malloch-Brown, and most other participants, this isn't about replacing national governments, or about disbanding existing institutions, rather it's about making them more effective by showing how collaboration can transcend boundaries and bring about better solutions. What's increasingly clear is that we've entered a brave new world where people won't show deference to organizations that fail to deliver. "An age of deference is being replaced by an age of accountability," says Malloch-Brown.

This model of networked problem solving also provides the world with one not-to-be underestimated quality: ongoing environmental and risk-focused scanning. By bringing together a large component of the world's expertise, no matter their organizational affiliation or geographic origin, the Global Agenda Initiative might just provide the world with a means of monitoring and informing national leaders on what's coming next. As Hu Jintao, president of the People's Republic of China, noted at a recent gathering of world leaders: "We lack a risk radar system, so even when the World Bank rang the bell [about the impending global financial crisis] no one heard."[13] An ongoing network of discussion about the key issues related to global governance might well offer the framework for this risk radar system.

It's one thing to identify risks, however. It's quite another to forge a collaborative approach when a new risk turns up on the radar. Changing today's model of global governance will require a transformation of the incentives for cooperation and action. Not an easy task given the domestic political issues most leaders prioritize. But with some creative thinking, the incentive problem may be solvable. Say global collaboration could yield a less costly and more immediate solution to climate change, wouldn't national leaders revel in the opportunity to claim credit, having instructed their national research institutes to assign their best people to the effort? Or say, through forums like the Global Agenda Initiative, we could assemble the global expertise to spot the next financial crises before it's too late? Tangible benefits like these would add to the reservoir of political will available to political leaders in their domestic constituencies. In other words, global cooperation would beget domestic success.

Rick Samans, managing director at the Forum, echoes our thinking on the topic, noting that the current system of global governance has failed to provide national leaders with the incentives to go beyond shorter-term, domestic self-interest. "International cooperative structures, whether they're formal institutions or networked communities, must provide national leaders with an extra justification for national leaders to look beyond immediate domestic political agendas," says Samans. "Bringing together broad groups of stakeholders to think ahead and identify challenges and opportunities can provide that extra justification, and could add legitimacy to the actions and mandates of national governments by way of their inclusivity." Experience shows that such inclusivity needs to be focused on the actors that are closest to the issues at hand. And that's one key reason why nongovernmental organizations and civil society networks are increasingly becoming focal points for global problem solving, as we discuss in the next section.

GLOBAL CITIZENS: NEW MODELS OF COOPERATION FROM THE BOTTOM UP

Many intriguing questions and possibilities are arising out of the Forum's global agenda process. But one series of questions comes up repeatedly: can individual citizens and organized citizen networks help solve some of the problems that big institutions have so far failed to address? Indeed, if the solutions we need won't be found in the world's national capitals, could networks of citizens supply the requisite creativity and capability? And if decisions made in distant international institutions seem illegitimate or disconnected from the realities and concerns of ordinary people, could civic organizations of all sizes and persuasions become conduits for greater citizen input—the kind of inclusivity that Samans spoke of?

You probably didn't realize this, but the NGO sector is now the eighth largest economy in the world—worth over $1 trillion a year globally![14] When Bretton Woods and other postwar arrangements were made, the concept of an NGO didn't even exist and there were only a handful of these entities in the world. Governments were the only players at the table as our systems of global cooperation were being established. Today the NGO sector employs nearly 19 million paid workers, not to mention countless volunteers. And NGOs spend about $15 billion on development each year, about the same as the World Bank.[15] Equipped with global reach, powerful brands, and a mission to ferret out public and private malfeasance, NGOs can powerfully influence the marketplace and the public sector—either as high-profile challengers or as partners in finding new solutions.

While organizations such as Greenpeace, World Vision, and Oxfam are perhaps the most visible manifestation of civil society, the sector as a whole has been exploding with legions of smaller organizations and networks becoming active around an increasingly wide range of issues and projects at local, national, and international levels. Basically, individual citizens are getting involved in change, in part because they like contributing to issues, and in part because networks like Twitter and Facebook make saving the world a social activity. Consider some recent evidence from Harvard political scientists Robert Putnam and Theda Skocpol. They independently registered increased levels of disillusionment with traditional political and civic institutions, but found growing interest in advocacy organizations attached to social causes like women's rights or the environment. They also found that citizens are now more likely to drop in and out of organizations and issues than they are to make a long-term commitment to membership in apolitical associations like the old Rotary Club.[16] In the meantime, the very fabric of social action is changing. Once the home of large, centralized nongovernmental organizations, today change is coming from everywhere. Be it individuals with time to spare while waiting for a doctor's appointment or senior executives with untapped ideas and unrealized dreams, social action is now everyone's business and the world is getting better as a result.

Ideagoras for Activists

Skills and expertise have become infinitely portable thanks to the Internet. And although people still cite "time scarcity" as the number one reason for not volunteering, even that problem can be tackled with a little ingenuity. Jacob Colker and his business partner, Ben Rigby, found, for example, that it's not that people don't have any time, it's that they don't have large chunks available to commit to larger tasks. "It's important to remember that as much as we don't think we have the time to volunteer," says Colker, "we spend 9 billion hours a year playing solitaire." Apparently, Americans spend 4.6 hours a week playing video games, 51 minutes riding public transportation to and from work every day, 18 minutes in an airport security lane, and half an hour on average standing in line at the post office.[17]

Colker and Rigby reckoned that all of this spare time could be harnessed and given a social purpose. So they designed The Extraordinaries, a microvolunteering platform that allows supporters to use their mobile phone to transform their spare time into social action. For example, someone with foreign language skills can help translate a nonprofit's Web site into another language, or someone with a passion for birds can help the Cornell Lab of Ornithology identify species in archived photographs. Call it an ideagora for activists.

Not only does the Web give rise to a broader spectrum of engagement options, it calls into question the role of large charities and advocacy organizations in a world where citizens can engage directly. Do citizens need to help fund an organization to do charitable or advocacy work for them? Not any longer. That doesn't mean everyone will choose to engage directly, but a growing proportion will. While writing a check still constitutes an integral aspect of "engagement," as does volunteering, concerned citizens can do everything from the relatively trivial expression of support (say, putting a logo or fund-raising application on your Facebook page) to organizing local meet-ups or even contributing your knowledge and expertise toward a specific, one-time project or problem.

As we write this, a new generation of marketplaces for social action is breaking down geographic and temporal boundaries and expanding the pool of candidates available to contribute to problem solving efforts. Think creatively and one can easily envision dozens of opportunities for collaboration between individual problem solvers and the organizations that need their expertise: from InnoCentive-like marketplaces that connect solution seekers with problems solvers to Digg.com-like forums where participants in the development community suggest and rank projects that require funding. Virtual worlds like Second Life could provide donors and recipients with the venue to build virtual mock-ups of their projects, while eBay-like development auctions could provide governments, communities, and individuals in the developing world with the ability to bid on the "development products" of aid agencies and NGOs. All of this NGO 2.0 activity would need to transpire in an environment where organizations worked harder to share knowledge and coordinate their activities through clearinghouses like Global Relief Web.

From Social Cause Marketing to Social Innovation

This new model of social innovation is not just a pipe dream. It's all happening today, just about everywhere around the world. Billy Bicket, for example, is the director of NetSquared, an online organization that seeks to match social entrepreneurs with social challenges through a series of funded contests. Bicket, who cut his teeth at Meetup (the online social networking platform made famous by Howard Dean's 2004 presidential nomination), is one among a growing cadre of social activists who believe that the traditional NGO model is broken. "Today's NGOs aggregate their networks in Washington, D.C., and essentially operate direct marketing organizations," he says. "Local affiliates are told by HQ what to focus on, and how to fundraise." When NGOs spend money on projects, the results are rarely measured and the outcomes are not always transparent. Moreover, in a competitive market for donors, the issue-focus can drift toward marketable causes rather than the real problems that need solving.

NetSquared is the exact opposite of this centralized approach: it's not about social cause marketing, it's about social innovation. Rather than push a predefined agenda out to a passive public and solicit donations to fund their work, NetSquared uses incentivized challenges to accelerate the work of social entrepreneurs and social causes with proven results. The fact that the challenges are run on the Web makes the whole process collaborative and transparent. The cash incentives motivate contributions and give winning entrepreneurs seed funding for their projects. Promising ideas are refined, rehashed, and rebuilt thanks to community input. It doesn't matter if the winners are looking to make money or not, what matters is that they achieve the results they promise.

When the United States Agency for International Development (USAID) was looking for new ways to use mobile phones to address development challenges in poor countries, for example, it partnered with NetSquared to launch a Development 2.0 Challenge and put up $20,000 in prize money.[18] The winners, announced at the start of 2009, highlight the breadth and abundance of capability available in the civic sector. First prize went to the Child Malnutrition Surveillance and Famine Response system designed by a team of six students at Columbia University. The mobile application transmits nutritional data from growth monitoring clinics in developing countries to government and UNICEF databases, while providing instant feedback to mothers on the changing status of their child's growth and nutritional needs. Second and third place prizes were no less transformative. The runner-up was a health diagnostics application that connects health care workers in underserved regions to medical specialists and collects real-time data for interventions in areas such as maternal mortality, cancer, and AIDS. Third place went to Ushahidi, the crisis reporting application we featured in chapter 1.

To date the NetSquared team has run nine such challenges, distributing almost $375,000 in cash to the winners, which Bicket sees as "the community writing a check for the innovator."[19] For some this is the democratization of funding, but it's actually more than that. This is a wholly self-organized network of solution-starved organizations and independent problem solvers that coalesces to satisfy the requirements of each party. Altruism is still part of the equation, but so too is the challenge of applying business principles and innovative technologies to social issues needing attention. This new generation of social innovators isn't bound by organizations, industries, or traditional causes—it's bound by a commitment to providing solutions, and increasingly that means leveraging the best global expertise on the planet.

DEMOCRACY BEYOND BORDERS

One thorny challenge for the new model for solving global challenges is that multistakeholder networks have power, but not formal power. In other words, they can't enact new laws or send human rights violators to jail. They can't raise taxes to finance a big project or intervene militarily in a crisis. Nor can they automatically claim the same legitimacy as a member of Congress or parliament who has been elected by citizens through a democratic process. These formal powers are reserved for nation-states and their governments. Even international institutions like the IMF, the UN, or the G20 must, to varying degrees, defer to the power of their national benefactors.

So given this reality, what can individual citizens and informal networks really contribute? Can they help pull the levers of state power so that massive financial and other resources can be brought to bear on global problems? Or put it another way: What if a government or an international organization wanted to tap the power of mass collaboration to address an international issue? How could it provide a platform for individuals, no matter their background, to join in and debate the future, and in the process set the agenda for a more granular group of experts and decision makers? And how would this change the role of national political representatives and policy makers that are so accustomed to holding the reins of power?

In 2005, the Canadian government, through its Minister of Labour and Housing, Joe Fontana, asked precisely these questions. In particular, it wanted to use the global reach of the Internet to drive ideas to action to help address the world's most challenging urban issues in a democratic manner without hierarchy, and allow people who would never have an opportunity to discuss the issues to actually have a voice. Of course, to do so would represent a big change. After all, in most countries, policy making is a top-down process and it conforms to national boundaries. Politicians study issues, seek counsel from a select group of advisers, deliberate, and enact laws on the population's behalf. Most citizens are on the periphery, playing no role other than casting a ballot every few years.

Fontana envisioned a different kind of democratic experiment altogether. Partnering with IBM, the Canadian government, through its sponsorship of the World Urban Forum Secretariat, decided to take the conversation about urban sustainability issues to the streets on a global scale. The idea was to bring thousands of participants from government, business, academia, and civil society together for a seventy-two-hour facilitated online discussion where they could strategize around how to provide access to clean water and sanitation, boost environmental sustainability, and improve local governance in the world's rapidly growing metropolitan regions. The engagement process, a digital brainstorm, was modeled after an event IBM first held in 2001, called the Innovation Jam, where IBM brought together its employees worldwide to explore solutions to global problems. In 2006, the same event was even more successful, bringing together 150,000 employees and dozens of thought leaders online to brainstorm new areas of opportunity for IBM in sectors such as health care, transportation, and energy. CEO Sam Palmisano believed so strongly in the concept that he committed up to $100 million to develop the ideas with the most social and economic potential.[20] The World

Urban Forum Secretariat didn't have that kind of money to throw at urban sustainability initiatives, but through the courage and support of hundreds of organizations and individuals from around the world, the Habitat Jam broke down the barriers of language, literacy, disability, poverty, war, and the digital divide to enable over 39,000 people from 158 countries to begin a conversation that some say will change the world.[21]

The diversity of the 39,000 was impressive. Slum dwellers participated alongside government ministers, who participated alongside schoolchildren, who participated alongside leading academics. The conversation ranged across issues of transportation, clean water, governance, poverty, and other issues of importance to people living in cities—especially those who are poor.

As perhaps one of the largest public consultation exercises ever attempted, the event proved that it is possible to reach out to thousands to discuss and deliberate about ideas that might be the source of new and more effective policies and services. Indeed, these loosely distributed networks can coalesce to provide focused advice from those most affected by an issue, or those in the best position to take action. This pioneering experiment was grounded in the belief that this would be the fastest way to innovation. The goal of the Jam was to get all participants working on the most pressing problems of urban sustainability from cities around the world. Six unique forums framed the most critical issues—improving the lives of people living in slums; sustainable access to water; environmental sustainability; finance and governance; safety and security; and, finally, humanity—the future of our cities. An invitation was open to anyone who had something to say about the cities in which they lived.

According to Fontana, "It was an unbelievable learning experience . . . This was not a policy conference anymore. It was a gathering of practitioners from civil society and the private sector exploring these questions: What things have worked? What have we learned? What mistakes have we made? How do we do things better?"[22] The Habitat Jam was successful in leveraging global networks to bring forward the experience and voices of people who would never have been able to attend the World Urban Forum (WUF). Their "actionable ideas" were the starting point for the conference, designed to build networks that would carry their ideas into implementation through improved policies and services—at a global, national, and community level. What is remarkable is the number of the actionable ideas from the Jam. More than four thousand pages of discussion were captured, six hundred ideas generated, and seventy actionable ideas researched and summarized in a workbook for the WUF meeting held in Vancouver in June 2006.[23]

Although the official meetings are now long over, the spirit of Habitat Jam lives on. Bill Tipton, project manager for Hewlett-Packard and contributing author at the Global Dialogue Center, wrote about what it meant to him, as a blind person, to participate in the Habitat Jam: "This is so exciting it makes my hair stand up on end to see and talk with all people with disabilities on-line."[24] Bill has gone on to lead an ongoing dialogue with seventy disabled people from slums around the globe. He helps to raise money that is used to directly provide services in these communities.

More and more of these digital conversations are helping to set agendas that may not have originally seemed significant to national governments. Global dialogue and research into climate change, hunger, and AIDS in Africa are current examples. They have achieved standing on governmental agendas primarily as a result of widespread public concern—from rock stars to citizens to public officials to not-for-profits and to those directly affected.

Local has increasingly become global. And while this global interconnectedness is not necessarily new, what is new is the amount of information and ease of access to it for today's information seekers, while Wikipedia, YouTube, and Twitter provide the tools for a truly global network of citizen problem solvers to emerge. Fortunately, Web 2.0 technologies make the process of engaging citizens in problem solving easier and less costly than ever before. As with Habitat Jam, international institutions can post background information on the Web and use online video conferencing to bring in expert testimony. Web-enabled forums can allow discussion and debate among hundreds, thousands, and even millions of geographically dispersed participants. Wikis provide a platform for collaborative editing of policy documents, while social networking technologies can connect citizens and organizations with common goals and interests.

The promise is that digital engagement will support global problem-solving approaches that integrate policy development and implementation into a seamless and flexible practice of continual engagement, improvement, and innovation that can reach across national borders. Depending on the issue, emerging problem-solving networks will draw participants widely from governments, international organizations, businesses and industry associations, think tanks, academic institutions, civil society organizations such as NGOs, associations, religious groups, and the general public. In doing so, we can better connect ordinary citizens to communities where conversations are happening and help build greater legitimacy for resulting decisions and projects. Indeed, if the first wave of democracy established elected and accountable institutions of governance but with a weak public mandate and an inert citizenry, the second wave should be characterized by strong representation and a new culture of public deliberation built on active citizenship.

However, while multistakeholder networks hold great promise and are already having a profound impact on the world, they pose a number of difficult questions too. Do these networks lack legitimacy because they are not democratically elected? In whose interests do they act? To whom are they accountable? The United Nations may have growing inadequacies as a vehicle for global cooperation, but at least it appears to be a representative and legitimate body and its delegates are accountable to the national governments of which the UN is composed. As multistakeholder networks claim larger roles in governance, national leaders and governments will need to find ways to respond and participate in these discussions. They will face choices about whether to remain reactive or whether to find new ways of entering the conversation in proactive and productive ways that lead to better policy, services, and outcomes. The Habitat Jam brings powerful testimony to bear in pursuing the latter and provides a potential blueprint for the creation of participative forums, made legitimate through their transparency and membership, and made vital by their ability to funnel ideas and innovative solutions to those in position to make them happen.

THE FOUR PILLARS OF SOCIETY: TOWARD NETWORKED PROBLEM SOLVING

For some, *only* national institutions can provide the type of governance we need to cope with a "runaway world." After all, only national governments, as we mentioned, possess the powers of taxation, rule making, and coercion. Columbia University scholar Saskia Sassen makes a good point: "If we don't also deal with the complex, and at this point, more effective environment of national states, we're nowhere . . . too much power and too much of the capacity to make something into law, regulation, or statute still runs through national state institutions." Sassen is right. Nation-states will remain important. But the disaggregated nature of power and authority in today's world means that other actors will have to be involved. Governments will not meet the challenges of global governance alone. We need all four pillars of society—businesses, civic organizations, governments, and individuals—to contribute ideas and collaborate around promising solutions.

The goal of networked problem solving is not, as discussed, to replace existing institutions and their well-entrenched leadership in areas of policy and governance. Rather, the aim is to create an underlying "motor" that moves ideas, foresight, and dialogue from a broad, stakeholder-based debating forum to the organizations that have the financing and human resources to turn them into actions. In their new role, governments should become the framework setters for a new era of global accountability. This means contributing leadership, democratic input, and state resources into a shared project of building more responsive institutions of global governance.

As more decision making shifts to the international arena, the role of elected officials and legislative assemblies in debating, refining, and resolving issues will change profoundly. But rather than diminish the authority of elected representatives, new forms of collaborative problem solving could actually empower politicians to claim more credibility to speak on the electorate's behalf. Indeed, elected officials must ensure that digitally enabled decision making at the national and international levels does not merely amplify the voices of organized interest groups that are already heard. In this new role, politicians will need to complement their skills in political strategy and communications with skills in listening, mediating, consensus building, and public outreach. Above all, politicians—as the only truly legitimate *political* representatives—will need to exercise good judgment in weighing inputs and making policy decisions in international arenas.

That said, perhaps we need a whole new model of legitimacy. Indeed, let's not forget that leaders in most democratic countries rule with the support of less than a quarter of the population (Conservative leader David Cameron took power in Britain with 36 percent of the popular vote and only 23.5 percent of eligible electoral voters since only 65 percent of the electorate actually voted).[25] In fact, governments across the industrialized world face falling voter participation, declining political engagement, and reduced levels of trust. So rather than assuming that the authority to act on behalf of a population is derived solely from a four-year electoral mandate, legitimacy could instead be earned by any organization (or combination of organizations) that is open to public participation and sufficiently transparent about its objectives, operations, and progress. The World Economic Forum process, for example, is inspired, but is it legitimate? Professor Schwab was among the first to address this issue. "Networks need to have legitimacy. These are not entertainment networks. We're talking about networks that have global impact on the state of the world," he says. So how do those networks become legitimate players to make, or at least prepare and shape, decisions that affect the lives of people outside the network? "With transparency your track record and progress is clear, in turn conferring legitimacy on the network," says Schwab, who hopes the Forum can help spawn multistakeholder networks that are welcomed broadly by society as legitimate, open, and accountable to the world.

Other smart organizations outside government should also exploit new technologies to build stronger relationships with their constituents, and in doing so become facilitators of political discourse and citizen engagement in an era of global decision making. Decades of participation in local and international development efforts have shown that civil society organizations like NGOs can be effective change agents and make important contributions to decision making with a blend of effective leadership and adequate access to information and resources. Governments have even come to rely on NGOs in many cases to help create and implement policies that better reflect the needs and aspirations of citizens. But as with the Forum, civic networks and other nongovernmental bodies will need to wrestle with legitimacy questions and operate with the level of transparency and accountability that society demands.

Like NGOs, companies have not always been viewed as legitimate formal players in most of our institutions for global cooperation. There are no firms in the G20, the UN, or the World Bank, for example. But clearly there is a role for the private sector not just in wealth creation, but in helping to advance social development and protect the environment. Indeed, we think corporate leaders with a combination of vision, energy, and communication skills can rise to the challenge. Forums like the World Business Council for Sustainable Development (a coalition of two hundred international companies) and the Global e-Sustainability Initiative (a partnership of technology companies and the United Nations Environment Programme and International Telecommunications Union) already facilitate industry dialogues where corporate leaders can better understand and address the evolving roles of business organizations in the twenty-first century.

As the fourth pillar of society, individuals must accept new leadership roles in addressing global challenges too. Some will opt to do this in their local communities, but choose to share their experiences and insights with the world. Others may plug into global conversations and communities hosted by the World Economic Forum or organizations like NetSquared. Regardless of which forum one chooses, now is the time to contribute your ingenuity to rebuilding our institutions for global governance. We have a small window to ensure that we put our best foot forward in solving global problems. The principles of wikinomics—openness, collaboration, sharing, integrity, and interdependence—provide the blueprint for getting it done.

Global governance has too long been the subject of expert organizational debate, off-limits to the various stakeholders who are affected by the decisions made on a small block of streets in central Washington, D.C., no matter the rate of success or failure. In contrast, a world driven by organizational concert and collaboration, built on shared values and the legitimacy of shared leadership, offers a promising approach to building a proactive, forward-thinking, and sustainable world. One might even say, a wikinomics world.

17. FIGHTING FOR JUSTICE: USER-GENERATED FREEDOM FROM TEHRAN TO RANGOON TO BEIJING

When millions of angry, disillusioned Iranians took to the streets of Tehran in June 2009 to dispute election results that saw incumbent Mahmoud Ahmadinejad retain power, the world got its first glimpse of a nascent perspective on a reform-oriented Iran. No longer was it solely part of the so-called Axis of Evil, with unclear nuclear ambitions and a firebrand president. Instead, thanks to modern technologies and their ability to circumvent government repression of free speech, the world began to see the rise of a new, young, and increasingly moderate Iran that might augur a new start in the world's relationship or at least a different perspective on the Persian country.

This new narrative on Iran was shaped in large part by millions of tweets during the election protests, including peak Twitter usage of 221,744 tweets an hour (3,695 per minute) during the height of the protests that gave the world a real-time perspective on what was happening on the streets of Tehran.[1] As hundreds of thousands of Iranians poured into the streets and a bloody crackdown ensued, graphic images displaying the violence were instantly uploaded to Twitter and other networks by protesters on the ground. A firestorm of concern and sympathy soon engulfed the Internet. And with few foreign journalists actually stationed in Tehran, updates on social media sites were the primary source of information for reporters. The BBC, for example, was forced to preface its news with the disclaimer that "verifying reports is difficult" and to quote journalists from its London-based Persian bureau, rather than staff on the ground.[2] Other Western media outlets faced similar restrictions.

With its control over the broadcast networks, Iran's clerical oligarchy—and the massive security apparatus that supports it—likes to portray itself as a genuine Islamic democracy in which the true interests of all Iranians are protected by a leadership with insight of divine origin. While few outside the establishment believe this to be the case, the regime's blatant control over communications and public discourse has left the government free to engage in years of wasteful and reckless practices that have seriously undermined Iran's welfare and security, despite the promise of its oil wealth and other advantages. These practices have also had serious consequences abroad, helping to destabilize much of the Middle East. And with nuclear technology now in the mix, what transpires in Iran has potentially grave implications for the world at large.

What happened following the elections, however, was a mass protest that escaped the control of all sectors of the Islamic ruling classes. It was an elementary expression of the thirst for democratic rights—one that was shared globally thanks to the fabric of social connectivity made possible by mobile phones and online services like Facebook and Twitter. Every social and political force, both nationally and internationally, has since tried to interpret and understand this new factor in Iranian politics.

FREEDOM AND ITS ENEMIES

It's easy to feel sympathy for the Iranians who want, more than anything, to share the values of freedom and openness, but must risk life and limb to do so. But the real reason for telling this story is that Iran is merely a flashpoint in a much bigger turn of events. The same forces that flattened the global economy are also helping to flatten and open societies around the world, even in countries where you would least expect it. And it turns out the same deep stirrings for individual expression and democracy that are evident in Iran are present in many other countries where the universal aspiration for freedom remains unfulfilled. Thanks in large part to the Internet, these stirrings are increasingly congealing into organized political movements that reach across national borders and effect change.

While this book has been largely about institutions that need rebuilding, the events in Iran and other countries suggest there is surely one human institution that urgently needs nurturing and expanding. That institution is human freedom: the right of peoples everywhere to determine their own destinies, to be free of oppression, and to enjoy bedrock liberties like the right to free speech, a free media, free association, and, most of all, the opportunity to participate equally in building a more prosperous, free, and sustainable global economy.

Unfortunately, according to Freedom House, a leading authority on tyranny and the promotion of democracy around the world, the current models for promoting freedom are broken. For over three decades, Freedom House has published comparative surveys and special reports focused on the state of democracy and human rights around the world.[3] Its most recent report concluded that political rights and civil liberties have suffered a net global decline for three successive years, the first such deterioration since the survey's inception in 1972. Meanwhile its global analysis of media independence revealed an even more prolonged, multiyear decline in the freedom of the press around the world. And unfortunately it gets worse. Countries that are politically bankrupt tend to be economically suspect as well. In democratic countries, individuals are generally free to work, produce, consume, and invest in any way they please, and these rights are steadfastly defended by the state. But in countries lacking political freedoms, economic opportunities are constrained by capricious leaders and a culture of corruption that makes life miserable for anyone trying to start or run a business on any scale. That means hundreds of millions of potential inventors, managers, engineers, educators, thinkers, and investors around the world are handicapped in their pursuit of opportunities that should be every citizen's birthright.

"But hold on," you say, "aren't people in countries like Iran, Russia, and China enjoying far more freedom than those who lived under the likes of Khomeini, Stalin, and Mao?" In many respects that is true. Ordinary citizens in these countries can access far more information than their parents ever could. Their growing middle classes can travel abroad for holidays, purchase a plethora of consumer goods, and enjoy other personal freedoms. Moreover, to varying degrees, all three countries have joined the global trading system, and international commercial relationships are flourishing. Indeed, it seems that today's despots have recognized that absolute control over information and economic activity is neither possible nor necessary.

While many of today's authoritarian states may appear to have little in common with the military dictatorships of the past, they are arguably even more adept at the black art of social control than their predecessors. In a recent report, "Undermining Democracy," Freedom House argues that authoritarian regimes are less dependent on brute military force and now prefer to exercise more subtle forms of manipulation. "These regimes have developed methods that allow them to 'guide' and 'manage' political discourse; selectively suppress or reshape news and information of political consequence; and squelch, co-opt, or parasitize the most important business entities," the report concludes.[4] In other words, the forms of social control in countries like Iran, Russia, and China have only become more insidious. Citizens and businesses—both domestic and foreign—can exercise some autonomy as long as they are prepared to acknowledge the supremacy of the ruling group and comply with its directives. And the extent to which citizens can exercise their rights depends not on the legitimate laws and due process, but on arbitrary and capricious decisions taken by an opaque and unaccountable establishment. Investors such as Warren Buffett have called Russia a "kleptocracy" and vowed never to invest there. Companies looking to operate in these markets must factor in a huge risk premium. And, unfortunately, some companies find themselves making ethical compromises that they wouldn't even contemplate in their home countries.

While national elites live lavish lifestyles, a sizable portion of their citizenries enjoy no such comforts. Instead, they stand perched precariously at the precipice of the global economy, unable to fully realize their potential or enjoy the prosperity that globalization has bequeathed to the citizens of free countries. While the historical trajectory points to more freedom and more democracy in the future, the battle is far from over and freedom could use more allies. In the new global interdependent world, human development, political openness, and economic success can and should go hand-in-hand. Not only are higher levels of economic freedom associated with higher per capita incomes and higher GDP growth rates, those higher growth rates create fertile ground for better governance and greater business investment and innovation.[5] Companies should anticipate and even encourage these transformations. Those that do will unleash powerful forces of choice and opportunity, help nourish other liberties, and improve their own competitiveness by gaining unique insights into the needs and aspirations of freedom-seeking people around the world.

When the Wiki World Meets International Diplomacy

All this highlights a new rule for the wiki world. Extinguish the flame of democracy and you dampen the spirit of economic creativity and innovation as well. Succeed, on the other hand, in extinguishing the dictators, and the world will liberate countless people to participate in wealth creation within every sector of the economy. After all, you can't have a wiki world without freedom. And it turns out that wikinomics presents our best hope for promoting freedom and defeating its enemies. In the old paradigm of international diplomacy, promoting freedom was the job of the UN, national diplomats, and a handful of NGOs. The methods of advancing freedom were slow and ineffective. Diplomats might make a few phone calls, activists would write letters, and if the media and activists complained enough, Western countries were occasionally cajoled into applying a smattering of sanctions. Social change took decades to take effect and sometimes change never happened at all.

In the wiki world, there is a new bottom-up model of international diplomacy where hundreds of millions of oppressed people use mobile phones and modern communications platforms like Twitter to advance their causes. This new generation of citizen activists is wired for social change, united in their aspirations, and equipped with a global platform for collaboration that has even the most ardent authoritarian freedom haters running scared. In fact, everyone can help promote freedom—not just oppressed people. And expectations are growing that the business community will play a major role too.

While the methods of authoritarians are sophisticated and finely calibrated, so too are the networks of pro-democracy supporters. Their networks thrive on transparency and leverage highly decentralized forms of organization to outwit and outmaneuver traditional political and military hierarchies. They collaborate with like-minded groups around the world, sharing critical resources and seeking common cause so as to acquire strength in numbers. They fight for integrity and justice, which gives them a moral edge in a battle that is increasingly about hearts and minds rather than bullets and roadside bombs. And they are always innovating—using the latest Internet-based technologies to amplify their impact.

Take WikiLeaks, a self-proclaimed "intelligence service of the people" with a mission to abolish official secrecy. The whistle-blowing site made headlines when, in April 2010, an anonymous tipster posted a video the Pentagon claimed to have lost of U.S. helicopter crews enthusiastically killing Iraqis on a Baghdad street in 2007. That event, in itself, might not have made headlines, except among the dead were two reporters with the Reuters news agency. The explosive leak, which sent shock waves around the world and had the Pentagon in a conniption, is merely the tip of the iceberg. In just a few short years, WikiLeaks has released more than a million confidential documents from highly classified military secrets to text messages of those killed in the 9/11 attacks. Not a great deal is known about its founders, except that WikiLeaks describes the group as comprising Chinese dissidents, hackers, computer programmers, lawyers, and journalists.[6] To protect their sources (and their own identities), they spread assets, encrypt everything, and move telecommunications and people around the world to activate protective laws in different national jurisdictions. But who watches WikiLeaks? Can it be fully trusted? The jury is out, but so far its track record is good. And when you put WikiLeaks together with the events in Iran and other places you can begin to see why dictators everywhere are wondering when this new citizen power is going to come knocking on their door.

Indeed, the Pentagon may wonder whether WikiLeaks can be held to account when sensitive information leaks to the public. But the question we're asking in this chapter is whether these developments are a prelude to an inevitable wave of democratization around the world—a wave generated by an explosive combination of youthful demographics, the spread of the Internet, and the lure of opportunity in the global economy. The evidence, to be sure, is mixed. "Tweets don't overthrow governments; people do," observes Evgeny Morozov in *Foreign Policy* magazine, noting that social networking sites can be both helpful and harmful to activists operating from inside authoritarian regimes.[7] For example, Morozov points out that secret police increasingly gather incriminating evidence by scanning the photos and videos uploaded to Flickr and YouTube by protesters and their Western sympathizers. "They might even serve as an early warning system for authoritarian rulers," says Morozov.

All true—the Internet is no panacea for freedom fighters. But on the other hand there is copious evidence to suggest that new freedoms of expression and networking enabled by the Web underpin a profound shift in attitudes in many countries, especially among youth. Will youth rise up to effect positive change, cementing new rights and freedoms not just through their tweets, but also through their courageous actions? Who can be certain. What we do know is that the stories we convey here inspire hope—hope that the Internet can empower freedom movements around the world to wrest open despotic regimes, even if decisive political revolutions have yet to materialize in some of the more authoritarian countries we examine in this chapter. So buckle up, all dictators, despots, and tyrants; it's about to get a whole lot rougher out there.

A NEW NARRATIVE ON IRAN

If you ask average Iranians, they will tell you that the protests that erupted on the streets of Tehran in the summer of 2009 were not simply about who got elected; it was a revolt against an election that many consider to have made a mockery of the country's secular institutions.[8] Indeed, the very mood of the nation appears to have changed and all manner of Iranians have been speaking out. "I was out in the streets 30 years ago protesting against the Shah because that regime was brutal and savage and today I'm out again, this time older, again seeking justice and standing against dictatorship," said one sixty-two-year-old retired accountant quoted by *The Guardian*.[9] "I might not see a free Iran in my lifetime, but I'm proud of the battle of today's youth against injustice," he adds. A forty-five-year-old nurse concurs: "Something has changed in this country. I don't think this is just about fraud in the election. It's about the blood that has been spilled and people who have been raped or tortured or harassed by this government."

The so-called Twitter revolution may not have brought an immediate end to this tyranny, but it's the beginning of the end for a bankrupt regime that is desperately clinging to power. For decades the Islamic Republic has suppressed Iranians' elementary yearnings for freedom with an authoritarian cocktail that combines blatant coercion and powerful intelligence agencies with a form of mass bribery in which the government spends an estimated $100 billion annually subsidizing things like bread, sugar, and gasoline.[10] Today, there are many signs that the regime's grand social engineering project has failed. Even windfall oil revenues have been unable to mask the regime's failed economic policies. The crippling effects of the global financial crisis have heaped even more pressure on this oil-dependent economy and on Iran's rulers. In the meantime, the Iranian democrats have a powerful new weapon with which to expose the regime's failings.

At long last the Internet and the rise of popular social networks have broken the Iranian regime's stranglehold on power, allowing new groups to communicate and organize outside the boundaries of control set by government and religious leaders. Unlike the older, hub-and-spoke architecture of the mass media, the peer-to-peer architecture of the blogosphere is more resistant to capture or control by the state. Despite repeated attempts to extinguish the flowering of independent online dialogue among Iranians at home and abroad, Iranians are enthusiastically blogging, twittering, and connecting in a way that would have been impossible just a few years previously.

This technology-enabled rebellion has evolved in step with an Iranian demographic revolution. An astounding 60 percent of the country's 70 million citizens are under thirty years of age.[11] These young Iranians are surprisingly global in their disposition, savvy in their use of the Web, and secular in their values and ideals. Their adoption of mobile technologies has seen the country's mobile penetration rate jump over 500 percent since 2005. Iranian communication and information technology minister Mohammad Soleimani estimated that by late 2008 there were over 40 million mobile phone users throughout the country, sending over 80 million text messages per day, and over 23 million Internet users.[12] In addition to growing text message use, some sixty thousand routinely updated blogs feature a rich and varied mix of Iranian perspectives. The "Persian blogosphere is a large discourse network, incredibly rich in the types of discussions taking place and the different groups of Iranians who are blogging," say John Kelly and Bruce Etling of the Berkman Center for Internet and Society at Harvard University. "Early conventional wisdom held that bloggers were all young democrats critical of the regime, but we found conversations including politics, human rights, poetry, religion, and pop culture."[13]

These Iranian social networks extend globally, linking up to an enormous diaspora located largely in developed countries such as the United States, France, England, Germany, Australia, Canada, and parts of Scandinavia. This diaspora began under the shah's dictatorship following the 1953 U.S.-engineered coup and continued to grow, especially during the first five years of the current Islamic republic when repression against the more liberal middle class became extreme and almost wanton. Now largely wealthy or middle-class, as well as university educated, the Iranian diaspora is almost completely wired and maintains contact with relatives, friends, and colleagues back home via the Internet.

Regime change in the Persian homeland may be elusive, but the final chapter is far from written. What happens next is anybody's guess, but what is verifiably true is that the world now has two potential images of the oft-maligned Persian country. One focuses on the country's nuclear ambitions, its outspoken president, and what the combination of the two may mean for peace in the Middle East and beyond. The other portrays a burgeoning demographic of young Iranians and their desire for change. And while there's no doubt that the first narrative will continue to be the focus of foreign policy so long as Ahmadinejad is in power, the rise of the latter presents the world with hope that a more accommodating and friendly Iran is just around the corner.

BLOGGING FOR FREEDOM: A NEW BOTTOM-UP DIALOGUE BETWEEN MUSLIMS AND THE WEST

If democracy is stirring Iran, what's happening in the broader Middle East? Do the same technological, demographic, and social forces underpinning Iran's freedom movement provide new hope for averting American political scientist Samuel Huntington's grim prediction that a devastating "clash of civilizations" is looming between the West, East Asia, and Islam? In 1992, Huntington stirred great controversy when he asserted that the great divisions among humankind in the twenty-first century would be cultural and religious. He predicted that the fault lines between civilizations will be the battle lines of the future and warned that the resulting conflicts will dominate global politics in the foreseeable future.

With America fighting wars in Iraq and Afghanistan, and other tensions simmering around the globe, it's easy to see why some people think Huntington's dire thesis may be coming to fruition. But our optimism stems from the fact that the wiki world offers vastly more productive options for smoothing over cultural fault lines and solving global issues peacefully. The root problem is a clash of ignorance, not a clash of civilizations. And a global, interdependent world based on openness and engagement provides the best antidote to ignorance that we can think of.

In his new book called *Forces of Fortune: The Rise of the New Muslim Middle Class and What It Will Mean for Our World*, Vali Nasr asserts that the rise of a business-minded middle class is reshaping societies across the Muslim world and that how the West engages this burgeoning middle class will provide the key to countering the threat from Islamic extremists. That alone represents a considerable paradigm shift from the West's longtime support of autocratic nations in the region that have failed to democratize and liberalize their economies and their societies. Nasr makes a compelling argument that the way to win over the Muslim world is to engage it over business, capitalism, and trade—not to fight it over religion.

Nasr is right, but why not take his thesis one step further? By all means, engage the Muslim middle class in business and trade, but why not also engage in an exchange of words and ideas? As the world seeks new models for smoothing relations between Islam and the West, the blogosphere is already carving out a space for a new, bottom-up dialogue between Muslims and citizens in Western countries. Although the region's despots are eager to stamp it out, an increasingly wired and liberal-minded cohort of young Muslims around the world are risking their lives and liberty to get engaged.

By early 2010 nearly 200 million bloggers were sharing their views with a worldwide audience of approximately 350 million readers.[14] And millions of online groups promote discussion and information sharing on nearly every imaginable subject. This growth has occurred in some strange places—Iran, Syria, Saudi Arabia, and China boast moderate Internet penetration growth rates, despite repressive regimes that are known to try to control what is said about their countries on the Web. Perhaps not surprisingly, countries like Iran, Burma, and China have been named "the worst places in the world to blog from."[15]

Rise of the Arabic Blogosphere

Most people in the West have no idea what is going on. And few would know it by observing the major news media. But, against all odds, the transnational Arabic blogosphere is exploding and liberating new channels for uncensored dialogue. The heart of this new movement is in Egypt, but it encompasses other countries such as Saudi Arabia, Kuwait, and Syria where criticism of the state is dealt with harshly and the Internet provides the only outlet for secular pro-democracy groups. Arguably, the blogger has become the new freedom fighter in a war that is as much about images, ideas, opinions, and aspirations as it is about tracking down terrorists in mountainous regions of Afghanistan.

In fact, the expression "blogger" has become almost a revered term in much of the Arab world, where it takes considerable courage to express one's views freely on the Internet. In Egypt's "soft dictatorship" there is relative freedom in the mainstream press to complain about the government and even criticize feared security services. Blogging, on the other hand, seems to cross the line from speaking (which is tolerated) to acting (which is not). This intolerance could be because few bloggers exercise the kind of restraint that is common among journalists working for major papers. Journalists tread cautiously when writing about President Hosni Mubarak and the state's religious institutions; bloggers tend to attack them head on. Many bloggers have been beaten and jailed for their writing. The Arabic Network for Human Rights Information in Cairo was handling cases for over one hundred bloggers facing criminal charges at the time of writing.[16]

Despite the high stakes, some 35,000 active Arabic language blogs in eighteen different countries have risked everything to add their voices to the growing democratic chatter.[17] These Arabic bloggers tend to be predominantly young and male, and they are more likely to link to YouTube videos and Wikipedia as news and information sources than to outlets such as Al Jazeera. Many give blogger Salam Pax credit for initiating the blogging concept in the Middle East; his vivid accounts during the Iraq war gained a worldwide following and were eventually turned into a movie, and his writings prompted other Iraqi bloggers to document their experiences. The result was an extremely vibrant tapestry of authentic Iraqi commentary on current events with contributors across the political spectrum. To be sure, the Arab blogosphere is nascent and it's not clear that it has the critical mass needed to be able to change the course of events in Arab countries or launch successful campaigns for human rights and pro-democracy causes. Yet the foundations are emerging to support the growth of free expression and a robust civic sector that stretches across national borders.

Youth Bulges: A Force for Freedom or Tyranny?

Imagine if the number of Arab bloggers were to grow ten- to twelve-fold within a few years. Consider the amount of pressure this could place on despots to embrace freedom and openness. Is it just wishful thinking? Isam Bayazidi, founder of the popular blogging portal Jordan Planet, doesn't think so. He cites increasing mainstream attention as a factor in drawing more bloggers out of the shadows and credits the fact that free open-source blogging software now incorporates templates that enable posting in Arabic with ease. While such developments are important, the Arab world's exploding youth cohort provides an even more powerful tipping point. Veteran Egyptian columnist and blogger Mona Eltahawy points out that young Arabs have been vilified in the Western media and they have a desire to be heard. "While Al-Jazeera and Al-Arabiya have smashed through old taboos such as criticizing governments and providing forums for dissidents," she says, "they are still out of bounds for young people."[18]

Of course, there is a risk that demographic factors could tip the other way, giving rise to large populations of disenfranchised young people who turn toward dangerous forms of nationalism and fanaticism. Hitler mobilized dissatisfied youth in Germany to become brownshirt thugs who came into the streets to attack labor unions and anti-Nazi protesters. Some point to the recent resurgence of radical Islam as evidence that the Middle East's youth bulge has created a similar powder keg that, if sparked, could explode into armed conflict, even riots and civil war.[19] Herbert Moller, one of the first academics to seriously consider the impact of youth bulges on political stability, concluded that the presence of a large contingent of young people in any population can provide an impetus for progress or instead intensify or exacerbate existing problems, depending on the circumstances. "[Youth bulges] may make for a cumulative process of innovation and social growth; it may lead to elemental, directionless action-out behavior; it may destroy old institutions and elevate new elites to power; and the unemployed energies of the young may be organized and directed by totalitarian rulers."[20]

Time will tell, but so far there is no question that demographics are favoring democracy. As compared with elder generations, young Arabs are better educated, more wired, more likely to favor democracy, and generally less hostile toward the West, although they still view the U.S. government with suspicion.[21] Meanwhile the rising din of debate in the blogosphere suggests that repression and hard-line tactics in their homelands may be fanning the flames of freedom. "The very same restrictions that were once merely irritating, such as dress codes and government censorship, have now become absolutely suffocating," writes Melody Moezzi, a young Iranian author and blogger. In the meantime, she argues that young Iranians are beginning to realize the power of being in the majority. "And they are growing up fast," she says.[22] If Moezzi is right, the Middle East's future role in the world will be in large part determined by whether secular pro-democracy movements can win the hearts and minds of the youth. For now, those struggling for democracy inside and across the Muslim world will hope that the Internet remains a forum for broadening democratic engagement and a refuge for openly discussing the most salient issues facing Muslims today.

BURMA AT THE CROSSROADS: A CAUTIONARY TALE

In August 2007, the domestic frustrations in Burma had reached a boiling point. Fuel prices had skyrocketed as much as 500 percent following the junta's decision to lift fuel subsidies.[23] Food and other commodities quickly followed suit, leaving millions of people across the country unable to perform basic functions like buying food, commuting to work, or paying for their children's education. Even Buddhist monks were engaging in rare acts of defiance. They started by refusing to receive alms from the Burmese generals. In other words, they stopped giving these generals Buddha's blessings. Then on August 19, all hell broke loose. Thousands of Burmese monks poured onto the streets of Rangoon and soon tens of thousands of teachers, nuns, and local residents were joining them in a call to bring an end to decades of military rule. The uprising was dubbed the Saffron Revolution, after the color of the monks' golden yellow robes.

By September 24, crowds had filled the streets of more than twenty-five cities across Burma, with an estimated 100,000 marchers in Rangoon alone.[24] The regime got increasingly nervous and, on September 26, the Burmese military government responded with violence, seizing and beating thousands of protesters. Sympathetic campaigners in thirty cities around the world showed solidarity by organizing local demonstrations against the crackdown on antigovernment protests. Little could be done to stop the bloodshed; but this was just stage one in the battle over Burma's future.

Like most pro-democracy movements today, Burmese activists are harnessing the Web to reduce isolation, build far-flung networks, and coordinate for collective action. So far, iron-fisted government control over telecommunications has not prevented individuals from posting numerous grainy, and sometimes gory, videos on YouTube. Human rights organizations like Witness have traveled deep into the jungles of eastern Burma to document the military's persistent attacks on ethnic minorities. Meanwhile, grassroots media outlets such as Indymedia.org have become a rallying point for local Burmese bloggers who upload photos and provide a running commentary on news and events on the ground.

Unfortunately, all of the protests and grainy photos in the world have so far done little to improve the plight of the Burmese people. Indeed, in the midst of prosperity and dynamic transformation in Asia, Burma seems to reel from one wave of suffering after another. The list of Burma's woes includes an ongoing civil war, an endless series of political conflicts that have killed several hundred thousand people on all sides, chronic rice shortages in the land once dubbed "the rice bowl of Asia," and an epidemic of HIV/AIDS and other, curable, tropical diseases such as malaria. It is a legacy of abysmal mismanagement inflicted by decades of dictatorship and military rule. "The army approaches politics as if it were a war, seeking unity at gunpoint," says Maung Zarni, a prominent figure in the pro-democracy movement. "It suffers from a sense of being under siege by the West. It is the public that bears the enormous cost of the country's conflict."[25] A succession of military governments has prioritized security and utterly failed to deliver any gains in public health, education, ethnic integration, or the economy. Burma ranks 138th out of 182 countries on the UN's human development index, despite the country's sizable stores of natural resources, including oil, natural gas, timber, hydropower, gemstones, cash crops, and a periodic table's worth of minerals.[26]

The Evolution of Conflict in Burma

Against this backdrop, the fight for freedom in Burma has been long in the making, stretching back, to the country's fight to gain independence from Britain in the 1940s. After decades of false starts and single-party rule, the summer of 1988 was a seemingly auspicious moment for thousands of pro-democracy activists who, like the monks, had taken to the streets to demand free and democratic elections and a more open society in general. The National League for Democracy (NLD), an opposition party led by Aung San Suu Kyi, was emerging as a popular movement throughout Burma. Suu Kyi is the daughter of General Aung San, a Burmese revolutionary, nationalist, and architect of Burma's independence from Britain. Her family history made her exceptionally popular and therefore particularly threatening in the eyes of Burma's military generals.

When the elections were held in 1990, the military government expected to win overwhelmingly because it controlled all media outlets and banned advertisements in favor of the opposition. To its surprise, the NLD won 80 percent of the seats in the proposed parliament. The military immediately denounced the election, refused to seat the parliament, and initiated a massive campaign to arrest all of the elected NLD representatives. Like the 2007 uprising, the scenes that ensued were ugly. A collection of Burma's highest officials dispatched troops who slaughtered an estimated three thousand pro-democracy supporters (mainly students). Some students were caught and executed summarily, others were thrown into prison for lengthy terms, and others escaped into the jungles and joined armed ethnic resistances against the government. Aung San Suu Kyi was sentenced to house arrest until 1995, although she remains there to this day as the military junta invents new reasons to justify her confinement.

Those who escaped into the jungle continued their struggles for several years, but eventually decided it would be more effective to leave Burma and focus on gaining overseas support. Burmese expatriates managed to get away to Japan, Europe, the United States, and Canada, where they now maintain an international network that stretches from Los Angeles to Tokyo to Paris. Once in Western universities, these students took advantage of the Internet to disseminate information and communicate among themselves—without fear of repression.[27] In a span of a few years, Burmese activists were able to turn an obscure, backwater conflict into an international issue; force multinational corporations to divest from Burma; and make the so-called State Law and Order Restoration Council (SLORC) one of the world's most vilified regimes.

Their tactics, which focus primarily on starving SLORC of foreign currency, were inspired by and modeled after the antiapartheid movement that freed South Africa from colonial rule. But while the South African boycotts took decades to take effect, the Internet was steadily increasing the effectiveness of activists and accelerating the speed of social change. In the United States alone, more than twenty states and local governments enacted laws to disqualify companies doing business in Burma from bidding on public contracts. Major companies such as PepsiCo, Eddie Bauer, Apple, ConocoPhillips, Motorola, Texaco, Heineken, and Carlsberg quickly pulled their operations to avoid the embarrassing and costly boycotts.

SLORC did not take all of this lying down. The army, doubled in size and might since 1988, continues to maintain its stranglehold on information. Prison terms of seven to fifteen years are now routinely doled out for individuals caught with unauthorized possession of a computer with networking capability. But even stiff penalties do little to deny access to information on the Internet to those who are willing to work hard enough to get it. Sophisticated encryption programs and the use of foreign service providers are two common means of evading the watchful eye of authorities.

Conscious of its pariah status, SLORC changed its name to the State Peace and Development Council (SPDC) and launched a worldwide PR offensive. To attract legitimate sources of foreign currency, the regime promoted Burma as a haven for tourists and aggressively pursued foreign investment from companies that are not bound by national sanctions or local purchasing laws. Most of these efforts were in vain. The business climate is still widely perceived as opaque, corrupt, and highly inefficient. The foreign investors coveted by Burma's rulers stayed away from nearly every sector except for natural gas and power generation. The country still suffers from pervasive government controls, inefficient economic policies, and stifling rural poverty. Rising inflation hurts even those with jobs in the urban centers.

Which Way Forward?

As is often the case, the international community has been largely impotent. On October 11, 2007, the UN Security Council issued a statement condemning the actions of the Burmese regime, and the United States and the European Union announced tighter sanctions. But with soldiers on the streets of every city and the monasteries largely emptied, mass demonstrations ceased. The Saffron Revolution was quelled, at least temporarily. Not much has changed since then. When U.S. Secretary of State Hillary Clinton visited Asia on her inaugural foreign trip in early 2009, she weighed in on the Burma question, acknowledging, "Clearly the path we have taken in imposing sanctions hasn't influenced the Burmese junta," and that "reaching out and trying to engage them hasn't worked, either."[28] Arguably, sanctions could work if countries like China, India, Thailand, Malaysia, and South Korea weren't so busy ignoring them in order to feed their own insatiable appetites for resources.[29]

While Burma's future hangs in the balance, the situation exposes the raw international political economy of freedom. When powerful and corrupt tyrants and their benefactors profit at the expense of the masses, it is never easy to dislodge them or to correct deep injustices. As the monks and the students before them know all too well, revolution almost always means violence and, in turn, great tragedy and human sacrifice. Nevertheless, the information-age tactics of increasingly innovative, agile, and powerful networks of Burmese activists and their supporters have made life more difficult for Burma's dictators and inspired hope among the pro-democracy movement. We certainly know much more about their plight than we would without the social networks that have become a key channel for communication and networking. Moreover, there is hope that Burma's generals may choose another path, freeing their people and joining the international community as respected members. In April 2010, three years after the monks took to the streets, Burma's generals declared that they would hold elections—the first since 1990—and chart a new path to democracy. Skeptics will say they have seen this movie before—and they have. But with genuine and vigorous support from democratic leaders and the business community, a growing international network of citizen activists can play an important role in bringing this small nation closer to a democratic future.

AWAKENING FREEDOM'S ALLIES

To the north of Burma lies China, home to a fifth of the world's population and a similar percentage of the world's Internet users.[30] China engages in trading relationships with all the world's major democracies. Most of the world's top companies already have major investments there and many more investments are planned. Every year, more than 200,000 Chinese students flock to liberal university campuses throughout North America, Europe, and Japan to further their education and career prospects.[31] And an impressive array of Chinese companies is shaking up the international business landscape in industries ranging from finance to automobiles to consumer electronics.

Despite all this, China remains somewhat of an enigma. It's certainly no democracy, and the indelible images of the bloody clashes in Tiananmen Square on June 4, 1989, still provide a stark reminder of the steep price pro-democracy advocates have paid for daring to fight for freedom. But on the other hand, many in the West assumed that as China grew richer it would also become more liberal, following essentially the same path of development as today's advanced democracies. Well, that didn't quite happen. On the contrary, China's phenomenal growth has only emboldened its Communist leaders. Over time they have grown increasingly steadfast in their determination that China will never embrace Western freedoms or the West's tradition of competitive elections. And while it was once common for foreign leaders to chastise China for its human rights record, such criticisms are barely audible today. High rates of economic growth and rapid integration into the global trading system have effectively pushed issues of democratic governance to the back burner.

The Dictator's Dilemma

It's easy to understand why China and other authoritarian regimes tend to see the Internet and the knowledge economy as a bit of a mixed blessing. On one hand, they have little choice but to count on information technology to drive economic growth in the twenty-first century. On the other hand, these regimes must contend with the Internet's democratizing effects as increased scrutiny by citizens, both at home and abroad, undermines their ability to act in isolation and without criticism from the international community. Increasingly pervasive mobile phones are now the freedom fighter's most effective weapon—in terms of both organizing on the ground and documenting instances of abuse. When cell phones are combined with platforms like Twitter and YouTube, democracy activists now have a permanent channel to get their message out to the rest of the world. And the rest of the world now has a way to find out, inform others, and organize in support of oppressed peoples around the globe.

Of course, the Internet can easily amplify both sides of the freedom equation. Just as the Internet provides a platform for democratic chatter, it is also a meeting place for freedom's enemies. In China, the Internet provides a forum for a young generation of tech-savvy nationalists who view the country's sovereignty on internal and external affairs as trumping the promise of a liberal democracy. Such nationalism is reportedly on the upswing thanks to social media tools that help unite like-minded bloggers who see democracy as but one more attempt by the outside world to influence China's internal progress. One young nationalist interviewed by *The New Yorker* puts it this way: "Chinese people have begun to think, one part is the good life, another part is democracy. If democracy can really give you the good life, that's good. But, without democracy, if we can still have the good life why should we choose democracy."[32]

Watchdogs like Freedom House have suggested that the government deliberately fosters these reactionary sentiments by enlisting paid commentators and provocateurs like the "Fifty Cent Party" to overwhelm or disrupt undesirable conversations. Incidentally, the term "Fifty Cent" is not a reference to the rap star; it comes from the fact that Communist Party–backed bloggers are paid fifty mao, or roughly seven cents, for each "positive" post they make.

Like Iran and Burma, China has modernized and adapted its authoritarianism for the twenty-first century. Rather than simply suppress news and information, it tries to influence online debate with tactics akin to those deployed by spammers and fraudulent marketers. By some estimates, there are as many as 300,000 Internet police who continuously monitor the Web for comments that run counter to the Party's singular definition of national interests.[33] When, in July 2009, the world's attention was turned to riots in Urumqi, the capital of China's Xinjiang region, the Chinese government began blocking social networking sites such as Facebook and Twitter, and many users reported receiving zero results when searching for information on the region on local search engines.[34]

The Ghosts of Tiananmen

The issue of China's Internet censorship came to a head, of sorts, when on January 12, 2010, Google declared that it would stop censoring search results, as required by Chinese law. Google has always been tentative about its operations in China, hoping, like others, that its services would do more good than harm. "But over the past three years," said Nicole Wong, Google's deputy general counsel, "We have endured intimidation of our employees, an increased trajectory of censorship, attempts to steal our intellectual property and attacks on our corporate infrastructure." In December 2009, for example, alleged Chinese government hackers tried to break into the Gmail accounts of suspected dissidents and democracy activists. And after investigation, Google claimed that the attack targeted at least twenty other large companies, including other Internet and technology companies as well as businesses in the financial, media, and chemical sectors.[35] Wong described the event as "the straw that broke the camel's back." In protest, Google moved its Chinese search engine to Hong Kong and promised to spare no effort to protect the identities of Chinese dissidents who use Gmail. Although clearly annoyed by Google's public rebuke, China's leaders appear ultimately unmoved on the central issue of Internet freedom.

Indeed, it's now been more than twenty years since the June 4 incident in Tiananmen, and political change has been, as Mao predicted, "like crossing a river, feeling for the pebbles one at a time." The question, over the long term, is whether the ghosts of Tiananmen will come back to haunt China in ways its leadership could not have predicted. Just as India aspires to be more than the world's back office, China aspires to be more than the world's workshop. Can China move to a knowledge-based economy and achieve the same creative alchemy that we've seen emerge from Silicon Valley without the equivalent freedom of thought and openness that characterizes life in the high-tech capital of California? Could it have been even better off today had its leaders embraced freedom and democracy in the wake of Tiananmen?

It's impossible to say with certainty. But Maria Cattaui, former secretary-general for the International Chamber of Commerce, argues that you can't discount the continued impact of economic growth on public expectations. The irony is that the rapid economic growth that many authoritarian countries desire triggers the very internal forces that will see despots and dictatorships crumble under the weight of their citizens' aspirations. "The experience of most countries is that the growth of economic capacities internally spurs . . . the rise of citizen demands and citizen responsibilities,"[36] she says. In China, rapid economic growth is giving rise to a significant middle class with purchasing power, and with time to articulate social concerns and demands. The expectations rise quickly as economic gains translate into demands for a better overall quality of life. This kind of citizen is becoming more vocal. Cattaui says economic growth engenders a paradox—namely, that the very government that makes improvements becomes the object of further criticism.

Let Freedom Ring

The question now is: what can the natural allies of democracy do—the diplomats and business leaders who ought to be freedom's champions? A good start would be a better understanding of the dangers these authoritarian forces pose to democracy, the rule of law, and the global economy as we know it today. It's easy enough to be complacent when modern authoritarian governments are integrated into the global economy and participate in many of the world's established financial and political institutions. On the surface, such tolerance for ideological differences seems largely benign, and you could even argue it's a refreshing alternative to the ideological tensions that drove the Soviet Union and the United States to the brink of mutual annihilation during the Cold War.

But such complacency could just as easily lead humanity down an even darker path. What would happen, for instance, should nuclear capabilities in Iran or Pakistan fall into the hands of radicalized forces, led by a large contingent of disaffected youth in those countries? Or, say, if Russian criminal syndicates, which have been growing in power and numbers, amassed the resources to consolidate control over the organized crime networks that already wreak major havoc in democratic countries around the world?[37]

Companies like Google are making tough calls as the fault lines over the future of Internet freedom are drawn. Will we build virtual borders around the Internet in the same way that physical borders now separate countries? In other words, will there be one Internet for the free world, and one for people in China, Iran, Saudi Arabia, Vietnam, and other places where they see only what their leaders allow them to see? Or will the Internet become a place where everyone is free to see everything, minus that limited set of things which clear, explicit global rules specify should not be available? Moreover, what would you do in Google's shoes? Would you stick to your principles (don't be evil), even when your share of the world's largest market for Internet services is at stake? Or would you compromise now and hope that China's leaders eventually loosen the reins on free speech and democracy? Tough questions. Some companies will argue that it is not their responsibility or indeed their prerogative to meddle in the domestic affairs of sovereign nations. To be sure, bad things can happen when powerful companies have too much influence over the rules or rulers that govern their conduct. But if done transparently and in partnership with other legitimate bodies, a principled stand on the open Internet and an open economy is consistent with both Google's mission *and* its business objectives. After all, one global open Internet is a much more fertile domain for Google's services than one balkanized into regional subnets, where only some people get access to all of the world's online information.

If other companies were to join in the fight for freedom, perhaps this reality could materialize more quickly. The Global Network Initiative, for example, was founded by Google, Yahoo!, Microsoft, and a number of human rights groups in China and the United States. The group is developing a code of conduct around privacy and Internet freedom. It hopes that other companies, and even governments, will sign on. But with only three American companies on board to date, the code has neither the critical mass nor the legitimacy needed to effect change in China or anywhere else. Nevertheless, Nicole Wong is hopeful. "I will always bet on the users," she says. "We have a system that was architected to ensure the flow of information—that's on our side." The issue now, she adds, "is how do we keep users ahead of the regimes that would build infrastructures to choke off those information flows?"

In the end, it is worth emphasizing that the strength and competitive advantage of democratic states and free-market economies lie in their rules-based, accountable, and open systems, and in the values and standards that support them. By extension, a global economy and an international system that are grounded in human rights and the rule of law are far more desirable than the opaque and capricious alternative being actively pursued by the regimes examined in this chapter. In the long run, it's in our best interests to safeguard and promote the very qualities that set us apart from the authoritarians. After all, it's the very qualities of freedom, openness, integrity, and collaboration that provide the essential raw materials for building a sustainable global economy in which everyone can participate.

VII
CONCLUSION

18. GROUND RULES FOR REINVENTION: MAKING WIKINOMICS HAPPEN IN YOUR ORGANIZATION

Macrowikinomics is the story of a world with two starkly contrasting realities. On one hand, many of the institutions that have served us well for decades—even centuries—seem frozen and unable to move forward. On the other hand, we see sparkling new possibilities as people with drive, passion, and expertise take advantage of new Web-based tools to get more involved in making the world more prosperous, just, and sustainable. It's a story of atrophy versus renewal, stagnation versus renaissance. And now every organization, and every budding leader within them, must grapple with a profound choice: participate in rebooting all the old models, approaches, and structures or sit on the sidelines and risk institutional paralysis or even collapse. We're not talking about tinkering at the margins, but about reinventing and overhauling crumbling institutions and outmoded ways of working. Indeed, virtually all examples cited in this book challenge the traditional M.O., where the way to win is to stick with traditional hierarchical structures, rely exclusively on the people inside your organization to build the business, ferociously guard your proprietary knowledge and intellectual assets, and think only of your shareholders' interests.

That's all well and good, you might say, but how can I make wikinomics a reality in my organization or my community? Where should I start? What should I do? How can I lead? And what will it take to succeed? After studying hundreds of organizations across a dozen sectors of society, we've identified six rules, or ways of working, that successful individuals and groups follow to enable wikinomics in their organizations and sectors. 1) Instead of creating something and guarding it aggressively, as most organizations do, turn your thing (be it a good or a service) into a platform where others can self-organize and create new value. In other words, don't just be a creator, be a curator too. 2) In order to collaborate, you're going to need to share some intellectual property and get your IP lawyers on board. So think about what parts of your business activities could benefit from being released, and what parts you'll keep inside. 3) To control your future in this volatile world, you also need to start with a paradoxically different attitude: You need to let go. Encourage people to organize themselves to help you solve problems and come up with new ideas. 4) Of course, even self-organization needs prodding, and the requisite leadership in any large-scale collaboration usually comes from a small group of enthusiasts in the vanguard. So strengthen that vanguard and spur them on by providing incentives, recognizing excellence, and promoting talented individuals to positions of leadership. 5) Broadening and deepening the culture of collaboration in your organization is essential to making a lasting change in the way you create value. Transcend the old-style hierarchy and instead create a dynamic meritocracy where ideas and information can flow freely through the organization. 6) Empower the Net Generation, as today's young people are the first to grow up with an innate understanding of the digital world and its possibilities. Collaboration comes naturally to them, and wise leaders can leverage this by empowering young people to help lead the process of reinvention.

1. DON'T JUST CREATE. CURATE A CONTEXT FOR PEOPLE TO SELF-ORGANIZE

Many organizations take the wrong approach to exploiting the Web. They might create a new technology or system that promises to improve the way people work. To do so, they'll assess what people say they need, they'll design a system, and then they'll implement it, doing their best to manage change. They'll use the same approach to create any other content—whether it's a Web site, a new product and service, or a new way of dealing with the customer. In each case, they'll think of themselves as *creators* of content.

This is the wrong approach. In order to succeed in a wiki world, you cannot just think of yourself as a content provider, or as someone creating an initiative, product, or service. Instead, become the curator, someone who creates a context or a platform that allows other people to self-organize and create things that are valuable, both for you and for them, and maybe even for the world. If you build a Web site, don't simply load it up with static content. Instead, create the framework and tools for others to create their own content and build communities. Say you're a newspaper. The old way was: We'll tell you what's important today. The new way is: We're a venue for conversation, a community builder. We create the context—political discussion for the willing and able, for example, or the juiciest gossip about celebrities. We provide stories too. But we've redefined our core business as a place where a community can have a conversation. Increasingly, all institutions must think this way: it's not just about what we create, but about the overall value we can orchestrate through collaboration with our broader ecosystem.

Because of the networked age, organizations can be much more than just organizations. They can be platforms for value creation. Creating a platform for innovation, as we've seen in this book, expands the power of a business or an organization because it harnesses the creative power of a larger, more diverse, and ultimately more capable network of contributors than you could ever find in a single organization. And as the manifold examples in this book suggest, you can be a curator in many different contexts. Building an open and intelligent network for transportation innovation could help convene communities around shared problems like developing more advanced electric car batteries or building the next generation of in-car services and apps. Likewise, open platforms in government can broaden input into policy or get citizens involved in helping government agencies design and deliver better services. As federal CIO Vivek Kundra put it: "Government doesn't have to build and create everything—people can actually help create, through the participatory process, some of the solutions that the public sector needs."

Perhaps the most powerful way to curate is to let your customers or stakeholders help you invent a new product or service from scratch. When Jay Rogers, the Marine veteran who started Local Motors, set out to design the next generation of cars, he didn't follow the traditional path by hiring a full-time design team to translate his vision into blueprints. Instead, he built a platform for the design and creation of cars that enthusiasts really want—like high-speed dirt racers, or cars designed to navigate a city's narrow streets. While Rogers is tapping the passion of auto enthusiasts, virtually any product can unleash user creativity. In fact, a growing cohort of customers, especially young digital natives, treat products as a platform for their own innovations, whether companies or other organizations grant them permission or not. Whether hacking iPhones or remixing music tracks, they invent new ways to create extra value by collaborating and sharing information. If you do not stay current with customers, they invent around you, creating opportunities for competitors. Inevitably, it is preferable to share control with users than to cede the game to a more adept, prosumer-friendly competitor.

The corollary is that if people are going to hack your products anyway, then you may as well get ahead of the game. Make your products modular, reconfigurable, and editable. Set the context for co-innovation and collaboration. Provide venues. Build user-friendly toolkits. Supply the raw materials that collaborators need to add value to your product. Make it easy to remix and share. But don't forget: your collaborators won't appreciate being treated like they are just part of "the crowd." You've got to carve out meaningful roles and foster a sense of community. As Seth Godin put it: if you have a loosely affiliated group of people, they can be called a crowd. If you give them a leader, then they can be a tribe.[1]

And finally, don't expect a free ride. Your collaborators will increasingly expect to share in the ownership and fruits of their creations. If you make it profitable for potential collaborators to get involved, you will always be able to count on a dynamic and fertile ecosystem for growth and innovation. Amazon, InnoCentive, and eBay have already done it, YouTube and Facebook are trying to figure it out, and many more are on their heels. The bottom line is that the opportunity to bring customers, suppliers, and other third parties into the enterprise as co-creators of value presents one of the most exciting, long-term engines of change and innovation that the world has seen. But innovation processes will need to be fundamentally reconfigured if businesses and other organizations are to seize the opportunity. Just as you can twist and scramble a Rubik's cube, customers and other collaborators will reconfigure and build on your products and services for their own ends. And whether we're talking about government, health care, education, or beyond, static, immovable, noneditable items will be anathema—ripe for the dustbins of twentieth-century history.

2. RETHINK THE COMMONS

In this book we have argued that all organizations should abandon their fortress mentality and open up, not only by communicating pertinent information to stakeholders, but also by sharing some of their assets, within their business network or beyond. As we explained in chapter 2, the wikinomics principle of "sharing" is much more than playground etiquette; it's about growth, innovation, and profit. Of course companies need to protect critical intellectual property. But they can't collaborate effectively if all of their IP is hidden. So to make wikinomics happen in your organization you'll need to selectively put intellectual property and other assets in the commons, thus allowing larger numbers of contributors to interact freely with larger amounts of information in search of new projects and opportunities for collaboration. This sounds like a potential threat to business. But, in fact, it's an opportunity for organizations to discover new pools of creativity, both inside and outside the walls of their organizations.

Smart firms use sharing in many ways: to strategically shift the locus of competition in their industry, to get to market faster, to reduce R&D costs, to generate valuable follow-on inventions, to boost demand for complementary offerings, and to develop relationship capital with a community of collaborators. Think of Nike and Best Buy's decision to swap green innovations through the GreenXchange, or Novartis's decision to publish its diabetes research in order to enlarge the community of researchers participating in the hunt for solutions.

However, opportunities to use the commons to foster wealth creation and social development go way beyond business and the economy. In virtually every institution most new innovation and knowledge today is the product of networked individuals and entities looking for new solutions to specific problems and needs. And the big opportunity for all leaders is to leverage shared infrastructures as a way to accelerate their transition to wikinomics models. Recall, for example, how some of the world's great universities, like MIT, are laying down the first planks of a global network for higher learning—an open platform of world-class course material that includes everything from lecture notes and lesson plans to videos of inspiring talks by the world's great thinkers. Some see MIT's efforts as a mere gesture of intellectual philanthropy. We see something bigger: an opportunity to move toward a new model of collaborative pedagogy that enriches the learning experience with the best educational content and communities the Web has to offer.

Of course not even universities ought to share everything—it makes sense to protect your crown jewels. Instead we advocate a portfolio approach where organizations diversify intellectual property holdings across a range of open and closed offerings. Start by taking a look at all of the assets in your organization and ask whether some of them could be more valuable if they were available for reuse or modification. And as you examine your own IP portfolio, consider these guidelines. First, try hitching a ride on public goods. For example, is there an open-source community working on products that could boost demand for your consulting services or lower your cost of doing business? Alternatively, if you have an asset that others may find valuable, ask yourself whether opening it up could help foster new opportunities or revenue streams. Take a page from *The Guardian*, which opened up its vast collection of digital articles, images, videos, and data sets to encourage third-party innovation. And as we explained in chapter 11, if a third party uses *The Guardian*'s IP to invent new content-driven services that expand its audience and/or make money through its proprietary ad network, then both parties stand to profit. The bottom line: if you're seeking to engage a community and/or enlarge the pool of talent addressing a particular problem, sharing some IP is often the best way to do it.

Perhaps the biggest long-term opportunity, however, is to use the commons to drive sustainability. Consider the challenges of building a green energy economy and tackling climate change. Could the advantages of pooling competencies and reducing R&D costs around green technologies exceed the benefits of having exclusive rights in the innovations produced—particularly if an open-source approach sped the development and adoption of solutions? Say the United States or the EU masters a specific carbon sequestration technology—such as injecting carbon dioxide safely underground or beneath the ocean floor—and then gives the blueprints to the Chinese. Wouldn't that change everything? China would grow greener and the whole world would benefit—but it will take a new way of thinking about intellectual property to bring changes like these about.

Finally, organizations need to think creatively not only about what they share but also about how they share it. Remember, there is a spectrum of options between open and closed. "The interesting model," says IBM strategy alumnus Dan McGrath, "is the one that falls in the middle ground. It's a model that says we're collaboratively building something that will be privately owned by our consortia, or maybe your shares of ownership are apportioned in proportion to how much you brought to the table." One could envision a "digital-age co-op" with peer-rating systems that dynamically apportions shares to contributors based on the community's assessment of the value added by individual contributors. Annual profits from sales and services could then be distributed across the community of contributors. Whatever the precise arrangement, it's clear that the future of collaborative innovation lies in hybrid models where participants both share and appropriate at the same time.

Sharing does come with challenges. It means less control and requires practitioners to learn and abide by the rules of scientific and creative communities. It means investing in infrastructures for collaboration, while carefully considering when and how to distribute rewards and profits within the community. To make it work, you'll need to reveal your IP in an appropriate network, socializing it with participants and letting it spawn new knowledge and invention. You'll need to stay plugged into the community so that you can leverage new contributions as they come in. You'll also need to dedicate some resources to filtering and aggregating contributions. It can be a lot of work, but these types of collaborations can produce more robust, user-defined, fault-tolerant products in less time and for less expense than the conventional closed approach.

3. LET GO

Leaders in business and society who are attempting to transform their organizations have many understandable concerns about moving forward. One of the biggest is a fear of losing control. *I can't open up, it's too risky. Our lawyers would go berserk. There are too many obstacles. I can't empower others to make decisions because I'll get all the blame if they get it wrong.*

To be sure, there are real dangers to shifting to the new open networked model. There can be unintended consequences. When Goldcorp's Rob McEwen decided to publish his biggest secret—his geological data—for the world to study, the results could have been disastrous if he had had no gold on his property. It turns out his data revealed enormous assets, and for $575,000 in prize money McEwen transformed his $100 million company into a massive $9 billion enterprise. Of course, McEwen was the CEO; many typical change agents are not in positions of absolute power. Sometimes the roadblocks they face appear insurmountable. It might be a hierarchical, top-down organization, filled with middle managers who fight to protect their turf within the existing power structure. Or, as in the case of climate change and green energy, there could be an entrenched incumbent lobby with deep pockets to fight change through regulatory barriers, legal roadblocks, and other tactics that undermine the growth of alternative solutions. Organizations may fear that by opening up and sharing IP, they will make themselves vulnerable, reveal future product plans, or end up getting clobbered by the competition. The impasse could be a psychological one—a deep-seated fear of losing control, either as an institution or as a single person.

The paradox of today's age is this: to be strong, to have control, to ensure your security as an organization or society, you have to let go. This letting go can take many forms. It could mean giving your employees more freedom and more flexibility to innovate and co-create with their peers. It could mean tapping your suppliers and partners for ideas and collaborating more closely on product design and manufacturing. It could mean sharing at least some of your assets with the public to expand your reach or help attract a larger network of contributors. It could mean letting a loyal and engaged community of enthusiasts help grow your brand using their own DIY marketing campaigns. Or, in the case of government, it could mean giving up a monopoly on policy creation and handing citizens a meaningful role in the process. In each case, letting go can entail some risk. There's always a chance that an ill-conceived strategy could blow up in your face. But done right, letting go will make you stronger, more robust, and better prepared to navigate an increasingly complex environment. Says Mark Dajani, CIO of

Kraft: "There are so many uncontrollable forces in our company: people blogging, sharing intellectual property, experimenting, using tools from outside like social networks, self-organizing to do all kinds of things. How do we achieve control, security, and keep everyone marching in the right direction? I've concluded that the best approach is to just let go—to open up, and empower people."

Of course, no one said letting go is straightforward. Indeed, as IBM discovered when it joined the Linux community, control is not something that large organizations give up easily. Dan Frye, vice president of open system development at IBM, said that letting go was one of the hardest lessons to learn. "There is nothing that we can do to control individuals or communities, and if you try, you make things worse," Frye told the Linux Foundation's Collaboration Summit. "What you need is influence. It goes back to the most important lesson, which is to give back to the community and develop expertise. You'll find that if your developers are working with a community, that over time they'll develop influence and that influence will allow you to get things done."[2]

To influence a process rather than control it represents a big change in mind-set for a company that is accustomed to planning and carefully executing every aspect of its work. Open-source software communities run on instantaneous, transparent back-and-forth communications and rapid product iterations. Conversations use instant messaging, e-mail, and real-time collaborative software—whatever is fast. By comparison, internal company communications, attentive to internal sensitivities, are frequently slow and measured. Frye says, "When we were responding slowly with canned answers we weren't fast enough or transparent enough. It was not a level of technical exchange that was attractive to Linux developers." Frye told his team: "I'm unplugging you from the network. You can only communicate about Linux through the Linux community." And from then on the team used the same bulletin boards and chat rooms as Linux developers. "It's easy to form a community around yourself," he continued. "It's much harder and more valuable to participate in a community that you do not control—it took us time to learn that."

It took IBM about five years to learn that engaging in collaborative communities means ceding some control, sharing responsibility, embracing transparency, managing conflict, and accepting that successful projects will take on a life of their own. According to Frye, the company is still learning today. But no doubt the journey can be awkward for organizations accustomed to command-and-control systems. It means learning new skill sets that emphasize building trust, honoring commitments, changing dynamically, and sharing decision making with peers.

4. FIND AND STRENGTHEN THE VANGUARD

To harness the power of people who organize themselves, you first need to build a platform or a context where they can self-organize. But that may not be enough. Collaborative communities never get off the ground without a core group of leaders who establish the vision and community values, help manage group interactions, champion the cause, and attract more people to the ecosystem. This small group of key participants does a disproportionate amount of the work, often providing the social capital and technical infrastructure that other participants build on. This is the Vanguard, and if you want to tap into the power of self-organization you need to strengthen it.

As the community grows, the Vanguard plays a crucial role. If you look closely at the most developed communities, you'll find a delicate balance between self-organization and hierarchical direction. All successful open-source communities, for example, deploy structured and hierarchically directed processes for managing the tedious, tiresome work of joining together all of the fragmented pieces and contributions. It allows these communities to harness an incredibly diverse talent pool while still achieving the tight integration required for something as sophisticated as an operating system.

The direction comes from the Vanguard. Although Linux relies on contributions from thousands of programmers, a core group led by Linus Torvalds makes strict judgments about which contributions of code make it into the kernel of the operating system. "I certainly support anybody's right to modify and publish their own version of Linux," says Torvalds, "but at the same time, almost all my efforts are spent on the actual joining back of the results, and that's what most of what you'd call 'core developers' end up doing: guidance and quality control at various levels." The importance of community coordination and integration cannot be overstated. If value is not integrated properly, users may become disenchanted and stop contributing.

This new collaborative thinking is starting to show up in the most unlikely places, including stodgy international organizations like the World Bank. Dennis Whittle was a longtime bureaucrat at the World Bank at a time when it was loudly criticized for its failure to doing anything useful for the poor. Then in 1997, he was handed a new job, to come up with new products and innovations. Whittle had no idea where to start, so he proposed the creation of an idea marketplace where anyone in the bank could put forward their ideas for new products and innovations aimed at any of the bank's stated goals. But as soon as the ideas were submitted the initiative ran out of steam—senior bureaucrats balked at the time and effort needed to vet and judge the submissions. Then Whittle tried again, this time with a carrot of several million dollars in prizes, and with a jury of both internal and external experts to pick the winners. Eventually over three hundred submissions were chosen as finalists and a dozen chosen for funding, starting what would become known as the World Bank Development Marketplace. Then Whittle was approached by a South African woman who had made it to the finals but didn't win. "Just because the World Bank didn't fund our idea doesn't mean that other funders in the world might not," she told Whittle, asking when a secondary market was going to start.[3] It was a good idea, so Whittle quit his job at the World Bank and started working full-time on an online marketplace meant to tap the (then) over $200 billion U.S. philanthropy market for funding innovators and entrepreneurs. The DevelopmentSpace Web site was launched in February 2002 and soon renamed GlobalGiving, a marketplace that has served to channel over $19 million in funding toward over 1,340 projects.[4] Perhaps more impressive, the site has tapped over 49,000 unique donors to make those projects come to life.

The people in the Vanguard, like Whittle, set the tone for the community. They can have a profound influence on the type of community that evolves, especially since newcomers tend to model their behavior on what they have already observed. This early impact is important because the communities that evolve—inside or outside organizations—can take on a life of their own. Communities that are creating things develop their own rules (written and unwritten) that govern issues such as communications, appropriation, and the form and manner of contribution. The Wikipedia community, for example, has so far resisted the idea of advertising, and it has become such a potent force that Jimmy Wales has backed away from adopting an advertising model, despite the potential for a windfall. In the Linux community, Torvalds is careful to respond constructively to criticisms from other developers. This is typical of many open-source communities. Issues are debated publicly on e-mail lists and Web sites, which helps build consensus for final decisions on, for example, which code and features to include in official releases of programs. Debates remain on the public archive. Given this power of the community, it's crucial for the Vanguard to steer the community culture from the start.

How do you strengthen the Vanguard to grow a community within your organization or in the broader world? Build them a platform where they can innovate. Share your intellectual property so that they have something to work with. Let go, so they have the freedom to flourish. And above all, provide incentives that will help motivate the Vanguard to be successful. Let them expand slowly, step by step. Treat collaboration as real work, not as a distraction. And understand that leadership in the group can be its own reward.

5. CREATE A CULTURE OF COLLABORATION

The hardest challenge for anyone who wants to transform their institution for the networked age is to deepen and broaden the culture of collaboration. To make any of this work, you, as an organization and as an individual, must assume a collaborative disposition. This means being genuinely open to new ideas, irrespective of whom or where they originate from, instead of jumping at the opportunity to undermine them. It means giving up the instinct to protect turf and exert control—creating a dynamic meritocracy where ideas and information can flow freely through the organization. You need to abandon the old-style hierarchy and encourage people within the organization to talk directly to one another—even if they're in different departments. CEOs and other organizational leaders must, above all, demonstrate and encourage these values in the way they lead their organizations.

How do you start? You might reach outside the enterprise, as Procter & Gamble did when it looked outside the company to InnoCentive for new ideas, instead of relying solely on its internal R&D. But the most obvious place to start is at home—in your own workplace, and in yourself. You might be tempted to simply buy a suite of collaborative tools, and there are plenty of them. But our experience shows that importing tools like wikis will not turn your organization into a hive of collaboration overnight. In fact, if your organization has a very rigid and hierarchical corporate culture, you probably won't use the tools at all. Instead, start small. If you spot an enthusiast, a potential member of the Vanguard, give him or her the backing to launch a pilot project. That will demonstrate the value of collaboration, and of the collaborative tools. You might also use reverse mentoring, and get young people to teach the older ones how to use the new tools to collaborate more effectively. The right incentives can be important too. Incentives for group performance rather than individual performance, for example, could encourage people to work together more productively. And if you're a leader, set an example through your own behavior, in the way you collaborate and share information.

Indeed, when it comes to transforming an organization around the principles of wikinomics, there is no substitute for executive leadership. No matter how inspired, well meaning, or determined others may be, no one else can institutionalize a wikinomics culture across an organization like the CEO—the person who ultimately defines and embodies an organization's culture. That said, Peter Senge, who coined the concept of organizational learning, argued that the person at the top, regardless of IQ, can't learn for the organization as a whole. The Lee Iacocca–type leader who creates a vision and sells it down into the organization is being replaced by the model of the leader who draws on the collective brainpower of employees and other stakeholders and curates a context whereby change can occur. "When you bring people together and you put them together in an effective network that allows them to collaborate, you actually get new, bigger and better things than you ever got before, and that's a catalytic effect that's pretty powerful" says Tim Brown, the CEO of IDEO, one of the coolest and most inventive design companies on the planet. It has to start with the right mind-set, though. "You've got to believe that by connecting people in a way that inspires them, that you will get better performance," he says. As a result of Brown's leadership about 90 percent of employees have embraced IDEO's collaborative platform, called The Tube, which is really astonishing when you consider that in most companies, collaborative tools are only used by one in ten employees.[5] If on the other hand you're a "hierarchical company where people are scared of putting things out there," says Brown, "this is likely to be a very long slow journey."

One of the best examples of this kind of gradual transformation may be the BBC. It's a professional communicator—with five radio brands, two TV brands, and a global Web presence. Yet a decade ago, the globally dispersed staff at the BBC had no way of sharing information among themselves. Then, in 1999, Euan Semple took over as head of knowledge management. He could have spent a fortune on conventional knowledge management tools, but instead he bought simple tools for $400 that enabled users to ask questions of one another. It was more important, Semple thought, to share knowledge among the staff than to store it. The resistance from some quarters, especially middle management, was predictable. How are managers and employees supposed to define the line between collaboration and wasting time? Was contributing to a wiki a useful way to work?

This is a question managers often ask. But at the BBC, the advantages of sharing knowledge soon became apparent. Executives found they could use the collaborative tools to garner support for a new initiative and gain influence. Journalists found they could share skills and knowledge, discover hidden talent, and leverage collective intelligence. Today, with over 25,000 staff around the globe, a single blog within the intranet attracts over 8,000 individual readers each month.[6] With that readership comes influence, which, in turn, leads to action. The readers who comment on others' blogs attract attention to themselves, and if they form a community around their thoughts, it becomes a further stimulus to action. Ideas are turned into action more readily as the socialization of a new venture and the questions that help formulate the action are rapidly and comprehensively brought to the surface, before spending and resources are mobilized. The BBC's hierarchy did evolve: it loosened up and became more flexible. But the BBC has not lost its hierarchy; on the contrary, it has been strengthened by allowing many issues that formerly would have required an "up-and-over" negotiation to be worked out quietly and quickly on a horizontal basis, leaving the hierarchy free to spend more time managing and developing the business. This is the power of a collaborative workplace.

To turn into a collaborative workplace, your organization's leadership will need to decide that collaboration is an essential part of your business or mission and direct the organization toward that common achievement. You will need to assign your most appropriate resources on a dedicated basis to enabling the essential collaboration endeavor. You must relentlessly communicate the need to make enhanced collaboration a success, sustain the momentum, and then describe the next steps. Unlike previous IT efforts, you'll find collaboration is not primarily about the tools or the cost to implement them on an enterprise basis. For successful organizations it's about tribes, sociology, and cultural change. That means tailoring the approach to the people and processes involved—collaboration for and to a purpose, not in and of itself.

Indeed, early adopters of the collaborative workplace generally start with one, or possibly two, of the tools. They build on one success after another. Then communities grow across the organization. Someone from a different office or group joins a community out of interest, and it spreads. No two paths of adoption are exactly the same: it's not about learning to live with the technology, rather it's about merging the culture of collaboration into the culture of the organization. Both will grow and change together.

6. EMPOWER THE NET GENERATION

After publishing *Wikinomics*, we visited executives of a U.S. auto company to brief them on the Wiki Workplace. "This sounds great," one of the managers said. "Let's get a pilot wiki going with our top one hundred executives." That would be a bad idea, we told him. The place to start is with the demographic who will understand and make it happen—young people. Fortunately, the leaders of the future, the people who will inherit the deep challenges we've described in this book, appear to be equipped for the job. We call them the Net Generation, the first generation in history to grow up digital. Since birth they've been immersed in an interactive digital environment that has profoundly affected their way of working, the way they behave, and even the way they think. Unlike their parents, they've grown up with a technology that assumes you talk back. They've been reared on social networking (while the TV is playing in the background), so collaborating is natural.

Globally, they represent over one quarter of the world's population, and will soon come to dominate the workforce, the marketplace, the university, and politics and society. They've already displayed their power in a spectacular fashion: in 2008 they played a major role in electing the president of the United States, and by 2015 they will represent one third of the U.S. voting electorate, becoming one of the largest and most influential voting blocs.[7] As they become fully engaged adults, the Net Generation is beginning to shake up the old order. Their distaste for hierarchy and preference for collaboration are forcing organizations to rethink how they recruit, compensate, develop, and supervise talent. As consumers, they're eager to help companies innovate products and services—they're used to having a digital voice and are not shy about using it. Companies can leverage this to seek out fresh insight to build better products, services, and even a stronger brand. They're simultaneously challenging the old-style teacher's lecture and calling for a new model that revolves around the student. They may even revolutionize the medium of education by demanding the use of new tools, new teaching styles, and a new model of student engagement. The Net Generation has grown up immersed in games and digital media, and multiplayer games like World of Warcraft have given young people skills useful for leading and organizing people, as well as sharing and managing digital information, such as within databases, wikis, and forums. As citizens, they're in the early days of transforming how government services are conceived and delivered and how we understand and decide what the basic imperatives of citizenship and democracy should be. They also may think more as world citizens than their parents did—this generation is the first to experience the world in a truly global sense. Their worldview has been shaped by global events, and the geographical and knowledge-based barriers that may have prevented past generations from learning about and participating in the wider world have all but dissolved. They're also the first to face truly global challenges like climate change, and this could cause the Net Generation to redefine what it means to be a citizen, and how they think about nationalism and their connection with other cultures. They may gain a whole new perspective as young people in emerging economies participate in the global economy, for the first time as equals.

This generation is, in other words, best adapted to play by the rules of wikinomics. Indeed, many of the social innovations described in this chapter were created and nurtured by young people. Think back to chapter 1, where we met Ory Okolloh and Patrick Meier of Ushahidi, the crisis-mapping application that revolutionized relief efforts in Haiti. In chapter 9, we met Kevin Schawinski, a PhD student who became the driving force behind Galaxy Zoo, one of the most exciting science projects on the Web. Matt Flannery and Jessica Jackley launched Kiva.org while still in their twenties. Steve Ressler built GovLoop, the leading social network for government employees, on a shoestring budget and has scooped up numerous awards for his efforts. What leaders in all organizations should do now is empower young leaders to help guide them through the twenty-first century. "They're plugged in and natural global citizens with high expectations and a new view of the possible," says Klaus Schwab. The danger if we don't do this, he adds, is that the gap between their hopes and reality will widen. "And if their dreams are not achievable," he says, "we could see a global generational crisis."

Unfortunately, most organizations and institutions do a pretty poor job of engaging young people. Rather than harnessing all this young energy, collaborative instincts, and enthusiasm to diffuse a new set of collaborative tools and new work practices across an organization, many firms restrict, ban, and even outlaw them. Many companies are banning Facebook and other social media. Smart companies, on the other hand, understand that these tools and platforms are becoming the new operating system for their business. And for a new generation these tools are natural, like the air.

Best Buy gets it. Former CEO Brad Anderson could see that the young people working in the consumer electronics giant's stores were the closest to the customer. Their youthful culture ought to be the culture of the company, he thought, but there was a problem: "Between them and me are layers and layers of managers trying to prevent them from changing the company. So my task is to get them a license to self-organize." He put it this way: "Rather than making decisions, my main job is to unleash the power of human capital at Best Buy." With this support, tens of thousands of young employees flocked to a social network called Blue Shirt Nation. This network turned out to be a beachhead for the Wiki Workplace within the company. It was like an electronic water cooler for eighty thousand people throughout the organization.

The culture of embracing the Net Generation goes deep at Best Buy. In 1994, the company also set up the Geek Squad, a service arm of youthful "agents" who would help customers tackle problems from all kinds of venues—in the store, on the phone twenty-four hours a day, in emergency on-site visits, and eventually over the Internet. The remarkable feature of the Geek Squad is that they're empowered to solve problems and come up with new ideas without waiting for orders from the chief. Geeks used wikis, video games, and all kinds of unorthodox collaboration technologies to brainstorm new ideas, manage projects, swap service tips, and socialize with their peers. Even though their official job descriptions didn't call for it, they were empowered to contribute to product innovation and marketing—the kinds of things that usually happen only at headquarters. And all this made Geek Squad a great place to work, and contributed to its stellar service record.

Since then, the Geek Squad has grown to 24,000 agents and its ethos has spread through the company.[8] Best Buy, for instance, is opening up its data and asking outsiders to improve the offering. This move is already saving time and money: a creative agency built an entire home theater experience, based on Best Buy's data, without spending months with the company's technical teams. Best Buy has also set up consumer forums in several languages, and in the English forum, 85 percent of the conversations are peer to peer. Customers, in other words, are helping other customers. It's a terrific early warning system. A customer shopping for an iPhone in Miami reported that a single buyer had scooped up the whole stock. Two weeks later, the region instituted a new policy to prevent bulk buying.

Moral of the story: engage young people. They are the natural champions of the transformation we've described in this book. Get going with them, build momentum, deepen transformation, break down old industrial models, and enable reinvention.

THE NEW AGE OF NETWORKED INTELLIGENCE CALLS FOR A NEW KIND OF LEADER

The six ground rules for reinvention provide a guide for how to make wikinomics happen in your organization, community, nation, and indeed the world. None of us should assume, however, that these institutional transformations will be easy or straightforward. Leaders will face a constant stream of dilemmas, hard choices, and tough trade-offs. Creating a platform for self-organization, for example, can generate copious innovation, but it can also unleash more competition and keep companies on their toes. "It's almost like you're taking down your borders and opening up for no tariffs, no tax competition," says Better Place's Shai Agassi. "You need to know that your core assets and your skill sets allow you to continue to innovate fast enough as a corporation." Likewise, creating a culture of collaboration can flatten the hierarchy and unleash creative potential, but taken to the extreme, such an organization might spin out of control or lack the focus to execute great ideas. Indeed, how do you let go and embrace youth culture when the tools and approaches young people use are deeply unfamiliar to you and everything you have learned about running an organization?

There are long-term challenges too as leaders are called upon not just to reinvent their organizations but to participate in a global reinvention. Should I cannibalize revenue from legacy products to kick-start a transition to the future? How can my company provide leadership when others in our industry are lagging behind in adopting enhanced social or environmental performance standards? Will your company suffer if it unilaterally takes action to reduce pollution or cut carbon emissions, handing competitors a short-term price advantage? In a world where your job may survive on the results of your last quarterly report, how do you take the steps to ensure long-term success and sustainability for your company?

There are many other dilemmas and tough choices to be made. As in all times of great change there are many naysayers and the path forward may be littered with obstacles. As we discuss in the next chapter, these challenges call for a new kind of leader and a new brand of leadership for the age of networked intelligence.

19. LEADERSHIP FOR A CHANGING WORLD

On cool winter evenings near Oxford University (or anytime on YouTube) thousands and thousands of starlings come together over a marsh after foraging all day over a twenty-mile radius. Before they settle down, the starlings put on one of the most spectacular shows in the natural world. It's called a murmuration, in reference to the beating of thousands of pairs of wings as the birds swirl together. But the visual spectacle is even more astounding. The starlings, flying at warp speed, create incredible shapes that change minute-by-minute, second-by-second. It can look like a menacing dark heaving cloud, rising and coiling like a tornado, darkening the sky as it surges up like a thunder anvil and then crashes down like a torrential rain. But then, suddenly, the starlings can change shape, converting themselves into an airy drawing, delicate and fine, before they change again in unexpected and mysterious ways.[1]

The dazzling display is not just for show. The murmuration protects birds from predators and warms them for the night. There is also information exchange about good feeding areas. A number of operating principles, analogous to those of wikinomics, drive their behavior. There is interdependence and a kind of collaboration. Despite their high-speed aerial acrobatics, they never seem to collide. The murmuration is the epitome of self-organization. Every bird participates, and yet there is an implicit structure, as the stronger birds are more influential in directing the mass. The cloud is never the same, as new shapes are constantly created in their quest to sustain their survival. Leadership within the group seems to change constantly and dynamically as individual birds somehow manage to take up the mantle. It's not "collective intelligence" or "collective consciousness" because individual birds are not intelligent or conscious. But there does seem to be some kind of emergent, shared brain of sorts—a loosely conjoined network of relationships and impulses.

Unlike the starlings, human beings do have intelligence and consciousness. So is it possible that, as everyone connects through the global digital platform, we can begin to share not only information but also the capability to remember, process information, and even think? Is this just a fanciful analogy or will we come to consider networking as the neural routes that connect human beings in a way that creates something fundamentally new?

Arguably it is now possible, with modern technology, for organizations and even societies to share a perspective and even some kind of consciousness, a state of being aware, informed, concerned, and intentional. If an organization becomes conscious as a group, then perhaps it can learn. Unconscious organizations, like people, cannot learn. Conscious ones can. It may turn out that networked intelligence is the missing link in creating so-called learning organizations. Indeed, this new consciousness and interconnectivity—this capacity to learn dynamically as organizations—could be key to making the transition from old industrial models of value creation to the new networked models described in this book. If networking can speed up the metabolism of our collaboration and enable shared thinking and learning, a new kind of collective consciousness could be applied within and between organizations to innovate better, create prosperity, and advance society.

THE DARK SIDE OF MACROWIKINOMICS?

Not everyone agrees with the vision described above or with the broader transformation programs and principles we have laid out in this book. Rather than a time of optimism and new possibilities, they see a dangerous erosion of bedrock institutions and values like privacy or they question the capacity of collaborative communities to rival the quality and originality produced by the closed model of innovation that dominated the twentieth century. Some worry about our ability to create jobs or get paid in an economy crowded with an abundance of free goods and services. Others fear that online social networking is ruining true relationships and destroying communities. To be sure, the digital age will herald a mix of good and bad. But in tackling each of these "dark side" issues, it is also clear that solutions for each are at hand and, if anything, wikinomics provides the right framework for deriving fresh thinking.

The Hive Mind, Collective Consciousness, and Collectivism

There are many critics of the digital age, but one of the most articulate is Jaron Lanier, who, unlike many pundits, has a lot of street cred. Being a forerunner in virtual reality, he can't be dismissed as a Luddite. In fact, in 2010 *Time* magazine chose him as one of the one hundred most influential people in the world.[2] His much-awaited first book *You're Not a Gadget* is certainly the most erudite discussion of the downside of the digital age to date. Lanier argues that the Web has created a "hive" mentality that emphasizes the crowd over the individual, and is changing what it means to be a person. "Anonymous blog comments, vapid video pranks, and lightweight mash-ups may seem trivial and harmless, but as a whole this widespread practice of fragmentary, impersonal communications has demeaned interpersonal interaction." Having grown up digital, "a new generation has come of age with a reduced expectation of what a person can be, and of who each person might become."[3]

As a result, we behave like gadgets. We are all suffering from a "digital reification" where the basic characteristics of underlying technology algorithms are now determining how we relate to one another. In particular, Lanier seems concerned about a new form of online "collectivism" that is suffocating authentic voices in a muddled and anonymous tide of mass mediocrity. He laments the idea that the collective is all-wise and compares mass collaborations to totalitarian regimes. This collectivist mentality is led by a subculture of "Digital Maoists," who are the "folks from the open culture, Creative Commons world, the Linux community, and the Web 2.0 people." To him, "online culture is filled to the brim with rhetoric about what the true path to a better world ought to be and these days it's strongly biased towards an authoritarian way of thinking."

To be sure, the long-term effects of the digital revolution on humans are largely unknown. But often Lanier's critique, while brilliant, seems misplaced. He writes that Internet-enabled collaborations produce mediocre outcomes when compared with the secretive, closed-shop approach to innovation that dominated the previous century. "When you have everyone collaborate on everything," argues Lanier, "you generate a dull, average outcome in all things. You don't get innovation." Lanier dismisses Linux, the open-source operating system, as "ordinary," and claims that the most sophisticated, influential, and lucrative examples of technology stem from proprietary development.

Lanier has some suggestions on how to move forward. If you want to foster creativity and excellence, you have to introduce some boundaries, he says. "Teams need some privacy from one another to develop unique approaches to any kind of competition. Scientists need some time in private before publication to get their results in order." Making everything open all the time creates what Lanier calls "a global mush." Of course, scientists, and the rest of us for that matter, need private time. Some of the best insights often come while lying in bed at night or when secluded in contemplation. Martin Luther himself conceived his treatise for the reformation of the church while sitting on a privy. But collaboration today is not about a bunch of people sitting in a room constantly brainstorming with no time to reflect. Nor is it about designing by committee or having "everyone doing everything." It's about tapping a broader talent pool and bringing together the complementary skills and knowledge required to create a superior product or solve a problem. Take Apple's iPhone, which Lanier erroneously singles out as the epitome of "closed shop" development. The iPhone is, in fact, the result of a massive network-based collaboration involving thousands of companies. Although one of Apple's core competencies is their in-house design capability, there are various partners who help design the product. A Taiwanese company does the technical design, specs, manufacturing, and assembly, collaborating with hundreds of their own suppliers. And most of the software—supposedly Apple's main source of competitive advantage—is developed not by Apple, but by an army of third-party developers who have created upwards of 185,000 applications for the App Store.[4]

Other examples cited throughout this book confirm that collaborative innovation does not produce a "dull, average outcome" if you follow the design principles described in the previous chapter. Quite the opposite is often the case. When Kevin Schawinski (one of the astronomers behind Galaxy Zoo) enlisted 250,000 citizen scientists in a quest to categorize galaxies across the universe, he didn't sacrifice his capacity to think independently or compromise the integrity of his research.[5] Rather, he vastly accelerated progress in his field and, by proving that citizen science works, he also provided the broader scientific community with a novel solution for scientists who are currently drowning in an immense glut of data. When Jay Rogers invited a large community of automotive enthusiasts to help him design next-generation cars, the result should have been mediocrity or worse, according to Lanier's theory. On the contrary, Local Motors now has a diverse lineup of cars that address the specific needs of car buyers who are underserved by traditional car companies.

The most disappointing aspect of Lanier's argument is his troubling equation of the collaborative communities on the Web with Stalinist-style collectivism. Mass collaboration and Soviet collectivism are really polar opposites. Collaboration is based on self-organization, decentralized power and knowledge, and freedom of action. Collectivism is based on coercion and centralized control. Whereas communism stifled individualism, mass collaboration is based on individuals and organizations working to achieve shared outcomes through loose voluntary associations. One produced the Gulag: the other Linux, Wikipedia, myriad large-scale scientific collaborations, and the Twitter-inspired Iranian youth mobilization for freedom and a secular society, among other things.

Surely the debates that Lanier and others have stimulated about the nature of the Web and its impact on how we work, learn, live, and think ironically belie the idea that we are all becoming mindless gadgets marching to some unitary, authoritarian collective voice.

What About Jobs in a Networked Economy?

The term "jobless recovery" is an oxymoron, especially if you are one of the 30 million Americans who are trying to find work.[6] As we've explained, if you're young, the picture is particularly dismal, with upwards of 20 percent of youth unemployed in the United States and up to 45 percent in countries like Spain.[7] Economists euphemistically refer to this as "structural unemployment," meaning in lay terms that it won't go away. A 2010 study found that only one sixth of the young people in the American workforce say they are earning enough to live comfortably. Nearly 60 percent of them are trying to pay off student loans or other debts. And about one fourth of those aged twenty-five to twenty-nine said they are still or once again living with their parents—often after losing jobs they hoped would enable them to achieve independence.[8] This is a problem of epic proportions. It's not just that recovery depends on people with jobs having income to spend. Rather, there is a real danger that the largest, most highly educated cohort of young people in history could become "the lost generation." Considering that they are also the most capable of organizing for change given their digital alacrity, this could become an explosion that would make the youth radicalization of the 1960s look like kid stuff.

Many pundits now blame the networked economy as the source of the problem. Large companies shed jobs during the recession to sustain profits, but now they're not hiring back. Many are saying they plan on outsourcing rather than hiring or otherwise further reducing their head count. *Personnel Today* reports that IBM plans to reduce its global workforce of 399,000 permanent employees to 100,000 by 2017, the date by which the firm is due to complete its HR transformation program. "Tim Ringo, head of IBM Human Capital Management, the consultancy arm of the IT conglomerate, said the firm would re-hire the workers as contractors for specific projects as and when necessary, a concept dubbed 'crowdsourcing.'"[9] Although a spokesman for IBM said Ringo's comments were pure speculation, the trend for many companies is clear: if we can do more with fewer, we will. Corporate boundaries are becoming more porous and market forces can be brought to bear on every business function and every person—on a global basis. Because the Web drops transaction and collaboration costs, talent can be inside and outside of firms. Companies will increasingly use ideagoras to find uniquely qualified minds, create platforms upon which partners can co-create value, and tap the ingenuity of prosumers to design new products. Aren't wikinomics business models the death knell for jobs?

But when it comes to the jobless economy, it is more appropriate to fault the sputtering industrial economy with its failing banks and outmoded approaches to innovation and value creation. Sure, new forms of collaboration could make some jobs redundant or lead companies to reduce their workforce. But we think that there is a stronger case to be made that wikinomics principles help bolster fledgling enterprises by supercharging their innovative capabilities and that small enterprises in turn are the most reliable job creators. As we mentioned in chapter 1, a landmark study by the Kauffman Foundation shows that jobs come primarily from new firms and thereby from entrepreneurship. In 2007 (the last year that data was available) 8 million out of the 12 million new jobs were created by start-ups less than five years old.[10] Government leaders are barking up the wrong tree when they look to a country's largest and most successful corporations to be a source of jobs. Rather, the Web enables a new era of entrepreneurship and new business designs. Small companies can have many of the same capabilities as large companies without the main liabilities—bureaucracy, legacy cultures, antiquated systems, and old ways of working—all of which can impede innovation. As more small firms exploit the Web for new resources, they can gain unprecedented access to global markets previously enjoyed by only the largest corporations. And as small companies expand, they create new jobs at a faster rate than their larger competitors.

So entrepreneurship creates jobs. Collaboration and networked business models enable competitive entrepreneurship. But there is a missing ingredient. If we are to tackle structural unemployment, we need government policies that create the context for this to occur. The opportunities offered by wikinomics are universal, and entrepreneurs in the United States or Europe have global competition too. An entrepreneur in India can just as well take advantage of global sourcing, international expertise, and access to customers using various new online platforms. However, the jobs will disproportionally go to places where the culture and institutions are conducive to bringing new innovations to market. Governments need to invest in education to create a highly skilled workforce. We need to avoid protectionism and ensure that global markets are not closed to entrepreneurs. Governments can encourage R&D through tax and other incentives and make it easier for start-ups to get access to venture capital and marketing support. Leaders in every country could be conducting digital brainstorms and challenges to engage citizens in thinking about how to promote entrepreneurship. If we are to create and retain jobs in the emerging global marketplace, governments need to stop relying on old-style large companies and become champions of entrepreneurship, in the schools, in the media, and everywhere else. Every country in the Western world needs a "Jobs Through Entrepreneurship" campaign, launched by multistakeholder partnerships.

How Will People Get Paid in the World of Macrowikinomics?

Many worry that mass collaboration may result in a volunteer economy. They point to peer-to-peer communities like Wikipedia and Linux where everyone seems to do work for free. In most innovation contests, only the winner receives compensation. Consumers contribute ideas to companies that make products and get no remuneration in return. Writers for publications like *The Huffington Post* are not paid, nor are most bloggers. Even the world's biggest community, Facebook, with hundreds of millions of members, doesn't have a sound monetization plan. How is a person (or company) to make a living in this crazy world?

A survey of wikinomics business models reveals a more nuanced story. Linux is made up of "volunteers," but the vast majority of these contributors are employed full-time by companies like IBM, Intel, and Novell. And as we pointed out in chapter 4, the Linux community produces billions of dollars of revenue from products and services that incorporate the free operating system. When P&G holds an InnoCentive challenge, everyone can benefit financially. P&G staff help construct the problem. The winners are compensated well, with prizes sometimes reaching seven figures. InnoCentive itself is a profitable company taking a piece of the prize. And P&G as a company gets a problem solved. Yes, the winners of the Goldcorp Challenge shared over $500,000 for finding billions of dollars' worth of gold. But the prize money awarded to the winner is only one aspect of the overall commercial benefit. Winning the challenge helped establish Fractal Graphics, the prize taker, as one of the world's leading companies in analyzing complex data sets. Similarly, when Amazon opened up its e-commerce platform, it enabled thousands of third-party developers to create new businesses selling Amazon products through new channels and services that they invent. Amazon benefits too; over a third of all their revenue now comes from their cut of the third-party sales.[11]

Cultural content is trickier. What happens to the incentives for writers, artists, filmmakers, musicians, and journalists in a world where too many think knowledge-based products are free? Fair enough. The world will be a poorer place if nobody pays directly for high-quality content anymore. But as we have argued in these pages, people will pay for noncommoditized value if you get the business model right. Moreover, there is some validity to those who argue making certain content "free" can help maximize exposure and reap rewards through some complementary avenue. Music revenue, as mentioned in chapter 12, has shifted from sales of recordings to live performances, rights for ring tones, and other products, and many musicians do very well. Films are, in part, ads for their spin-off products. Authors of books (like us) make most of their revenue through speaking and consulting fees and related services, even when the book is a bestseller.

Journalism, as we've explained, is a tougher case. Many writers will not be employed by traditional publications and thousands have lost their jobs already. Their future is unclear and that's a big problem, for them and for society. Get ready for a lot of conflict too. Adam Pagnucco, who blogs under the name of Maryland Politics Watch, received a letter from *The Washington Post* inviting him to be an official blogger. He declined, saying: "If bloggers fill their functions for free, the *Post* will inevitably phase [paid columnists] out. In the labor movement, we have a term for workers who undercut other workers and threaten their jobs: *scabs*. As a labor guy for sixteen years, I have no intention of blogoscabbing."[12] Journalism can shift to the networked model we've described in this book, where journalists can make a good living and where society gets accurate news, good investigative reporting, and great writing. But the transition is already getting ugly.

Some truths are becoming self-evident. People will not pay for commodities, whether it's news or other content. But they will pay for compelling, differentiated value, as *The Economist*, rock concerts, Thomson Reuters, and many other examples show. iTunes competes with "free" every day and it does a very good job by providing quality, convenience, and value-added services like Genius mixes that generate automatic playlists. The same could be said of Netflix, the popular online DVD rental and video streaming service. All of the videos provided by Netflix are readily available on BitTorrent networks for those who put in the effort to find them. But loyal customers stick with Netflix because of the flat-fee business model and its innovative customer-driven recommendation and rating service. So, companies can compete with free. The key will be to develop new business models, offer distinctive value, and not get too hung up on trying to defend a legacy business that has been killed by the digital age.

Is Online Collaboration Killing Privacy?

This is a big topic, and rightly so. It's pretty clear that everyone gives away too much of their personal information on Facebook, Twitter, and other social networks. There are probably thousands of new graduates this year who won't get that dream job because the employer did a "reference check" online and found them doing something inappropriate.

But this is just the beginning. In this book we've argued for collaborative learning, health care, science, and media. We've made the case for citizens to create value on open government platforms and for customers to collaborate more closely with the companies they buy products from. We've also shown how companies that adopt social networks, microblogging, and other social media tools in the workplace perform better as a result. But in every case individuals are giving away information, often personal, to various institutions. As the Net becomes the basis for commerce, work, entertainment, health care, learning, and much human discourse, and people participate more deeply in the institutions of society, each of us leaves an ever-increasing trail of digital crumbs in our wake. Indeed, everything gets recorded digitally these days: the books, music, and stocks you buy online, your pharmacy purchases and groceries scanned at the checkout, your child's research for a school project, or the card reader in the parking lot. Just as every click online is filed away in your Web history, your journeys by car are increasingly tracked via satellite. To put this in perspective Google CEO Eric Schmidt notes that between the dawn of civilization and 2003 there were five exabytes of data collected (an exabyte equals 1 quintillion bytes). Today five exabytes of data gets collected every two days!

Online or off, our digital footprints are being gathered together, bit by bit, megabyte by megabyte, terabyte by terabyte, into personas and profiles and avatars—virtual representations of us, in a hundred thousand simultaneous locations. This digital shadow is used to provide us with extraordinary new services, new conveniences, new efficiencies, and benefits undreamt of by our parents and grandparents. But there are also great risks when very little about our lives is truly private, and that's understandably a big problem for many.

In the past we only worried about Big Brother governments assembling detailed dossiers about us. But now the threat also stems from Little Brother:

the myriad individual corporations that collect data from their customers. Intense competition is making marketing departments look for any edge they can get. Companies can't afford to squander marketing dollars on people who have no intention of ever buying their product. That means companies want to know more and more about what makes each of us tick—our motivations, behavior, attitudes, and buying habits. The good news is that companies can give us highly customized services based on this intimate knowledge. The bad news is that as these profiles are compiled they are rarely, if ever, deleted. There are weak safeguards to prevent unauthorized people from snooping. And sometimes datasets are sold off to third parties for dubious purposes.

At the same time, novel risks and threats are emerging from this digital cornucopia. Identity fraud and theft are growing threats in a networked world, along with new forms of discrimination and social engineering made possible by the surfeit of data. Personal information, be it biographical, biological, genealogical, historical, transactional, locational, relational, computational, vocational, or reputational, is the stuff that makes up our modern identity. It must be managed responsibly. When it is not, accountability is undermined and confidence in our evolving networked society is eroded.

The net result is the potential end of privacy as we know it—a trend now amplified by the meteoric rise of social networking. Ironically, we are increasingly willing accomplices in dissolving our own privacy rights. In 2005, who would have predicted that hundreds of millions of people would be voluntarily giving up detailed data about themselves, their activities, their likes/dislikes, etc., online every day. This situation has turned traditional privacy laws and regulations upside down. Privacy and data protection laws emphasize the responsibility of *organizations* to collect, use, retain, and disclose ("manage") personal information in a confidential manner. But collaborative networks, in contrast, encourage *individuals themselves* to directly and voluntarily publish granular data about themselves (tagged photos, preferences/settings/likes, friends' lists, groups joined, etc.), short-circuiting the obligations of organizations to seek informed consent and to manage this data responsibly according to defined criteria. The integration of personal profiles on networks such as Facebook with myriad other sites, communities, and applications on the Web further undermines privacy. Toss in the emerging "augmented reality" tools where you point your mobile device at the street and it gives you real-time information about the world around you—everything from recognizing the faces of people nearby to letting you know about all the people on Twitter in your vicinity—and we can be sure that a ton of personal information about most of us is deeply and irrevocably embedded into the fabric of the Internet and instantly available to the world.

Unfortunately some confuse this issue with transparency. But transparency is the opportunity and even the obligation of institutions to communicate pertinent information to their stakeholders. Individuals have no such obligation. In fact, to have a secure life and self-determination, individuals have an obligation to themselves to protect their personal information. And institutions should be transparent about what they do with our personal information. Transparency and privacy go hand in hand. It makes sense to be a privacy advocate and also advocate transparency, as we are.

Clearly there is some role for government in fixing this situation, but more will be required. Networking sites have become the subject of regulatory scrutiny in the United States and Canada, the European Union, and elsewhere. Regulators have focused on the lack of transparency that social networking sites provide about their information management practices, including the collection, use, and disclosure of personally identifiable information. However, everyone, not just governments, has a role to play in managing privacy. Given that personal data is effectively available everywhere on the Web today, there is a movement among businesses to seek common privacy rules that target acceptable or unacceptable *uses* of personal data. Increasingly companies understand that good privacy policies are good business practices as the precondition for trust and customer relationships.

One emerging principle is that personal information "belongs to the individual." As such, individuals should be empowered to see exactly which entities have data about them and they should be able to control how that data is used. To the extent possible, the individual should also be free to remove or "port" their personal data to another platform or service provider should they choose. Leading collaborative platforms may even be setting new privacy standards and raising the bar for privacy's future. Despite the ongoing controversy over specific settings changes, Facebook's highly granular controls over one's profile and activity info may provide a vital learning curve for consumers and a model for other industries to follow. But this is not enough. User-friendly and intuitive controls must be backed with an organization's commitment to allow users to preserve, as well as maintain, their own privacy, rather than forcing their data into the open. There is an emerging firestorm on the issue of privacy, and increasingly people everywhere understand the implications of leaving an irrevocable digital footprint. Many are becoming more vocal about privacy rights and are prepared to use the self-serve tools made available to them as customers, citizens, and patients. Companies that attempt to exploit their personal data inappropriately will be met with fierce resistance.

Ultimately, in order to protect privacy, all of us will need to change our own online behavior. Impossible assignment, you say? Once again look to young people of the Net Generation to show the way. Recent research suggests that youth are already more diligent than older adults in taking steps to protect themselves. In a 2010 study, the Pew Internet Project has found that people in their twenties exert more control over their digital reputations than older adults, more vigorously deleting unwanted posts and limiting information about themselves.[13] This supports our findings and those of others who have argued that young people who have grown up digital are confronted with the privacy issue at an earlier age and naturally come to grips with it earlier.[14] Helen Nissenbaum, an NYU professor of culture, media, and communication, came to the same conclusion in her book *Privacy in Context*, explaining that teenagers naturally learn to be protective of their privacy as they navigate the path to adulthood.[15]

Managing the Dark Side

The list of concerns goes on and the skeptics are ascendant. Media critic Nicholas Carr posed the question "Is Google making us stupid?" in an *Atlantic Monthly* cover story[16] and followed it up with a book called *The Shallows: What the Internet Is Doing to Our Brains.* His conclusion is that the Web is taking us from the depths of thought to the shallows of distraction, fostering ignorance and changing the very conception of ourselves.[17] Writer James Harkin says social networking is killing human relationships: we've all ended up in "Cyburbia: a peculiar no man's land, populated by people who don't really know each other, gossiping, having illicit encounters and endlessly twitching their curtains."[18]

One thing is for sure. We are entering an age of enormous leaps in technology, and in its pervasiveness in our lives—the way we spend our time and interact with others, learn, plan, and even think. Because of this we should take heed. The Age of Networked Intelligence demands that we step back and consciously design our lives. We need to decide explicitly what we stand for and whether we are the slave or the master of the new technologies. Today, smart companies and organizations are taking initiatives to help their employees cope with the new technology-rich world. They give their employees training in time management and help create a healthy work-life balance. They ensure that integrity is part of their corporate DNA. They design business models, structures, and business processes to ensure that work systems best serve the organization and maximize collaboration and the effectiveness of its people. But on the personal front, most of us muddle through this new networked and open world, stumbling from decision to decision or crisis to crisis without an overarching strategy. There is some truth to Nicholas Carr's assertion that with the myriad technologies interrupting us, it's tougher to focus deeply on a task or to read and analyze a long piece of text. Do we rely too much on Google to remember things for us? Sure, our kids are able to perform multiple tasks, including light homework, at once, but do we ensure that they have balance in their lives?

Certainly we all need to consider how we manage our time and make sure we have balance in our lives, cherishing face-to-face engagement with those for whom we truly care. And it's good for all of us to reflect on how to design our lives to ensure that the digital experience is enriching. All of us should be applying principles of design to our families and lives, making conscious choices about how our families will function and what we believe in. It makes sense to adopt a values statement—and constantly revise it as the world and conditions change. Don't complain about technological overload. Harness the power of new technologies and transparency for the good—design them rather than have them control you.

LEADERSHIP FOR TRANSFORMATION

The list of potential fallouts from the digital age is long, as evidenced by this growing storm of concern about where the *macro* wiki world is taking us. "Many of the transnational networks fostered by the Internet arguably worsen—rather than improve—the world as we know it," writes Evgeny Morozov in *Foreign Policy* magazine. He notes that at a recent gathering devoted to stamping out the illicit trade in endangered animals, the Internet was singled out as the main driver behind the increased global commerce in protected species. And "today's Internet is a world where homophobic activists in Serbia are turning to Facebook to organize against gay rights, and where social conservatives in Saudi Arabia are setting up online equivalents of the Committee for the Promotion of Virtue and the Prevention of Vice." He concludes: "Sadly enough, a networked world is not inherently a more just world."[19]

That said, few people, including the digital cynics and proponents alike, would argue that technology itself deterministically creates our future. The networked world simply enables new possibilities. But its potential for unleashing both good and evil can only be realized by people who act with free will and who have choices. Past media revolutions—the printing press, broadcast media—were hierarchical, immutable, and centralized. As such, they carried the values of their powerful owners. But as we have explained, the new Web is interactive and has no gatekeepers over the content services that are added to its backbone. As such the Internet has an awesome neutrality as a platform for self-determination. Ultimately, it will be what we want it to be. It will do what we command of it. It will achieve what we as humans can achieve.

In this book we have taken a journey through a vast swath of contemporary institutions that are both failing and being renewed on a networked model. But there are others undergoing the same process as the digital revolution extends into every nook and cranny of society. Each of these is a "digital chapter" and discussion group at MacroWikinomics.com. For example, there is growing agreement in the defense community that the military is organized on an industrial model and designed to fight yesterday's wars, involving advanced countries with massive military apparatuses. Our enemies are being effective with a more distributed, networked approach to combat. Primary education, K–12, needs a transformation similar to our discussion of the university. There are exciting developments in the arts today, such as how arts organizations are embracing collaboration through the invention of new art forms. Our faith and religion are moving into a period of change and smart leaders know this. Sunlight has exposed Catholic priests and bishops around the world and churches everywhere are coming to grips with concepts of transparency and integrity. The world's cities are also under stress, with megacities such as São Paulo and Johannesburg paralyzed by population influx, lack of infrastructure, traffic congestion, pollution, and crime. In the United States and elsewhere, many cities built up since the Second World War are dysfunctional and getting worse as the industrial economy collapses. Detroit has lost more than half its population from its heyday, with large swaths of the city now wasteland, populated by wild animals.[20] Yet everywhere there are bold new collaborative initiatives for reinvention.

Food and water crises are looming. According to the OECD, 2.8 billion people, or 41 percent of the world's population, already live in high water stress areas. This will soar to 3.9 billion by 2030.[21] Droughts and floods caused by climate change will threaten our food supplies. Much of the world is hungry and our methods of mass food production are making people and the environment sick. One billion people in the world are starving and another billion are obese. Half of the food produced in the United States is thrown away.[22] Yet everywhere multistakeholder networks are mobilizing. The Global Water Challenge and Ashoka Changemakers have created an online competition to bring together experts and entrepreneurs to innovate ways to solve the water access and sanitation crisis. Online communities like Hyperlocavore and Landshare have popped up in the United States and United Kingdom to redesign our food practices along the new model. In a similar vein, the U.S.-based design firm IDEO is partnering with Acumen Fund and the Gates Foundation to redesign water distribution and sanitation in India and Africa. Called the "human centered" design process, it engages the public in the process of designing solutions.[23] One innovation, the Aquaduct Concept Bicycle, transports and sanitizes water at the same time through a pedal-powered filtration system. It seems everywhere you look there is atrophy contrasted with renaissance.

In *Wikinomics* we argued: "There has probably never been a more exciting time to be in business, nor a more dangerous one. The wikinomics genie has escaped from the bottle, wreaking havoc on some and bestowing long-term success on others who embrace it."[24] Today we make the case that there has probably never been a more exciting time to be a human. Nor has there been such a systemic danger to most of the institutions of society. We also argued, "A new kind of business is emerging—one that opens its doors to the world; co-innovates with everyone, especially customers; shares resources that were previously closely guarded; harnesses the power of mass collaboration; and behaves not as a multi-national but as something new: a truly global firm." Today a new kind of *institution* is emerging in communities, cities, countries, and societies around the world.

This is a time of uncertainty, confusion, and calamity. Some people will receive the ideas of macrowikinomics with coolness, even hostility. Vested interests are of course opposing this change. Leaders of the old paradigm are having the greatest difficulty embracing the new, even when they understand that change is upon us. Stefan Stern, until recently a longtime columnist for the *Financial Times* in London, is one of the most respected business writers in the world. He was candid to us about the problem in his own industry: "Maybe the old media are not up to the challenge of exploring the troubling realities of this age. We are built on a model of authority. We produce it. You pay for it. We'll shield you from the really scary stuff because we're not sure you can take it." He adds: "And we're not sure it's commercially a good thing to push out challenging analyses. Innovation is at the edge. It's not in the mainstream."

Understandably, there is a crisis of leadership emerging. To succeed, we need a new kind of leader in industry—one who sees improving shareholder value as complementary to improving the state of the world and who understands that businesses must operate with a new set of principles. We need leaders in government who see citizens as active shareholders in the democracy rather than passive spectators. We need civil society leaders willing to open up the old bureaucratic structures of the nonprofit world to forge a new model of social innovation.

The defining challenge for all leaders in the decades ahead will be to ensure a sustainable future. We are being called upon to undertake a historic act of stewardship—to take responsibility and care for a world that will soon be owned by our children. Given the evidence before us today, it is unacceptable that we continue to defer the real costs of today's affluence and waste to future generations. Goods and services must be priced to account for their true costs to the environment and society, both now and in the future. We must also build new multistakeholder networks with the capacity to solve global problems—networks that can reach across disciplines, institutions, nations, and cultures to tap into the collective ingenuity of diverse communities. There is no room to delay, procrastinate, or obfuscate—we need wikinomics more urgently than ever. In 2050 there will be 9 billion people on this planet.[25] If we don't make fundamental changes we risk the possibility of catastrophic conflicts erupting over dwindling resources. The idea of being able to apply all the brains on the planet to a time-urgent situation is something that we are going to look back on and be really glad that we figured out, because otherwise it's going to be too late.

The leaders profiled in this book understand that this is a time of transformation, not for tinkering with old models and structures. We need to get beyond the tired debates and reach beyond the parameters of a failing framework. CEOs like Local Motors' Jay Rogers, VenCorps' Sean Wise, Better Place's Shai Agassi, Zipcar's Robin Chase, and Ushahidi's Ory Okolloh intrinsically get this and are now in positions of influence with next-generation enterprises. Other fields have leaders who are embracing wikinomics too: Vivek Kundra in government, Kevin Schawinski and Jean-Claude Bradley in science, Ben and James Heywood in health care, and Gord and Susan Fraser in green energy, to mention just a few. But just as significant are the tens of millions of individuals who have assumed positions of influence and leadership in their own communities. Whether contributing to new institutions for media, education, science, health care, or other fields, they are showing that an informed and engaged citizenry can make a big difference. As Brookings scholar Ann Florini once remarked, "There's Jefferson's old quote about 'refreshing the tree of liberty with the blood of patriots'; well, it doesn't have to be blood, but it does, at least, have to be attention."

This process of transformation is proving to be challenging, exhilarating, and sometimes agonizing. But given the stakes we have no choice but to forge this new future. Three hundred years ago Martin Luther called the printing press "God's highest act of grace." With today's communications breakthroughs we have a historic occasion to reboot business and the world using wikinomics principles as our guide. Because each of us can participate in this new renaissance, it is surely an amazing time to be alive. Hopefully we will have the collective wisdom to seize the time.

ACKNOWLEDGMENTS

This book is much more than a publishing project for both of us. In truth, it has become somewhat of a mission. Our research has revealed an abundance of opportunity for innovation and renewal in the many sectors and institutions covered in this book. But we have also come to appreciate that the stakes are high and a lot could go wrong if we fail to mobilize new solutions for stalled institutions. The principles of wikinomics provide guidelines for reinvention, but the real work is happening in the trenches where people with passion, drive, and expertise are getting involved in making the world more just, sustainable, and prosperous. So first and foremost we would like to acknowledge the countless educators, entrepreneurs, community enablers, doctors and health practitioners, artists and activists, public servants, scientists, social innovators, journalists, technologists, and thought leaders who are showing the world a way forward.

Like *Wikinomics*, this book was inspired by several multimillion-dollar research programs directed by nGenera and funded by companies and governments around the world. We are thankful to nGenera CEO, Tom Kelly; chairman and founder, Steve Papermaster; and the director of nGenera Insight think tank, Joan Bigham, for their support and encouragement.

CORE RESEARCH TEAM

We also pulled together a team of professionals to conduct research for the book itself. Celina Agaton tackled health care and led the development of macrowikinomics.com. Paul Artiuch contributed to several chapters and pushed our thinking on climate change, energy, and transportation. Jude Fiorillo painstakingly fact-checked and corrected every sentence in the book. Bill Gillies brought his usual clear thinking to the chapters on newspapers, media, the music industry, and health care. Dan Herman tackled one of the most expansive jobs—new models of global problem solving and social innovation. Moritz Kettler researched the issues of food and water. Sarah Scott helped achieve clarity in several sections and took on the challenge of making the chapter on making it happen, happen. Jody Stevens lent her usual rigor and clarity to administrative matters and to the production of the manuscript. Last but not least, Bob Tapscott brought fresh thinking and the wisdom of his many years in the financial services industry to help us crack that tough nut.

NGENERA COLLEAGUES

Many colleagues from nGenera and its circle of collaborators contributed to the insights in the book. These include Laura Carrillo, Lisa Chen, Ian Da Silva, Jeff DeChambeau, Mike Dover, Tammy Erickson, Denis Hancock, Naumi Haque, Daniela Kortan, Ming Kwan, Alan Majer, Sean Moffitt, Derek Pokora, Deepak Ramachandran, Bruce Rogow, Antoinette Schatz, Roberta Smith, Nick Vitalari, and Tim Warner.

CONTRIBUTORS

The book was also a mass collaboration. There were literally hundreds of others who had a hand in crafting the core ideas of the book, either through conversations with the authors or by commenting on the manuscript. If a quotation appears in the manuscript without a citation, the source is an interview.

These contributors include: Euan Adie, Nature Publishing Group; Shai Agassi, Better Place; Timo Ahopelto, Blyk; Brad Anderson, formerly at Best Buy; Michele Azar, Best Buy; Jitendra Bajpai, World Bank; Matt Barton, St. Cloud State University; Irving Wladawsky-Berger, Spencer Trask Collaborative Innovations; Billy Bicket, NetSquared; Peter Binfield, *PLoS One*; Alf Bingham, InnoCentive; Jean-Claude Bradley, Drexel University; Diane Brady, *BusinessWeek*; Chuck Bralver, Open Models Company; Cody Brown, Kommons; John Seely Brown, author; Tim Brown, IDEO; Willms Buhse, CoreMedia; Tony Burgess, U.S. Army; Kay Carson, MassRIDES; Fred Carter, Office of the Information and Privacy Commissioner of Ontario; Joel Cawley, IBM; Robin Chase, Zipcar; Bob Chen, Columbia University; Calvin Chin, Qifang; Aneesh Chopra, federal chief technology officer, U.S. government; Jacob Colker, The Extraordinaries; Peter Corbett, iStrategy Labs; Marilyn Cornelius, Stanford University; Jim Cortada, IBM; Robert Crandall, formerly at American Airlines; Duane Dahl, EarthLab; Mark Dajani, Kraft; Ron Dembo, Zerofootprint; Sean Dennehy, CIA; John De Souza, MedHelp; Peter Diamandis, X-Prize; Paul Dickinson, Carbon Disclosure Project; Teddy Diggs, EDUCAUSE; Frank DiGiammarino, executive office of the president; Cory Doctorow, BoingBoing; Chris Dorobek, FederalNewsRadio; Jodi Echakowitz, Echo; Ken Eklund, World Without Oil; Derek Elley, Ponoko; Dr. Michael Evans, University of Toronto; Niki Fenwick, Google; Ariel Ferreira, Local Motors; Matt Flannery, Kiva; Ann Florini, Brookings Institution; Maryantonett Flumian, Institute On Governance; Joe Fontana, former member of parliament in the government of Canada; Gordon and Susan Fraser, the Ravina Project; Tory Gattis, Houston Strategies; Ian Gee, Nokia; Laura Gillies; Dan Gluckman, BBC; Heather Green, Twilight; Robert Greenhill, World Economic Forum; Bill Greeves, Virginia county government; Jim Griffin, Pho; Peter Gruetter, formerly with the Swiss Federal Department of Finance; Simon Hampton, Google; Lisa Hansen, Twilight; Rahaf Harfoush, author; Craig Heimark, Open Models Company; Kim Henderson, Ministry of Citizens' Services, government of British Columbia; Ben Heywood, PatientsLikeMe; Paul Hodgkin, Patient Opinion; Paul Hofheinz, Lisbon Council; Mathew Holt, Health 2.0; Rob Hopkins, Transition Towns; Steve Howard, Climate Group; Lee Howell, World Economic Forum; Arianna Huffington, *Huffington Post*; John C. Hull, University of Toronto; Tara Hunt, HPC; Larry Huston, inno360; Jessica Jackley, Kiva; Jeff Jarvis, BuzzMachine; Chris Johnson, Virtual Alabama; Jason Karas, Carbonrally; Jason Kelly, Ginkgo BioWorks; Kevin Kimberlin, Spencer Trask; Andrew King, MapEcos; Amanda Kistindey; Lakshmi Krishnamurthy, iCarpool; Vivek Kundra, U.S. federal chief information officer; A.G. Lafley, formerly of Procter & Gamble; Jonathon Landman, *The New York Times*; Andrew Lang, Oral Roberts University; Leo Laporte, TWiT; Kelly Lauber, Nike; David W. Lewis, IUPUI; Dave Llorens, 1BOG; David Lowy, SAP; Jane Lucy, Landshare; Hugh MacLeod, formerly of the Climate Change Secretariat, government of Ontario; Lord Mark Malloch-Brown, World Economic Forum; Douglas Marston, Open Models Company; Elliott Masie, Learning TRENDS; Terry McBride, Nettwerk Records; Jessica McDonald, formerly with the Office of the Premier, government of British Columbia; Jacqueline McGlade, European Environmental Agency; Dan McGrath, IBM; Rhonda McMichael, government of Ontario; Patrick Meier, Ushahidi; Ann Mettler, The Lisbon Council; Jerry Michalski, Sociate; Jonathan Miller, News Corporation; Dan Mintz, formerly of the U.S. Department of Transportation; Molly Moran, U.S. Department of State; Aaron Naparstek, Livable Streets; Steve Newcomb, Virgance; Gord Nixon, RBC; Beth Noveck, White House Office of Science and Technology Policy; Hilda Ochoa-Brillembourg, YOA; Ory Okolloh, Ushahidi; Toby Oliver, Path Intelligence; Meaghan O'Neill, TreeHugger; Molly O'Neill, U.S. Environmental Protection Agency; Dara O'Rourke, University of California, Berkeley, GoodGuide; Suzanne Pahlman, WWF; Daniel Palestrant, Sermo; Joe Paluska, Better Place; Dennis Pamlin, WWF; Mark Parker, Nike; Kal Patel, Best Buy; Jonah Peretti, *Huffington Post*; Christian Pesch, CoreMedia; Robert Pothier, Twilight; Luis M. Proenza, University of Akron; Jordan Raddick, Johns Hopkins University; Saad Rafi, Ontario Ministry of Energy and Infrastructure; Chris Rasmussen, U.S. National Geospatial-Intelligence Agency; Ben Rattray, Change.org; Michael Reinicke, Rideshare; Steve Ressler, GovLoop; David Rich, WRI; Ben Rigby, the Extraordinaries; Jay Rogers, Local Motors; Mechthild Rohen, European Commission; Hans Rosling, Trendalyzer; Adam Rothwell, Intelligent Giving; John Gerard Ruggie, Harvard University; Charles Sabel, Columbia University; Rick Samans, World Economic Forum; Saskia Sassen, Columbia University; Kevin Schawinski, Yale University; Eric Schmidt, Google; Henrik Schuermann, CoreMedia; Brent Schulkin, Virgance; Klaus Schwab, World Economic Forum; Eddie Schwartz, Songwriters Association of Canada; Zuhairah Scott;

Euan Semple, formerly at the BBC; Peggy Sheehy, Suffern Middle School; George Siemens, Connectivism; Anne Højer Simonsen, Danish Ministry of Climate and Energy; Larry Smarr, Calit2; Marco Smit, Health 2.0; Kirsi Sormunen, Nokia; Dwayne Spradlin, InnoCentive; Soren Stamer, CoreMedia; Tom Steinberg, mySociety; Robert Stephens, Best Buy; Unity Stoakes, Organized Wisdom; Susanne Stormer, Novo Nordisk; Val Stoyanov, Cisco; Tomer Strolight, Torstar Digital; Anant Sudarshan, Stanford University; David Ticoll, author; Bill Tipton, Hewlett-Packard; Michael Toffel, MapEcos; Linus Torvalds, Linux Foundation; Lena Trudeau, NAPA; Mike Turillo, Spencer Trask Collaborative Innovations; Wood Turner, Climate Counts; Gentry Underwood, IDEO; Jim Walker, Virtual Alabama; David Wheeler, CARMA; Dennis Whittle, GlobalGiving; John Wilbanks, Creative Commons; Sean Wise, VenCorps; Dave Witzel, Environmental Defense Fund; John Wonderlich, SunlightFoundation; Nicole Wong, Google; Jon Worren, MaRS; Doug Wright, RiffWorld; Nick Yee, Palo Alto Research Center; Jim Zemlin, Linux Foundation.

Thank you.

The title of the book came from a challenge we conducted online that showed there is indeed wisdom in crowds. We had over 150 suggestions, and we didn't see that the best term for wikinomics when applied to the macro situation was macrowikinomics until the thoughtful Eryc Branham pointed out that it was just staring us in the face. We encourage everyone to continue the collaboration and conversation at Macrowikinomics.com.

We are also thankful to Adrian Zackheim, our publisher at Portfolio Penguin, for believing, and for assigning the brilliant Adrienne Schultz to be our editor. We are also fortunate to have our publicity efforts managed by Allison McLean. Having worked her magic on *Wikinomics*, we are hopeful for an even greater impact with this book. We are also indebted to Bill Leigh, Wes Neff, and the agents of The Leigh Bureau, who keep us in front of audiences around the world.

As always, our respective wives, Ana P. Lopes (Don) and Michelle Williams (Anthony), kept us honest and provided valuable ideas and support. We owe them our highest gratitude. Don would especially like to thank Alex and Nicole Tapscott for their deep insights and encouragement along the way. Anthony would like to thank his sons, Immanuel and Evan Williams, for their patience and understanding while Daddy spent endless hours glued to his computer, and for providing fun and inspiration when he needed it most.

We are enormously thankful to all. Having said all this, we as the book's authors take full responsibility for its content, as well as for any errors or omissions.

NOTES

Chapter 1: Rebooting the World

1. Groups such as Médecins Sans Frontières, for example, complained that the U.S. Army prioritized getting military personnel and equipment into the country over humanitarian supplies. One aid worker quoted by the BBC claimed logistical disagreements between the Army and the UN were causing a "situation of utter chaos." Of course, it has not helped that infrastructures for transport and communications were severely damaged or that the Haitian government was ill equipped for such a crisis. But it appears that much of the blame for the lack of coordination lies with the relief agencies themselves. See http://news.bbc.co.uk/2/hi/americas/8472670.stm.
2. For example, the report found that efforts to rebuild communities were more successful when those affected by the tsunami were involved in planning the reconstruction efforts. "The only way to redeem that kind of loss is to empower and dignify those people who have suffered," said President Bill Clinton, who was the UN special envoy for the tsunami. "The Tsunami Legacy: Innovation, Breakthroughs and Change," International Federation of Red Cross and Red Crescent Societies (April 24, 2009).
3. Whereas 2 percent of the population owns a computer, 50 percent of Kenyans have a cell phone. "Quarterly Sector Statistics Report, 2nd Quarter Oct-Dec 2009/2010." Communications Commission of Kenya (March 2010).
4. Jeanne Sahadi, "America's hidden debt problem," CNN Money (March 1, 2010).
5. Christopher Hayes, "Twilight of the Elites," *Time* magazine (March 11, 2010).
6. "Where Will the Jobs Come From?" Kauffman Foundation Research Series: Firm Formation and Economic Growth (November 2009).
7. Ken Terry, "Health Spending Hits 17.3 Percent of GDP In Largest Annual Jump," B Net (February 4, 2010).
8. "Life Expectancy at Birth," CIO World Factbook. See https://www.cia.gov/library/publications/the-world-factbook/rankorder/2102rank.html.
9. D. U. Himmelstein, D. Thorne, E. Warren, et al., "Medical bankruptcy in the United States, 2007: Results of a national study," *The American Journal of Medicine* (August 2009).
10. Karen Pallarito, "Government to Pay for More Than Half of U.S. Health Care Costs," *U.S. News* (February 4, 2010).
11. Geoffrey Lean, "Water scarcity 'now bigger threat than financial crisis,'" *The Independent* (March 15, 2009).

12. See http://www.globalissues.org/article/75/world-military-spending.
13. Peter Voser, "Energy transition: not for the faint-hearted," *The Globe and Mail* (September 17, 2009).
14. Susan Kraemer, "China Now Spending $9 Billion a Month on Renewable Energy," CleanTechnica (December 1, 2009).
15. The links between poverty and extremism are contentious, according to Ömer Taşpinar, a professor of national security studies at the National War College and an adjunct professor at the Johns Hopkins University's School of Advanced International Studies. "In one camp, the center-left maintains that the struggle against the root causes of terrorism should prioritize social and economic development. Inspired by modernization theory, this camp sees social and economic development as the precursor of democratization. It also considers educational and economic empowerment as the best antidote against radicalization and terrorist recruitment. Since poverty and ignorance often provide a breeding ground for radicalism, socioeconomic development appears compelling as an effective antidote. This correlation between socioeconomic deprivation and terrorism is strongly rejected by a second group of analysts. Their logic is simple: most terrorists are neither poor nor uneducated. In fact, the majority seem to come from middle class, ordinary backgrounds. Terrorism is therefore perceived almost exclusively as a 'security threat' with no discernible socioeconomic roots or links with deprivation." However, these two camps are not mutually exclusive. After all, educated, middle-class terrorists are rebelling against what they perceive to be Western-induced injustices in the Muslim world and they act because they have the education and financial means to do so. Ömer Taşpinar, "Fighting Radicalism, Not 'Terrorism': Root Causes of an International Actor Redefined," *SAIS Review* (Summer 2009).
16. Quoted in Damian Carrington, Suzanne Goldenberg, Juliette Jowit, Jonathan Watts, Alok Jha, James Randerson, David Smith, David Adam, and Tom Hennigan, "Global deal on climate change in 2010 'all but impossible,'" *The Guardian* (February 2, 2010).
17. See http://www.ibm.com/ceostudy.

Chapter 2: Five Principles for the Age of Networked Intelligence

1. Gutenberg was not the first to invent printing. Asian cultures had previously developed printing but it was not as sophisticated or flexible as Gutenberg's wonder.
2. The ideas of Francis Bacon and Isaac Newton, which defined the scientific method, set the tone for much of what would follow in the century. Bacon and Newton believed that true science called for axiomatic proof to be fused with physical observation in a coherent system of verifiable predictions. For scientific theories and predictions to be verifiable, science needed to be open.
3. The long decline of collaborative scientific work and invention after the triumph of Christendom across the Old Continent broke down historical continuity with classical Greek, Egyptian, Persian, and even Roman science. Those classical experiences were driven by the exchange, critical debate, and research carried out by associated scientists and students located in schools, some very famous even today. The main channel to recover classical scientific achievements was the Mediterranean Arab world. Córdoba (Spain), under Moorish control, became a major cosmopolitan cultural center bridging Europe and the Arab world. By far

the biggest share of translations of the day were from Arabic into Latin, and more important, into French, German, Yiddish, and English. The "European" romance with Greek science and culture was chaperoned by translators and the printing industry. They came to believe that the most important scripture was found in the book of nature, which, when necessary, trumped anything written in stone. This brought them closer to the spirit of Classical Greek and Alexandrian-Arab science, whose main fascination was the nature of things, and natural processes—the great tradition of Epicurus and Lucretius.

4. George Sarton, one of the fathers of the history of science, writes that the discovery of printing "changed the very warp and woof of history, for it replaced precarious forms of tradition (oral and manuscript) by one that was stable, secure, and lasting." George Sarton, *Six Wings* (Bloomington: Indiana University Press, 1957).
5. Clay Shirky, "Newspapers and Thinking the Unthinkable," (March 13, 2009). See http://www.shirky.com/weblog/2009/03/newspapers-and-thinking-the-unthinkable/.
6. Mark Fishman quoted in "New genomic tool for diabetes," Broad Institute press release (February 12, 2007).
7. Denise Campbell, "Trust in politicians hits an all-time low," *The Observer* (September 27, 2009).
8. Scott Rasmussen, "Deficit of Trust: Most Voters Don't Believe President's Assertions About Economy," Rasmussen Reports (January 30, 2010).
9. Although many civic activists may feel they speak for the public good, single-issue NGOs are often myopically focused on their own agendas. They are not always interested in balancing different visions of the public good, or acknowledging the central role the private sector plays in creating wealth and fueling innovation in modern societies. Having a role in setting a broader policy agenda will carry with it a requirement to think and act beyond narrow interests and to act with integrity.
10. Elizabeth Duke quoted in "Regulators: accounting changes need global reach," *USA Today* (September 14, 2009).
11. James Kroeker quoted in ibid.

Chapter 3: Opening Up the Financial Services Industry

1. Matthew Bishop and Michael Green, *The Road from Ruin: How to Revive Capitalism and Put America Back on Top* (Crown Business), and Richard A. Posner, *The Crisis of Capitalist Democracy* (Harvard) are two well-written books that attempt to look at regulatory and structural changes to the financial services industry that might fix what ails it (and capitalism).
2. Fannie Mae and Freddie Mac, the Community Housing Act, and others all had vast programs to encourage lending money to low-income individuals. Banks could unload their mortgages on these institutions, guaranteed by the U.S. government.
3. "FT Global 500 2010," *Financial Times* (May 29, 2010).
4. There's another factor, too. Canadians can't write off their interest payments, so home buyers are more cautious. In the United States the mortgage tax deduction encourages people to stay in debt, and when they can't make payments they simply walk away. Not so in Canada, where homeowners must personally guarantee their loans.

5. "Congress is pressed for bailout with dire warnings," Reuters (September 24, 2008).
6. The call for greater transparency in financial services has a long backstory. From its early days in the nineteenth century, the New York Stock Exchange was the financial center of U.S. capitalism. Backroom deals, gambling, fraud, and self-dealing were rampant. Exchange members enjoyed lower trading rates than nonmembers. Share prices were rarely made known to the public or the press. Until Dow-Jones founded *The Wall Street Journal* in 1889—where the Dow-Jones Index ran on a daily basis from 1896—most financial newspapers were paid mouthpieces for stock promoters. This practice ended only after the 1929 crash. It took the worst (so far) business collapse in modern history—the Great Depression—to force transparency into the broader financial marketplace. The Securities Act of 1933 was the first piece of national securities legislation passed by Congress. During the previous two decades, some twenty states had passed a patchwork of so-called blue-sky laws to regulate the issuance of securities, but these were rife with loopholes. U.S. financial markets, in both banking and securities, operated pretty much free of regulation and visibility until President Franklin D. Roosevelt stepped in.

   Business leaders fought transparency tooth and nail. After World War II, at a time when conflict between management and labor was hitting the boiling point, the United Auto Workers staged a huge strike against the nation's biggest company, General Motors. As Don Tapscott and David Ticoll tell in their 2002 book *The Naked Corporation*, UAW leader Walter Reuther insisted that GM could afford to increase wages without increasing prices. He challenged GM to "open the books" to prove him wrong. The suggestion outraged GM vice president Harry W. Anderson: "We don't even open our books to our stockholders!" This was standard practice back then. Companies provided almost no information to shareholders.
7. Massimo Calabresi, "Geithner vs. the Regulators: A Time for Swearing," *Time* (August 6, 2009).
8. The most obscure financial products—the ones that are hardest to see and evaluate—are derivatives. These financial instruments are derived from some other asset, index, or event. In other words, they are one or more steps removed from the actual asset. Derivatives might be CDOs, or any number of other things, like credit default swaps (CDS)—a contract in which the buyer of the CDS makes a series of payments to the seller and, in exchange, receives a payoff if a credit instrument (typically a bond or a loan) fails to pay (defaults). A standard CDS insures a bond that you own against default. A "naked" CDS is a bet that a bond someone else owns will default. It's like buying fire insurance on your neighbor's house, where you only get paid if it burns. (These were a core business of the failed company AIG.)
9. "Number-crunchers crunched," *The Economist* (February 11, 2010).
10. Rick Bookstaber, *A Demon of Our Own Design: Markets, Hedge Funds, and the Perils of Financial Innovation* (John Wiley & Sons, 2007).
11. Rick Bookstaber, "Why Do Bankers Make So Much Money?", Rick Bookstaber Blog (October 23, 2009).
12. The company is beginning at Level 3, and plans to continue to other nongovernment credit securities and contracts, be they CDOs, CDSs, or other derivatives.

13. Roger Martin, "The business of fleecing others," *The Washington Post* (April 26, 2010).
14. "Venture Impact: The Economic Importance of Venture Capital Backed Companies to the U.S. Economy," Global Insight (2009).
15. VC-backed companies also have a disproportionate impact on innovation, as measured by patents filed. *BusinessWeek* reported on the work of academics researching the link between the filing of patents and the funding. Two researchers, Josh Lerner from Harvard, and Luigi Zingales from the University of Chicago, found that venture capital investments result in exponentially (almost ten times) more patent filings (often seen as a leading indicator on innovation) than monies invested by corporate R&D. Lerner found that although venture capital investments were equal to less than 3 percent of the monies spent by corporate America on R&D, the VC-backed companies were responsible for more than 15 percent of the total number of patents filed over the same time period. It should be noted, however, that other researchers are less optimistic on the link between innovation and venture capital. See Masako Ueda and Masayuki Hirukawa, "Venture Capital and Innovation: Which Is First?", Social Science Research Network (September 14, 2008).
16. "Breaking Through the Broken," North Venture Partners (2009).
17. Claire Cain Miller, "Do Web Entrepreneurs Still Need Venture Capitalists?", *New York Times* (May 14, 2009).
18. Rafe Needleman, "Marc Andreessen launches new venture fund," CNET (July 5, 2009).
19. Paul Kedrosky, "Right-sizing the U.S. Venture Capital Industry," Ewing Marion Kauffman Foundation (June 10, 2009).
20. See http://www.youtube.com/watch?v=jGC1mCS4OVo&feature=player_embedded#!.
21. Julianne Pepitone, "YouTube credit card rant gets results," CNN Money (September 20, 2009).
22. "Predicts 2010: Executive Decisions in Banking and Investment Services Demand a Longer View," Gartner (November 12, 2009).
23. Peter J. Brennan, "Peer-to-Peer Lending Lures Investors With 12% Return," Bloomberg.com (July 16, 2009).

Chapter 4: Bootstrapping Innovation and Wealth Creation

1. Tesla Motors press release. See http://www.teslamotors.com/media/press_room.php?id=2220.
2. "Electric Car Startup Fisker Buys GM Plant to Build Midsize Car," *U.S. News* (October 28, 2009).
3. Yoni Cohen, "Coda Automotive to launch all-electric vehicle in the fourth quarter," *Los Angeles Times* (March 1, 2010).
4. Don Tapscott and Anthony D. Williams, "Hack This Product, Please!", *BusinessWeek* (February 23, 2007).
5. SEMA Show. See http://www.semashow.com/main/main.aspx?ID=/content SEMA SHOWcom/HomePage.
6. Yochai Benkler, *The Wealth of Networks* (Yale University Press, 2006).
7. Larry Huston and Nabil Sakkab, "Connect and Develop: Inside Procter and Gamble's New Model for Innovation," *Harvard Business Review* (March 2006).
8. "GE adds green 'Odyssey' to its India center," *Business Standard* (March 13, 2009).

9. The term "infostructure" dates back to the 1980s with no clear origin. Nothing so powerful as an idea whose time has come?

Chapter 5: Reversing the Tide of Disruptive Climate Change

1. Rebecca Moore, "Seeing the forest through the cloud," Google.org (December 10, 2009).
2. "Stern Review on the Economics of Climate Change," HM Treasury (October 30, 2006).
3. Focus the Nation. See http://www.focusthenation.org/.
4. "Climate Change: Global Risks, Challenges & Decisions," International Alliance of Research Universities (2009).
5. To escape the worst effects of climate change global temperatures cannot rise by more than 2 degrees from preindustrial levels according to scientific consensus. The observed temperature today has already risen by 0.7 degrees. That may not seem like much of an increase, but for Mother Nature it's a lot. Tom Friedman makes a helpful analogy with body temperature. "If your body temperature goes from 98.5 to 100.6, you don't feel so good. If it goes from 100.6 to 102.6, you go to the hospital. So does Mother Nature." To put the challenge in perspective, an average American "produces" about 20 tons of CO_2 every year through normal activities such as driving, heating their home, and consuming food. Limiting temperature rise to below 4 degrees will require the average to fall to below 2 tons of CO_2 per year, an undertaking that is practically inconceivable given our current lifestyles and infrastructure. See Darrel Moellendorf, "Treaty Norms and Climate Change Mitigation," Carnegie Council (September 11, 2009).
6. To be fair, a binding international treaty on climate change is a tough ask. After all, a single text would have to be acceptable to rich and poor nations, democratic and not, left and right, binding them not just for now, but for decades to come. Even the developed countries cannot agree among themselves how carbon emissions should be counted, let alone reduced. What's more, elected leaders will have to be sure that whatever they sign at the international negotiating table will be accepted back home. This need for domestic ratification alone is the stuff of nightmares for international negotiators. One European diplomat quoted in *The Guardian* despairs at the ". . . headbangers who cannot resist a chance to damage Obama, believe global warming is based on junk science and regard action on climate change as ungodly because it will delay the second coming." See Jonathan Freedland, "If Obama can't defeat the Republican headbangers, our planet is doomed," *The Guardian* (September 15, 2009).
7. Tim Palmer, "Climate change uncertainty is no reason for inaction since we can't rule out risk," *The Guardian* (March 22, 2010).
8. Ibid.
9. "Livestock impacts on the environment," Food and Agriculture Organization of the United Nations (2006).
10. Bryan Walsh, "Sizing Up Carbon Footprints," *Time* (May 15, 2008).
11. See Vermont School Carbon Challenge, http://blog.carbonrally.com/2008/10/31/vermont-students-kicking-co2.
12. In February 2009, a NASA satellite carrying instruments to produce the first map of the Earth's carbon emissions crashed near Antarctica only three minutes after lift-off. The satellite would have measured carbon emissions at 100,000 points

around the planet every day, providing a wealth of data compared with the 100 or so fixed towers currently in operation in a land-based network. The head of the U.S. National Oceanic and Atmospheric Administration (NOAA), Professor Jane Lubchenco, has warned that the gathering of satellite data is now at "great risk" because America's aging satellite fleet was not being replaced. Even before her warning, scientists were saying that America, the world's scientific superpower, was virtually blinding itself to climate change by cutting funds to the environmental satellite programs run by NOAA and NASA. See Suzanne Goldenberg and Damian Carrington, "Revealed: the secret evidence of global warming Bush tried to hide," *The Observer* (July 26, 2009).

13. "How the Energy Sector Can Deliver on a Climate Agreement in Copenhagen," International Energy Agency (October 2009).
14. Daniel Goleman quoted in "On Web and iPhone, a Tool to Aid Careful Shopping," *New York Times* (June 14, 2009).
15. Paul Hawken, "The Next Reformation," *In Context* (Summer 1995).
16. "Climate Solutions 2: Low-Carbon Re-Industrialization," Climate Risk (2009).
17. You may be thinking, "Well, the WWF is a crusty old environmental organization; why should we heed what they say?" The truth is that the WWF is not your uncle's environmental organization. In addition to environmentalists, it employs a legion of scientists and MBAs. Not to mention the fact that some of the world's largest companies, including Coca-Cola, HP, IBM, Johnson & Johnson, Nokia, and Sony, now seek the WWF's advice on how to go carbon neutral.
18. "Stern Review on the Economics of Climate Change," HM Treasury (October 30, 2006).
19. Paul Hawken quoted in "Understanding Sustainability: Quality-of-life and opportunities for design," International Council of Societies of Industrial Design (January 2008).
20. Ray Anderson, Interface, quoted in "Green-biz pioneer Ray Anderson says sustainability literally pays for itself," Grist.org (October 19, 2009).
21. If you read our last book, you'll know that this was done before, during the Human Genome Project, a quintessential example of mass collaboration and certainly among the most important scientific endeavors of our time. When efforts to map the human genome began back in 1986, scientists had barely an inkling of how this fundamental part of our existence works—and to a large degree they still don't! But thanks to massive, distributed collaboration across institutions, countries, and disciplines that took over fifteen years to complete, society is at the crest of a new wave of medical and biological innovation that could see scientists reprogramming nature in much the same way open-source programmers code operating systems like Linux today.

    For us the Human Genome Project is important for an additional reason. It represents a watershed moment, when a number of pharmaceutical firms abandoned aspects of their proprietary human genome projects to back open collaborations. By sharing basic science and collaborating across institutional boundaries, these companies challenged a deeply held notion that their early-stage R&D activities are best pursued individually and within the confines of their secretive laboratories. As a result they were able to cut costs, accelerate innovation, create more wealth for shareholders, and ultimately help society reap the benefits of genomic research more quickly.

22. Even the pricing can be flexible. Say a company wants to charge their direct competitors but would happily waive the fee for firms with annual revenues below $10 million a year. "That opens up essentially all entrepreneurial use," says Wilbanks. "And it opens up the vast majority of developing world uses in a simple and scalable way too."

Chapter 6: Wikinomics Meets the Green Energy Economy

1. Ian Sample, "Oil: The final warning," *New Scientist*, Issue 2662 (June 25, 2008).
2. "The International Energy Outlook 2010," U.S. Energy Information Administration (May 25, 2010).
3. Tom Friedman likes to show a graph in his presentations illustrating a direct correlation between high oil prices and calamitous occurrences in the world. When oil prices were high, for example, the Soviet Union was strong and belligerent; when they fell, the Soviet Union collapsed. High oil prices emboldened Iraq, Iran, and other extremist countries. When oil prices are low, countries are more willing to cooperate with their neighbors. To be sure, long-term oil prices have only one direction to go, which spells more trouble for the world.
4. Patrick Mazza, "The Smart Energy Network: Electricity's Third Great Revolution," Climate Solutions (June 2003).
5. "The International Energy Outlook 2010," U.S. Energy Information Administration (May 25, 2010).
6. "China's Low Carbon Development Pathways by 2050," Energy Research Institute (September 2009).
7. "Who's winning the clean energy race?", Pew Charitable Trusts (2010).
8. Consultancies such as McKinsey have estimated that China will need to invest closer to US $200–300 billion each year in the development and large-scale use of renewable energy and nuclear power between now and 2030—and all that just to maintain its emissions at about 10 percent above 2005 levels. According to the report (REF), China will also need to be a leader in energy-saving technology, build environmentally friendly homes, develop more public transportation, and dramatically curtail car usage.
9. The Northeast Blackout of 2003. See http://en.wikipedia.org/wiki/Northeast_Blackout_of_2003.
10. "Berkeley Lab Study Estimates $80 Billion Annual Cost of Power Interruptions," Lawrence Berkeley National Laboratory (February 2, 2005).
11. Leonard Gross, Hydro One, quoted in "Utilities, government charged up about high-tech power distribution systems," CBC News (March 12, 2009).
12. Jeff St. John, "8.3M Smart Meters and Counting in U.S.," greentechgrid (July 17, 2009).
13. Maria Hattar, Cisco, quoted in "Cisco: Smart grid will eclipse size of Internet," CNET News (May 18, 2009).
14. The Digital Environment Home Energy Management System (DEHEMS). See http://www.dehems.eu/about.
15. David Miliband, U.K. secretary of state for environment, quoted in "Carbon emissions: Now it's getting personal," *New York Times* (June 20, 2007).
16. Richard MacManus, "IBM and the Internet of Things," ReadWriteWeb (July 22, 2009).
17. "World electricity: The smart grid era," *The Economist* (June 5, 2009).
18. "SMART 2020: Enabling the low carbon economy in the information age," The Climate Group (2008).

19. The argument in favor of radically decentralizing energy production is also subject to the specifics of geography. The southwestern United States, for example, has both sun and land in abundance, which makes desolate areas like the Mojave Desert ideal for large-scale solar power generation. In more northerly latitudes, where land and sun are more sparing, the economics are different. On the other hand, convincing millions of homeowners to install solar panels on their roofs hardly sounds like a light feat of social engineering either. And yet, countries like Germany—arguably now the world's leader in solar technology—have succeeded in large part due to the government's willingness to encourage households to install rooftop systems by subsidizing their efforts. Germany has also declared that all new buildings will be either energy neutral or net energy producers by 2018.
20. According to Gord Fraser, $30K accounts for all things related to the addition of solar power, including the movable support structure. Another $15–20K was spent on the boiler, insulation, windows, doors, and insulated siding that encapsulates the second story of the house and the back porch. Since this project attempts to track changes in household thermodynamic efficiencies, the Frasers have rolled out these changes over several years during the summer months and used the winter months to gather data. Each year the new data allows them to crunch the household efficiency numbers to see what gains, if any, have been made. This is basically a "bang for the buck" calculation.
21. Ontario's Green Energy Act. See http://www.mei.gov.on.ca/en/energy/gea/.
22. "The Danish Example—the way to an energy efficient and energy friendly economy," Danish Ministry of Climate and Energy (June 16, 2009).
23. For instance, if the typical retail price for fossil-fuel generated electricity is 15 cents per kWh, power generated by solar or wind is compensated at 40 cents per kWh.
24. "Case Studies in American Innovation: A New Look at Government Involvement in Technological Innovation," Breakthrough Institute (April 2009).
25. "Falling Panel Prices Could Bring Solar Closer to Grid Parity," Green Econometrics (October 27, 2009).
26. "The Danish Example."
27. Linda J. Bilmes and Joseph E. Stiglitz, "The Iraq War Will Cost Us $3 Trillion, and Much More," *The Washington Post* (March 9, 2008).
28. Jesse Jenkins, "National Institutes of Energy Needed to Fill Energy R&D Gap," Breakthrough Institute (October 9, 2009).
29. Juan Pablo Pérez Alfonzo, Venezuela's oil minister in the early 1960s and one of the founders of OPEC, referred to oil as the devil's excrement. Oil, he said, was not black gold; it was a curse, referring to the observation that poor but resource-rich countries tend to be underdeveloped not despite their hydrocarbon and mineral riches but because of their resource wealth. More often than not, large deposits of the devil's excrement lead to concentrated power, corruption, and the ability of elite rulers to ignore the needs of their populations. Since then, Pérez Alfonzo's insight has been rigorously tested and confirmed—by a slew of academic studies. Even when resource-fueled growth takes place, it rarely yields growth's usual full social benefits. So it turns out oil is not just bad for the environment, it's bad for economic development, bad for democracy, bad for human health, and bad for the world.

Chapter 7: The Transportation Revolution

1. "Oeko-bilanz eines autolebens," Umwelt-und-Prognose-Institut Heidelberg (1993).

2. Car-loving nations can take credit for much of this pollution. The United States, home to 5 percent of the planet's population and 30 percent of the world's automobiles, contributes 45 percent of the world's automotive CO_2 emissions. In fact, Americans burn through 20 million barrels of oil while driving more than 5 billion miles a day. See John DeCicco and Freda Fung, "Global Warming on the Road," Environmental Defense (2006).
3. "How the Average U.S. Consumer Spends Their Paycheck," Visual Economics (April 2009).
4. P. Wiederkehr and N. Caïd, "Transport troubles," Organisation for Economic Co-operation and Development Environment Directorate (August 2002).
5. "The World Factbook 2009," Central Intelligence Agency (2009).
6. M. Peden et al., "World Report on Road Traffic Injury," World Health Organization (2004).
7. Ariana Eunjung Cha, "China's Cars, Accelerating a Global Demand for Fuel," *Washington Post* (July 28, 2008).
8. Jonathan Watts, "China's powerhouse vision for 2050," *The Guardian* (February 10, 2006).
9. Phil Gott, "Is Mobility As We Know It Sustainable?", Global Insight (2008).
10. By comprehensively rolling out electric vehicles over the next two decades, China could cut its projected demand for imported oil by up to 30 to 40 percent by 2030, according to McKinsey. This wouldn't address all of China's environmental problems, but it's a start and it's easy to sell given China's automotive ambitions. Largely uninhibited by huge legacy costs, its fledgling domestic industry is well positioned to leapfrog over the industrial past and into the electric future, with the full weight of the Chinese government behind it. Indeed, if you want further evidence that the economics have now tipped in favor of electric transportation, look no further than the commitment made by China's most senior leaders to make the country the largest producer of electric vehicles in three years. See Zhao Tingting, "China can build 'green economy' by 2030," *China Daily* (February 26, 2009).
11. "Global Environment Outlook," United Nations Environment Programme (2007).
12. The numbers look something like this: Assume you drive 12,000 miles a year, gas costs $2 a gallon, and electricity is priced at 12¢ per kilowatt, about what most Americans pay. A gasoline-powered car that gets 20 miles to the gallon—say, a Chevy Impala or a BMW X3—will have annual fuel costs of $1,200 and generate about 6.6 tons of carbon dioxide. Equip those cars with electric motors, and fuel costs drop to $400 a year and emissions are reduced to about 1.5 tons (or zero if car batteries are recharged using renewable energy). Over a ten-year lifespan the savings add up to $8,000, using today's battery technology and today's gas prices. See Marc Gunther, "Warren Buffett takes charge," *Fortune* (April 13, 2009).
13. To illustrate, an average European car costs 12,000 euros to acquire, yet over its twelve years of life will require approximately 30,000 liters of fuel costing roughly 35,000 euros (assuming fuel prices do not continue to increase even further). In other words, we now have a container for energy built into the car—the fuel tank—costing $100 to build; yet our energy costs three times the price of the car. Contrast that with the electric vehicle where the container for energy, in this case a battery, costs roughly 7,000 euros, yet the electricity to run the car costs 2,000 euros for the entire life of the car. In the aggregate, energy to drive an electric vehicle has now crossed under 10,000 euros. Historic trend lines for batteries over

the last twenty-five years show a 50 percent price per kWh improvement every five years, stemming from technological and process improvements. Projecting forward to 2015, we should see the cost of the battery and solar generation sufficient for a car reaching a combined cost of 5,000 euros. See Shai Agassi, "Projecting the Future of Energy, Transportation, and Environment," Better Place (October 5, 2007).

14. As noted above, historic trend lines for battery production over the last twenty-five years show a 50 percent price per kWh improvement every five years, stemming from technological and process improvements. Similar effects are well documented in the chip industry, where Moore's law predicted chip improvements amounting to a doubling of performance at minimum cost every two years. See Shai Agassi, "Projecting the Future of Energy."
15. "From Home to Work, the Average Commute is 26.4 Minutes," U.S. Department of Transportation (October 2003).
16. Critics of Agassi's scheme have raised doubts over the feasibility of building a planet-wide network of exchange stations and recharge spots. Hundreds if not thousands of these would have to be built in the United States alone before this idea could work—an expensive endeavor and the same kind of problem that derailed the advent of hydrogen as a fuel source for autos. It's certainly not going to happen overnight and that's not the only technical challenge facing operators like Better Place. Stocking every kind of battery for every electric vehicle will be a nightmare, unless, in an unlikely scenario, international standards are agreed on quickly. And actually switching out batteries will prove difficult since every car will be designed differently and batteries weigh thousands of pounds. Will advances in battery technology (faster charging, longer duration) eliminate the need for exchange stations before they're even built? Time will tell.
17. Progressive Automotive X Prize. See http://www.progressiveautoxprize.org/.
18. Orteig Prize. See http://en.wikipedia.org/wiki/Orteig_Prize.
19. Jeremy Korzeniewski, "Chevy Volt will cost GM $750 million," autobloggreen (December 9, 2008).
20. See Zipcar Press Release: http://zipcar.mediaroom.com/index.php?s=43&item=8.
21. "Case Study: Zipcar," District of Columbia—Department of the Environment. See http://ddoe.dc.gov/ddoe/cwp/view,a,1210,q,499698.asp.
22. Ibid.
23. See Zipcar Corporate Overview: http://zipcar.mediaroom.com/file.php/61/corporate_overview.pdf.
24. Paul Keegan, "Zipcar: The best new idea in business," *Fortune* (August 27, 2009).
25. Bill Ford, quoted in ibid.

Chapter 8: Rethinking the University

1. "Education at a Glance 2009," OECD (2009).
2. Cindy P. Veenstra, "A Strategy for Improving Freshman College Retention," *Journal for Quality and Participation*, vol. 31, no. 4 (January 2009), p. 19, citing A. W. Astin and L. Oseguera, *Degree Attainment Rates at American Colleges and Universities*, rev. ed. (Los Angeles: Higher Education Research Institute, UCLA, 2005).
3. Anya Kamenetz, "How Web-Savvy Edupunks Are Transforming American Higher Education," *Fast Company* (September 1, 2009).
4. Elaine Allen and Jeff Seaman, "Staying the Course," Babson College and The Sloan Consortium (November 2008).

5. Kevin Carey, "What Colleges Should Learn from Newspapers' Decline," *Chronicle of Higher Education* (April 3, 2009).
6. Jeffrey Gangemi, "Do Online MBAs Make the Grade?", *BusinessWeek* (August 18, 2005).
7. Allen and Seaman, "Staying the Course."
8. Ben Terris, "The Latest File-Sharing Piracy: Academic Journals," *Chronicle of Higher Education* (October 30, 2009).
9. Mark C. Taylor, "End the University as We Know It," *The New York Times* (April 26, 2009).
10. "Letters: Do Our Universities Need a Makeover?", *The New York Times* (May 3, 2009).
11. See, for example, Taylor, "End the University as We Know It"; Barbara Leigh Smith and Jean T. MacGregor, "What Is Collaborative Learning?", Washington Center for Improving the Quality of Undergraduate Education (1992); and Mark C. Taylor, "Useful Devils," *EDUCAUSE Review*, vol. 35, no. 4 (July/August 2000).
12. Of course the lecture is more than this. Many faculty work hard to engage students and the broadcast model is enhanced in some disciplines through essays, labs, and even seminar discussions.
13. The notion of collaborative learning has been around for a long time, predating the World Wide Web. But it had a very limited scope. Barbara Leigh Smith and Jean T. MacGregor wrote in 1992 about a shift away from the typical teacher-centered or lecture-centered milieu in college classrooms. "In collaborative classrooms, the lecturing/listening/note-taking process may not disappear entirely, but it lives alongside other processes that are based in students' discussion and active work with the course material." Their spirit was right: "Teachers who use collaborative learning approaches tend to think of themselves less as expert transmitters of knowledge to students, and more as expert designers of intellectual experiences for students as coaches or mid-wives of a more emergent learning process." But the bottom line was a simple one: professors should spend more time in discussion with students. As educator Jeff Golub pointed out in 1988: "Collaborative learning has as its main feature a structure that allows for student talk: students are supposed to talk with each other . . . and it is in this talking that much of the learning occurs."
14. John Seely Brown and Richard P. Adler, "Minds on Fire: Open Education, the Long Tail, and Learning 2.0," *EDUCAUSE Review*, vol. 43, no. 1 (January/February 2008).
15. Richard J. Light, *Making the Most of College: Students Speak Their Minds* (Harvard University Press, 2001).
16. Seymour Papert, "New Theories for New Learnings," *School Psychology Review*, vol. 13, no. 4 (fall 1984), 422–28.
17. Papert's notion of "constructionism" should not be confused with constructivism, a theory of knowledge developed by Jean Piaget, that argues that humans generate knowledge from their experiences.
18. Scott Carlson, "The Net Generation Goes to College," *Chronicle of Higher Education* (October 7, 2005).
19. GoodQuestions at Cornell. See http://www.math.cornell.edu/~GoodQuestions/.
20. Warren Baker, Thomas Hale, and Bernard R. Gifford, "From theory to implementation: The Mediated Learning approach to computer-mediated instruction, learning and assessment," *Educom Review*, vol. 32, no. 5 (September/October 1997).

21. Charles M. Vest, "Open Content and the Emerging Global Meta-University," *EDUCAUSE Review*, vol. 41, no. 3 (May/June 2006).
22. MIT OpenCourseware at MIT. See http://ocw.mit.edu.
23. Susan Hockfield, "President's Message," MIT OpenCourseware Web site. See http://ocw.mit.edu/OcwWeb/web/about/president/index.htm.
24. Paul Hofheinz, "EU 2020: Why Skills are Key for Europe's Future," The Lisbon Council (2009).
25. Peter Drucker, "Seeing things as they really are," *Forbes* (March 10, 1997).
26. James J. O'Donnell notes: "Institutions and parts of institutions that go on thinking they're in the youth camp business will increasingly be seen as failing at their core mission." James J. O'Donnell, "To Youth Camp: A Long Farewell," *EDUCAUSE Review*, vol. 36, no. 6 (November/December 2001).

Chapter 9: Science 2.0

1. A project like Galaxy Zoo requires using actual people instead of a computer algorithm, Schawinski explains, because computers aren't very good at pattern recognition. Humans, on the other hand, are exceptionally good at picking out different patterns and shapes, such as being able to distinguish between people's faces at a quick glance, he notes. The same turns out to be true when it comes to distinguishing different types of galaxies.
2. The fact that ordinary people are getting involved in science has other benefits as well. The increasingly tight-knit community's members range from individuals with no astronomy background, to schoolteachers and students, to parents who participate with their children as a sort of family activity. They share experiences, solve problems together, and help educate new members as they join. Some community members have contributed improved user interface solutions, while other Zooites arrange regular meet-ups in places like New York, London, and Amsterdam. One could argue that citizen science has become a genuine social movement, complete with a shared sense of identity, shared goals and accomplishments, and a social fabric that binds them. On top of all that, broad participation in projects like Galaxy Zoo helps boost the public's general understanding of science, a nice side effect at a time when some degree of scientific literacy is required just to understand, let alone solve, some of our biggest public policy issues.
3. The ideas of Francis Bacon and Isaac Newton, which defined the scientific method, set the tone for much of what would follow in the century. Bacon and Newton believed that true science called for axiomatic proof to be fused with physical observation in a coherent system of verifiable predictions. For scientific theories and predictions to be verifiable, science needed to be open.
4. The average number of authors per scientific paper is up too, increasing steadily over the past sixty years from an average of slightly over 1 to averages of 2.22 in computer science, 2.66 for condensed-matter physics, 3.35 for astrophysics, 3.75 for biomedicine, and 8.96 authors for high-energy physics. See M.E.J. Newman. "Who is the best connected scientist? A study of scientific co-authorship networks," *Working Paper*, Santa Fe Institute (2000).
5. Kevin Kelly, "Speculation on the Future of Science," *Edge*, vol. 179 (April 7, 2006).
6. Fred Pearce, "Climate Wars," *The Guardian* (February 9, 2010).
7. As data sets continue to grow, scientists will need to hand off more and more of the routine tasks to machines. The problem in the interim is that machines won't necessarily spot the unusual or the unexpected, the kind of one-offs that can lead to game-changing breakthroughs. Galaxy Zoo cofounder Chris Lintott claims

citizen science data sets naturally provide large and powerful training sets for machine learning approaches to classification problems. By doing citizen science today, researchers like Lintott can help train them.

8. Quoted in "Science 2.0: New online tools may revolutionize research," CBC News (January 13, 2009).
9. Gigi Hirsch, MD, "NEWDIGS' New Drug Development Paradigms," MIT Center for Biomedical Innovation (January 2010).
10. MIT press release announcing the creation of the Center for Biomedical Innovation, "MIT launches Center for Biomedical Innovation," April 29, 2005.
11. Traditional journals aggregate academic papers by subject and deploy highly structured systems for evaluating and storing the accumulated knowledge of a scientific community. Each paper is peer reviewed by two or more experts, and can go through numerous revisions before it is accepted for publishing.
12. No doubt these problems are hangovers from a world of physical distribution and a much more limited volume of publishing. The current publishing regime emerged in seventeenth-century Europe, when the pace of discovery was glacial by twenty-first-century standards. Scientific journals provided the primary infrastructure for scholarly communication and collaboration. Apart from annual academic symposiums, journals were *the* place where scientists could find out about, engage with, and carefully critique each other's work. Publishing journals was expensive, entailing significant capital and operational costs.

Chapter 10: Collaborative Health Care

1. Jonathan Weiner, *His Brother's Keeper: A Story from the Edge of Medicine* (HarperCollins, 2004).
2. "About Us," PatientsLikeMe (accessed May 15, 2010).
3. Erica Westly, "50 Most Innovative Companies in the World: #23 PatientsLikeMe," *Fast Company* (February 17, 2010).
4. Ibid.
5. "Health care spending in Canada to exceed $180 billion this year," Canadian Institute for Health Information (November 19, 2009).
6. Robert Kelley, "Where Can $700 Billion In Waste Be Cut Annually From The U.S. Healthcare System?" Thomson Reuters (October 27, 2009).
7. D. Himmelstein, D. Thorne, E. Warren, et al., "Medical bankruptcy in the United States, 2007: results of a national study," *American Journal of Medicine* (August 2009).
8. J. Lubitz and G. Riley, "Trends in Medicare payments in the last year of life," *New England Journal of Medicine* (April 1993).
9. Greg Keller, "US Tops World in Health Care Spending, Results Lag," ABC News (December 8, 2009).
10. "Country Comparison: Life Expectancy at Birth," Central Intelligence Agency World Fact Book (2009).
11. "Country Comparison: Infant Mortality Rate," Central Intelligence Agency World Fact Book (2009).
12. "World Health Organization Assesses the World's Health Systems," World Health Organization (2000).
13. Barbara Starfield, "Doctors Are the Third Leading Cause of Death in the United States" *Journal of the American Medical Association* (July 26, 2000).
14. Ibid.
15. Stephen Pincock, "The great Australian loneliness; One in three adults in the

prime of life find loneliness a problem. What does this mean for their health, and our society?" *Australian Doctor* (March 5, 2010).

16. Ibid.
17. Those are the findings of the latest Gallup-Healthways Well-Being Index, which shows that close to two thirds (63.1 percent) of adults in the United States were either overweight or obese in 2009. See also: Bill Hendrick, "Americans Are Eating Poorly, Exercising Less, and Getting Bigger, Survey Finds," WebMD Health News (February 10, 2010).
18. Just over 59 percent of obese Americans exercised at least one day per week, compared to 70 percent of overweight people, and 74 percent of healthy-weight people. Obese people are less likely than people in every other weight category (overweight, normal weight, underweight) to have eaten five servings of fruits and vegetables on at least three days of the past seven. Obese Americans also are less likely to say they ate healthy "all day yesterday."
19. Thomas Goetz, *The Decision Tree: Taking Control of Your Health in the New Era of Personalized Medicine* (Rodale Books, 2010), p. 238.
20. M. Kwan and N. Haque, "Sermo and PatientsLikeMe: A Revolution in Collaborative Healthcare," Enterprise 2.0 Lighthouse Case Study, nGenera Corporation (2008).
21. Ibid.
22. Ibid.
23. Ibid.
24. "Social Networking May Benefit Patients with Common Skin Disease," Center for Connected Health (January 21, 2009).
25. Ibid.
26. "Testimonials," WeAre.Us (accessed May 15, 2010).
27. JoNel Aleccia, "Docs seek to stifle patients' rants on Web sites," msnbc (January 13, 2010).
28. RateMDs.com is just one of a host of sites operated by ratingz.com, a firm started by the founder of RateMyProfessors.com. Drugratingz.com is another of its sites.
29. D. Fallows and S. Fox, "Internet Health Resources," Pew Internet & American Life Project (July 16, 2003).
30. S. Ponder and J. T. Skyberg, "E-Prescribing: Is it Just What the Doctor Ordered?", HCT Project (July 17, 2004).
31. Ibid.
32. Ibid.
33. "President Bush's IT Doctor," *BusinessWeek* (March 28, 2005).
34. Matthew Holt, "The Past and Future of Health 2.0," Health2Advisors.com (January 2010).
35. "Flu Trends," google.org (2009).
36. J. Kruger, H. M. Blanck, and C. Gillespie, "Dietary and physical activity behaviors among adults successful at weight loss maintenance," *International Journal of Behavioral Nutrition and Physical Activity* (July 19, 2006).
37. Stefania Viscusi, "Weight Watchers: Mobile and Web Strategy Keeps Customers Engaged," TMCnet.com (April 19, 2010).
38. S. Woolhandler, T. Campbell, and D. Himmelstein, "Costs of Health Care Administration in the United States and Canada," *New England Journal of Medicine* (August 21, 2003).
39. Toni Johnson, "Healthcare Costs and U.S. Competitiveness," Council on Foreign Relations (March 23, 2010).

40. Ponder and Skyberg, "E-Prescribing."
41. Jay Parkinson, "All Physicians Are Not Created Equal: How to Fix Medicine's Two-Party System," *Fast Company* (April 30, 2009).
42. Chuck Salter, "The Doctor of the Future," *Fast Company* (May 1, 2009).
43. J. Parkinson, "Jay Parkinson's Medicine 2.0," PopTech! Conference (October 2008).
44. Ibid.
45. Ford Vox, "When Doctors Talk," *Newsweek* (March 10, 2009). See also: Occupational Outlook Handbook, 2010–11 Edition, United States Department of Labor, Bureau of Labor Statistics (accessed May 15, 2010).
46. James C. Robinson and Paul B. Ginsburg, "Consumer-Driven Health Care: Promise and Performance," *Health Affairs* (January 27, 2009).
47. "About Us," MedHelp (accessed May 15, 2010).
48. "Healthcare News," *Harris Interactive Survey*, vol. 4, issue 13 (August 2004).
49. Ponder and Skyberg, "E-Prescribing."
50. Peter Waegemann correspondence with Dan Herman (April 21, 2005).
51. "Personal Health Working Group Final Report," Connecting for Health (July 1, 2003).
52. Ibid.
53. Ibid.
54. Dave deBronkart, "What e-Patients Want from Doctors, Hospitals and Health Plans," Presentation at ICSI/IHI (May 2010).

Chapter 11: The Demise of the Newspaper and the Rise of the New News

1. Amanda Ernst, "Huffington Post's Traffic More Than Doubles Year Over Year," mediabistro (January 19, 2010).
2. Huffington Post Company Profile, CrunchBase (April 28, 2010).
3. Erick Schonfeld, "The Huffington Post Starts to Give Out Badges to Readers," TechCrunch (April 29, 2010).
4. Jonah Peretti as reported in "The Death and Life of the American Newspaper," *The New Yorker* (March 23, 2008).
5. Andrew Lipsman, "Huffington Post Defies Expectations, Reaches New Heights Post-Election," Comscore (June 4, 2009).
6. Sam Stein, "Bailed-Out Firms Distributing Cash Rewards: 'Please Do Not Call It A Bonus,'" *The Huffington Post* (February 11, 2009).
7. Jeff Jarvis, "Arianna Huffington is saving journalism," *The Guardian* (April 6, 2009).
8. M. J. Stephey, "White House Press Corps Dean Helen Thomas on 'Listen Up,'" *Time* (October 19, 2009).
9. GigaTweet Tweet Counter (accessed May 6, 2010). See http://popacular.com/gigatweet/.
10. Chris Putnam, "Faster, Simpler Photo Uploads," The Facebook Blog (February 5, 2010).
11. Chad Hurley, "Y,000,000,000uTube," The Official YouTube Blog (October 9, 2009).
12. "State of the Blogosphere 2009," Technorati (2009).
13. "Press Accuracy Rating Hits Two Decade Low," Pew Research Center for People and the Press (September 13, 2009).
14. David Carr, "Papers Try to Get Out of a Box," *The New York Times* (April 12, 2009).

15. Rupert Taylor, "Newspapers Are Closing Down," Suite 101: Newspaper Industry (July 22, 2009).
16. "The 10 Most Endangered Newspapers in America," *Time* (March 9, 2009).
17. "The State of the News Media 2010," The Pew Research Center's Project for Excellence in Journalism.
18. Ibid.
19. Michael Mandel, "The Journalism Job Market: Part I, Looking Back," *BusinessWeek* (September 16, 2009).
20. Taylor, "Newspapers Are Closing Down."
21. The American public's growing distaste for print on paper is also pulling down the magazine publishing industry. Sure, *Vogue* and *InStyle* are bursting with ads. But although there are a handful of very successful magazines, since 2000 single copy sales (think airport magazine racks) have fallen almost 30 percent. Iconic *Reader's Digest* filed for Chapter 11 in August 2009. In the last dozen years, *Newsweek*'s circulation has dropped from 3.5 million to a little over one million. One of the largest magazine publishers in the United States, Condé Nast, has a raft of famous titles, including *Vanity Fair*, *Vogue*, *The New Yorker*, *Wired*, and *Architectural Digest*. But in 2009 when advertising revenue dropped a whopping $1 billion, they began shedding magazines like a German shepherd in springtime, laying off two hundred staff and closing *Gourmet*, *Elegant Bride*, and *Modern Bride*. Sources: Averages calculated by the MPA (Magazine Publishers of America) from Audit Bureau of Circulations (ABC) statements for the first and second six months of each year. Domestic titles audited by ABC; annuals, international editions, and comics have been excluded. Totals may not add up exactly, due to rounding of averaged numbers.
22. Interview with Bob Garfield on *Talk of the Nation*, NPR (August 6, 2009).
23. Clay Shirky, "Newspapers and Thinking the Unthinkable," Edge (March 17, 2008).
24. Josh Young, "If News Is That Important, It Will Find Me," *The Huffington Post* (October 1, 2009).
25. Harold Innis, *The Bias of Communication* (University of Toronto Press, 1951).
26. Ibid.
27. David Leigh, "Guardian gagged from reporting parliament," *The Guardian* (October 12, 2009).
28. Noam Cohen, "Twitter and a Newspaper Untie a Gag Order," *The New York Times* (October 18, 2009).
29. "Twitter can't be gagged: online outcry over Guardian/Trafigura order," *The Guardian* (October, 2009).
30. Some, such as the *Cincinnati Post*, *Kentucky Post*, and *Rocky Mountain News* have gone out of business, while others have seen their share values plunge. A recent study suggested that one out of every five journalists working for newspapers in 2001 is now gone, and 2009 may be the worst year yet for job losses.
31. Tim Arango, "Fall in Newspaper Sales Accelerates to Pass 7%," *The New York Times* (April 27, 2009).
32. "The New York Times Company Reports 2009 Second-Quarter Results," iStockAnalyst (July 23, 2009).
33. Andrew Keen, *The Cult of the Amateur: How Today's Internet Is Killing Our Culture* (Broadway Business, 2007).
34. Cody Brown, "A Public Can Talk to Itself: Why the Future of News Is Actually Pretty Clear," http://codybrown.name (October 25, 2009).

35. "State of the Blogosphere 2009," Technorati (2009).
36. Peter F. Drucker, *The Practice of Management* (Harper & Brothers, 1954).
37. Hoaq Levins, "Martin Sorrell: Newspapers/Magazine Contraction Must Continue," *Advertising Age* (November 6, 2009).
38. Sinclair Stewart and Grant Robertson, "Media: The Zero-Paper Town," *The Globe and Mail* (March 14, 2009).
39. But again, let's keep this in perspective. Many media organizations were big, monolithic, and unapproachable corporations. They felt accountable to absolutely no one. Journalists were the gatekeepers of the facts and opinions, and fiercely guarded their independence. Good luck trying to get an editor of *The New York Times* on the phone twenty years ago. Readers were largely bystanders to the newsgathering, filtering, and publication process. The notion of newspapers appointing a public editor to represent readers' interests and respond to reader feedback and criticism was seen as so radical that many publications laughed at the idea. It took the Jayson Blair scandal to finally convince *New York Times* management to appoint a public editor in 2003.

    In fact, many journalists, especially the very senior ones, enjoyed being members of a cozy and exclusive club and the perquisites that came with it. They were "in the know." Publishers, columnists, and senior editors were often confidants of celebrities and politicians reaching all the way up to the president. Many scandals never saw the light of day. Press club dinners were annual events where politicians and journalists shared many drinks, many jokes, and all of it off-the-record. All this caused thoughtful people to be skeptical of the mainstream press. As John le Carré wrote: "until we have a better relationship between private performance and the public truth . . . we are absolutely right to remain suspicious, contemptuous even, of the secrecy and the misinformation which is the digest of our news."
40. James Madison, Virginia Resolutions (December 21, 1798).
41. "About Us," ProPublica. See http://www.propublica.org/about/.
42. "Pulitzer Prize in Investigative Reporting: Deadly Choices at Memorial," ProPublica.
43. However, a greater number, 68 percent, agreed with the statement that "Old-style, traditionally objective and fair journalism is dead." Fully 84 percent said national news media organizations were very or somewhat biased. Close to 46 percent said they have permanently stopped watching a news media organization, print or electronic, because of perceived bias. http://www.editorandpublisher.com/eandp/news/article_display.jsp?vnu_content_id=1004015442.
44. Michael Andersen, "Four crowdsourcing lessons from the Guardian's (spectacular) expenses-scandal experiment," Nieman Journalism Lab (June 23, 2009).
45. "EDITORIAL: No nonsense, no embellishment, just the news," Peoria *Journal Star* (July 21, 2009).
46. Bill Bishop and Robert G. Cushing, *The Big Sort: Why the Clustering of Like-Minded America Is Tearing Us Apart* (Houghton Mifflin, 2008), as quoted in Scott Stossel, "Subdivided We Fall," *The New York Times* (May 18, 2008).
47. Don Tapscott, *Grown Up Digital: How the Next Generation Is Changing Your World* (McGraw-Hill, 2008).
48. "1981 primitive Internet report on KRON," YouTube. See http://www.youtube.com/watch?v=5WCTn4FljUQ.
49. "MediaGuardian 100 2009," *The Guardian* (July 2009).
50. Jarvis used data from Borrell Associates, extensive interviews with media compa-

nies both legacy and start-ups, and a survey of more than 110 hyperlocal news sites.

51. "New Business Models," The City University of New York Graduate School of Journalism (September 2009).
52. "State of the Blogosphere 2008," Technorati (2008).
53. Michael Hirschorn, "The Newsweekly's Last Stand: Why *The Economist* is thriving while *Time* and *Newsweek* fade," *The Atlantic* (July/August 2009).
54. Jim Cooper, "4A's: 'Wired' Chief Says iPad Will Rescue Magazines," *Adweek* (March 2, 2010).

Chapter 12: Inside the Future of Music

1. Amra Alirejsovic, "Musical Group Wants to Unite World One Song at a Time," Voice of America (November 11, 2009).
2. Ian Cuthbertson, "Changing lives, one song at a time," *The Australian* (August 24, 2009).
3. "Orchestra: Planting Seeds," Youth Orchestra of the Americas (accessed May 9, 2010).
4. "About: Mission," Youth Orchestra of the Americas (accessed May 9, 2010).
5. Ed Christman, "Digital Bytes; Growth in Track Downloads Slows Sharply, Accentuating '09 Sales Woes," *Billboard* (January 16, 2010).
6. International Federation of the Phonographic Industry Digital Music Report 2009 (January 16, 2009).
7. Tom Vander Beken, *Organized Crime and Vulnerability of Economic Sectors: The European Transport and Music Sector* (Maklu Uitgevers N.V., 2005).
8. Don Tapscott, "Some Advise 'Everywhere Internet Audio'," *The New York Times* (September 15, 2003).
9. Big Champagne in California has attained high accuracy and does it in a way that maintains the anonymity of the file sharers. There are other technologies that give accurate results as well and a combination may be the way to go.
10. As explained by Steve Gordon in *The Future of the Music Business: A Guide for Artists and Entrepreneurs* (Backbeat Books, 2005).
11. "A Better Way Forward: Voluntary Collective Licensing of Music File Sharing," Electronic Frontier Foundation (April 2008).
12. In a recent University of Hertfordshire study, "What does the MySpace Generation really want?", over 80 percent of those surveyed said they would welcome the opportunity of a legal paid file sharing system. The Swedish Performing Rights Society conducted a similar study in Sweden (a country with relatively relaxed attitudes toward copyright infringement and home to notorious BitTorrent Web site The Pirate Bay) where 86 percent of respondents were interested in paying a voluntary fee for legal downloading. See Glenn Peoples, "Study: 86% Would Pay For Legal P2P," Billboard.biz (April 2009).
13. See "Since 1972 $75,000 package," joshfreese.com
14. Interview with OurStage founder Ben Campbell on indie-music.com (November 3, 2007).
15. Interview with Ben Campbell on Hypebot.com (2008).
16. Interview with Campbell, indie-music.com (November 3, 2007).
17. OurStage Facebook Page (accessed May 9, 2010).
18. Jake Coyle, "Radiohead's experiment rocks music industry," msnbc (October 24, 2007).
19. hitRECord.org (accessed May 9, 2010).

20. "hitRECord & Tumblr," hitRECord Tumblr Page (October 2009).
21. RegularJOE and Jared, "hitRECord: The New Deal," hitRECord.org (January 1, 2010).
22. hitRECord.org Terms of Service (accessed May 9, 2010).
23. See hitRECord YouTube Video "NewDeal" (January 20, 2010).
24. Amanda Lenhart, Mary Madden, Aaron Smith, and Alexandra Macgill, "Teens and Social Media," Pew Internet & American Life Project (December 19, 2007).
25. Andy Holloway, "Can Nettwerk save the record industry?", *Canadian Business* (May 22, 2006).
26. Andy Greenberg, "Free? Steal It Anyway," *Forbes* (October 16, 2007).
27. "France passes controversial anti-piracy bill," CBC News (May 12, 2009).
28. Charles Arthur, "Digital economy bill rushed through wash-up in late night session," *The Guardian* (April 8, 2010).
29. Cory Doctorow, "Digital Economy Act: This means war," *The Guardian* (April 16, 2010).
30. Victor Keegan, "Let's dance to a new tune," *The Guardian* (November 23, 2006).
31. Joseph A. Schumpeter, *Capitalism, Socialism and Democracy* (Harper, 1975).

Chapter 13: The Future of Television and Film

1. Leo Laporte, "What's TWiT Worth To You," TWiT Blog (December 3, 2009).
2. Clyde Bentley and Donald Reynolds, "ONA Keynote: 'Lunch with Leo,' Leo Laporte, Host, This Week in Tech (#ONAkey)," Donald W. Reynolds Journalism Institute (October 2, 2009).
3. "Americans Watching More TV Than Ever; Web and Mobile Video Up Too," Nielsen Wire (May 20, 2009).
4. "3-D movies help boost movie box office worldwide to record $30 billion," Associated Press (March 10, 2010).
5. Alex Dobuzinskis, "Film box office overtakes 2009 DVD, Blu-ray sales," Thomson Reuters (January 4, 2010).
6. Ryan Junee, "Zoinks! 20 Hours of Video Uploaded Every Minute!", Official YouTube Blog (May 20, 2009).
7. Peter Farquhar, "Fans pay the ultimate tribute with Casey Pugh's Star Wars Uncut: The New Hope," news.com.au (April 14, 2010).
8. See starwarsuncut.com (accessed May 15, 2010).
9. "Hollywood hits restart as more filmmakers venture into video games," *Los Angeles Times* (June 1, 2009).
10. "World of Warcraft Subscriber Base Reaches 11.5 Million Worldwide," Blizzard Press Release (December 23, 2008).
11. See WoWJutsu Guild Rankings (accessed May 15, 2010).
12. Heath Brown et al., *Diary of a Camper*, United Ranger Films (1996).
13. Adam Pasick, "HBO buys film made in Second Life," Thomson Reuters (September 4, 2007).
14. See YouTube video "Episode #1: My Second Life" (accessed May 15, 2010).
15. "About Us," SnagFilms.com (accessed May 15, 2010).
16. Ibid.
17. Wayne Friedman, "Nielsen: More TV in Homes," *Media Daily News* (April 28, 2010).
18. "Americans Watching More TV Than Ever; Web and Mobile Video Up Too," Nielsen Wire (May 20, 2009).
19. "Special Report on Television," *The Economist* (May 1, 2010).

20. Michael Dance, "Network TV Is Dead," *Co-ed Magazine* (April 29, 2009).
21. "List of most-watched television broadcasts," Wapedia (accessed May 15, 2010).
22. Gary Levin, "Nielsens: 'Idol' opens strong for its ninth season," *USA Today* (January 20, 2010).
23. David Bauder, "'American Idol' Ratings Drop 10% Still Wins," *The Huffington Post* (January 14, 2009).
24. Marisa Guthrie, "The Ruling Class of TV News," *Broadcasting & Cable* (September 21, 2009).
25. Adrian McCoy and Rob Owen, "Americans so caught up in the 'Net, they would set TV free: For the first time, more people would do without television than the Internet," *Pittsburgh Post-Gazette* (April 9, 2010).
26. "Lady Gaga hits 1 billion video views," TheBigTop40.com (March 26, 2010).
27. Jefferson Graham, "Hulu cues up for next step; Free site hits milestone: 1 billion videos viewed," *USA Today* (February 3, 2010).
28. Ibid.
29. Michael Starr, "Almost 'Idol'.Com—'AI' Creator Pushes Past Cowell To Next Big Thing," *New York Post* (March 2, 2010).
30. Ibid.
31. "Super Bowl attracts record-setting 106.5 million viewers," CBS Sports Press Release (February 8, 2010).
32. The 2009 World Series, for example, which saw the New York Yankees defeat the Philadelphia Phillies in six games, had only 22 million viewers for the final game. The NFL, NBA, MLB, and NHL operate their own television networks, streaming games over the Internet for fans who can't get enough sports action through cable or satellite television.
33. The Grammys were up 35 percent in 2010 over 2009. See "Grammy Viewership Went Up 35% This Year, Social Media Was Part of the Reason," mediabistro.com (Feburary 4, 2010).
34. Misty Harris, "Social media boosts power of TV; Boob-tube is benefitting from the digital realm as viewers go online with their thoughts," *The StarPhoenix* (March 17, 2010).
35. "Research: viewers are more engaged with what they watch," Entertainment Marketing Letter (April 1, 2010).
36. Stephen Shankland, "Google to test ultrafast broadband to the home," CNET News (February 10, 2010).
37. Richard Lai, "Google receives 'more than 1,100 community responses' for gigabit fiber network," Engadget (March 28, 2010).
38. Marguerite Reardon, "Verizon bets big on network infrastructure," CNET News (October 6, 2008).
39. "Cisco Visual Networking Index: Forecast and Methodology, 2008–2013," Cisco Report (June 9, 2009).
40. Ibid.
41. James Quinn, "Google sets its sights on television dominance," *Daily Telegraph* (March 10, 2010).
42. "From goggle to Google: TV meets the Internet," *The Independent* (March 10, 2010).
43. Chad Hurley, "Y,000,000,000uTube," The Official YouTube Blog (October 9, 2009).
44. Brian Stelter, "YouTube Videos Pull In Real Money," *The New York Times* (December 10, 2008).
45. Ibid.

46. "BBC scoops Digital Emmy Award for The Virtual Revolution," BBC Press Release (April 13, 2010).

Chapter 14: Creating Public Value

1. "Home," AppsForDemocracy.org (accessed May 19, 2010).
2. "Forums," TheStudentRoom.co.uk (accessed May 19, 2010).
3. "About Us," Netmums.com (accessed May 19, 2010).
4. "Money Saving Expert," thegoodwebguide (accessed May 19, 2010).
5. See "Country clean-up project 'Let's Do It 2008' Teeme Ara 2008," YouTube video (February 17, 2009).
6. David Mardiste, "Estonians scour country for junk in big clean up," Thomson Reuters (May 3, 2008).
7. Jeffrey M. Jones, "Trust in Government Remains Low," Gallup (September 18, 2008).
8. "About Us," mySociety.org (accessed May 19, 2010).
9. Dan Jellinek, "Focus—Participatory Budgeting: Who Wants to Spend a Million Dollars?", Headstar e-government bulletin (October 2, 2008).
10. Chris Elmendorf and Ethan J. Leib, "Budgets by the People, for the People," *The New York Times* Op-Ed (July 27, 2009).
11. Dan Jellinek, "Special Focus—Participatory Budgeting, Part 2: Sympathy for the Devil?", Headstar e-government bulletin (November 24, 2008).
12. Henry R. Nothhaft and David Kline, "The Biggest Job Creator You Never Heard Of: The Patent Office," Harvard Business Review Blog (May 6, 2010).
13. Liz Allen, "Your chance to participate in Patent review—Peer to Patent needs you," Public Library of Science (November 11, 2007).
14. Tom Steinberg, Ed Mayo, "Digital Engagement," The Power of Information Task Force (2007).
15. Ibid.
16. Professor Hans Rosling, "New insights on poverty and life around the world," TED (March 2007). See http://www.ted.com/talks/view/id/140.
17. "About," MapEcos.org (accessed May 19, 2010).
18. In most countries, the public sector is encumbered by complex institutional legacies that encompass hundreds of separate departments across multiple levels of government. The resilience of government institutions is profound, and their historical role has created a sense of permanence that has proven difficult to shake. Their complexity makes even the task of deciding on a starting point for change difficult—in one government, among several, in one department, at the organizational level, in the legislative arena, big or small? And, unlike the private sector, there is no single CEO in charge who can marshal resources on an enterprise-wide basis to accomplish the necessary changes.

    The inertia of old models is compounded by the political nature of government bureaucracies. Civil servants and their political masters are conditioned to avoid mistakes that could be used by opponents or the media to embarrass the party in power. Rather than develop flexible systems that give civil servants discretion, rigid systems of public administration make discretion virtually impossible. Even simple travel arrangements require endless forms and numerous signatures. Straightforward purchases take months; larger ones take years. This emphasis on process steals resources from the real job of serving the customer and engaging the constituent.

19. "About Us" and "Chapters," YoungGovernmentLeaders.org (accessed May 19, 2010).
20. "Members," GovLoop.com (accessed May 19, 2010).
21. Karoline Piercy, "Igniting innovation and collaboration," NetworkedGovernment (June 2009).

Chapter 15: The Rise of the Citizen Regulator

1. Eric Paulos, Ian Smith, and R. J. Honicky, "Participatory Urbanism," urban-atmospheres.net (accessed May 18, 2010).
2. Gardiner Harris, "Salmonella Was Found at Peanut Plant Before," *The New York Times* (January 28, 2009).
3. Moni Basu and Michelle E. Shaw, "Anatomy of the peanut salmonella outbreak," *The Atlanta Journal-Constitution* (March 15, 2009).
4. Michael Moss, "Peanut Case Shows Holes in Safety Net," *The New York Times* (February 9, 2009).
5. Regulators claim that they plan to try to change the law to require greater disclosure of food safety tests. But some officials in the FDA view such disclosures as a "double-edged sword" that might inhibit some companies from testing in the first place. On the other hand, in examining Peanut Corporation of America's records, federal investigators discovered that company tests had found salmonella twelve times since 2007. The inspectors claimed they only got the records by invoking a bioterrorism law.
6. Lisa Shames, "Federal Oversight of Food Safety," United States Government Accountability Office (June 12, 2008).
7. Robert Cooter, "Decentralized Law for a Complex Economy: A Structural Approach to Adjudicating the New Law Merchant," *University of Pennsylvania Law Review*, vol. 144 (1996).
8. Ann Florini, *The Right to Know: Transparency for an Open World* (Columbia University Press, 2007).
9. Julia Finch, "Twenty-five people at the heart of the meltdown," *The Guardian* (January 26, 2009).
10. Zachary A. Goldfarb, "SEC Chief Strives To Rebuild Regulator: Scrutiny Intensified In Financial Crisis," *The Washington Post* (June 4, 2009).
11. The prototype is showcased on Nokia's Web site. See http://www.nokia.com/environment/we-create/devices-and-accessories/future-concepts/eco-sensor-concept.
12. In a parallel experiment, graduate students in Paulos's newly created Living Environments Lab at Carnegie Mellon have loaded households with sensors to sample tap water and indoor-air quality. Results are uploaded to a Web site where participants can compare them with other people's contributions.
13. Abby Tan, "Cell phones may be key to cleaner air in Philippines," CSMonitor (July 19, 2002).
14. Ibid.
15. Since 2001, GlobeScan, a market research and public opinion polling firm, has been asking consumers across twenty-five countries to rate the extent to which large companies should be held responsible for a variety of different actions. These actions include operationally oriented responsibilities—in other words, actions that are directly related to a company's operations and that are seen as the standards companies should achieve in their normal course of business (e.g., not

harming the environment, ensuring a responsible supply chain, treating employees fairly, having the same high standards wherever it operates). They also include citizenship responsibilities: socially oriented actions that companies need not undertake in their normal business operations, but which can have a positive impact on people's impression of the company (e.g., improving education and skills in communities where they operate, increasing global economic stability, reducing human rights abuses, helping to solve social problems). Overall, CSR expectations are universally high, with more than eight in ten people across all countries surveyed saying that companies should be held at least partially responsible for all fourteen of the social, environmental, and economic actions tested. See Femke de Man, "Tracking the Gap between Societal Expectations of Companies and Perceived CSR Performance," Center for Corporate Citizenship, Boston College, Carroll School of Management (September 2007).

16. "About Us," Crocodyl.org (accessed May 18, 2010).
17. "CrocTail: making government data useable," *The CSR Digest* (July 8, 2009).
18. One enthusiast built a custom Google search engine that users can use to quickly browse information on a given corporation. Type in a company name and it spits out a list of recent pages, prioritized from a list of Web sites that focus on corporate scrutiny. Hit the "Controversy" link, and one can narrow the results using a list of keywords such as "human rights," "lawsuit," "labor violation," "superfund," and "abuse."
19. "Annual Report on the OECD Guidelines for Multinational Enterprises 2008: Employment and industrial relations," Organisation for Economic Co-operation and Development (March 23, 2009).
20. See the California Institute for Technology and Carnegie Mellon experiments where low-power intelligent wireless sensors measure everything from temperature to movement to chemical composition and report that information back in real time.
21. In some cases, the most up-to-date data can be found on the Web sites of national chapters such as Canada's. See http://www.globalforestwatch.ca.
22. Although the costs of GIS data are declining, accurate assessments of ecosystem health cannot be provided by GIS alone; they rely on government data about logging concessions (among other things) and require local researchers to perform extensive on-the-ground research that can take years to complete.
23. Matthew Steil and Jean Sylvestre Makak, "Working Towards Greater Forest Sector Transparency in Gabon," World Resources Institute (June 11, 2009).
24. The Interactive Forest Atlas for Gabon was officially launched in Libreville in May 2009. The Atlas can be accessed at: http://www.wri.org/publication/interactive-forestry-atlas-gabon.
25. The countries include: Brazil, Cameroon, Canada, Central Africa, Democratic Republic of Congo, Indonesia, Peru, Russia, and Venezuela.
26. Wade-Hahn Chan, "4 studies in collaboration; Case 3: Puget Sound Information Challenge," *Federal Computer Week* (February 29, 2008).
27. The current membership includes: Barry Callebaut, Cadbury Schweppes, Education International, European Cocoa Association, Ferrero, Free the Slaves, Global March Against Child Labour, Hershey Foods, International Confectionery Association, International Trade Union Confederation (ITUC), International Union of Food, Agricultural, Hotel, Restaurant, Catering, Tobacco and Allied Workers Associations (IUF), Kraft Foods, Mars Incorporated, Nestlé, National Consumers League, WAO Afrique, and Toms. Mr. P. Gillioz (Swiss lawyer) is also

a member of the ICI. See http://www.cocoainitiative.org/structure-membership-and-financing.html.
28. Leslie Kaufman and David Gonzalez, "Labour Standards Clash with Global Reality," *The New York Times* (April 24, 2001).
29. These last two recommendations are in tension. The ability to exert influence often depends on the capacity to mobilize narrow interests. However, the legitimacy to make public policy depends on the willingness of these same groups to think more broadly about the public interest.
30. Roger Martin, "The Virtue Matrix," *Rotman Management Journal* (Spring/Summer 2003).
31. John Braithwaite and Peter Drahos, *Global Business Regulation* (Cambridge University Press, 2000).

Chapter 16: Solving Global Problems

1. Kiva Facebook Page (accessed May 20, 2010).
2. "First loan he gave was $27 from own pocket," *The Daily Star* (October 14, 2006).
3. "1,500 farmers commit mass suicide in India," *The Independent* (April 15, 2009).
4. "Nobel Laureate Muhammad Yunus: Changing the System, One Social Business at a Time," *University of Virginia News* (September 21, 2009).
5. Kiva Presentation at TiEcon 2009 (2009).
6. "Facts & History," Kiva.org (accessed May 20, 2010).
7. Carol Realini, "Obopay CEO Shares Strategy to Transform Mobile Payments," The Official Obopay Blog (March 9, 2010).
8. "Global Agenda Partnership Concept Paper," World Economic Forum (April 2010).
9. "The State of Consumption Today," Worldwatch Institute (accessed May 20, 2010).
10. Ray S. Baker and William E. Dodd, *The Public Papers of Woodrow Wilson*, Authorized Edition, Vol. 1 (Harper, 1924).
11. Dominic Wilson and Anna Stupnytska, "The N-11: More Than an Acronym," Goldman Sachs Global Economics Paper No. 153 (March 28, 2007).
12. "Global Redesign Summit 2010," World Economic Forum (accessed May 20, 2010).
13. Francis Matthew, "Experts seek to plug holes in top global institutions," *Gulf News* (November 20, 2009).
14. Peter Hall-Jones, "The rise and rise of NGOs," Public Services International (2006).
15. Ibid.
16. Harvard professor Theda Skocpol, for one, has documented these transformations in civil society and worries that too many valuable aspects of the old membership-based civic tradition are not being reproduced or reinvented in the world of "memberless organizations." Theda Skocpol, "Associations Without Members," *The American Prospect* (July/August 1999).
17. Anis Salvesen, "Philanthropy: Change for Change Part II," RedGage (August 2009).
18. "USAID Development 2.0," NetSquared.org (2009).
19. "Challenges," NetSquared.org (accessed May 20, 2010).
20. Heide Malhotra, "Companies Embrace 'Open Business Model,'" *The Epoch Times* (August 4, 2009).
21. "Habitat Jam: Summary Report" (2005).
22. Debbe Kennedy and Joel A. Barker, *Putting Our Differences to Work: The Fastest Way to Innovation, Leadership and High Performance* (Berrett-Koehler, 2008).

23. "Habitat Jam: Summary Report" (2005).
24. Kennedy and Barker, *Putting Our Differences to Work.*
25. "Election 2010 National Results," BBC (May 2010).

Chapter 17: Fighting for Justice

1. Incredibly, there were also more than 19 million blog posts on the topic both in and outside of Iran (2.25 million in twenty-four hours); YouTube: 184,500 videos listed on the site (3,000 posted in twenty-four hours, representing 6,000 minutes of protest footage); Ben Parr, "Mindblowing #IranElection Stats: 221,744 Tweets Per Hour at Peak," Mashable (June 2009).
2. To be sure, the updates on Twitter were fragmented and devoid of a contextual narrative (except that conveyed by Western media outlets) that might have portrayed a more complete picture of what is really happening in the political battles over Iran's future. Journalists and other interested observers were left to flesh out the skeleton of agency copy and social media updates, on one hand, while attempting to pick apart content from the local Iranian media, which is widely known to be subject to government manipulation.

   Careful analysis reveals that the mainstream media misconstrued or glossed over the nuanced nature of political support on the ground. Reva Bhalla, director of analysis for Stratfor, a Texas-based strategic intelligence and forecasting company, observes that "You can get the notion that Ahmadinejad is very unpopular and that [Mir Hossein] Mousavi has this groundswell of support, but we don't have data that shows that . . . Ahmadinejad has real support, but his supporters don't have smartphones. There is a real risk of amplifying [one side]."

   Ahmadinejad supporters who did use social media got far less coverage than opposition supporters. For example, of the hundreds of stories logged regarding the use of blogs and Twitter related to the 2009 election, you can count on one hand the number that highlight the use of the Internet by Ahmadinejad supporters despite activity on this front. Their posts and stories, however, tend to be written in Farsi rather than English, thus limiting their accessibility outside of Iran.
3. Its annual *Freedom in the World* report is a respected global benchmark assessment of the state of political rights in 192 countries and 14 related and disputed territories.
4. "Undermining Democracy: 21st Century Authoritarians," Freedom House (June 2009).
5. James Gwartney and Robert Lawson, "Ten Consequences of Economic Freedom," National Center for Policy Analysis (July 30, 2004).
6. We do know, however, that WikiLeaks was founded by an Australian (Julian Assange) who lives in East Africa.
7. Evgeny Morozov, "Think Again: The Internet," *Foreign Policy* (May/June 2010).
8. Experts suggest that the differences between Mir Hossein Mousavi and Mahmoud Ahmadinejad are far less significant than what unites them. In the aftermath of their failed attempts to overturn the election results, many pro-democracy activists started to question whether Mousavi truly embraces their ideals.

   Mousavi has been described by many as an unlikely hero—a convenient vehicle for a larger reform movement long in the making. After all, more seasoned pro-democracy activists will recall that Mousavi was a top leader of the Islamic Republic during the most repressive years, when tens of thousands of left-wing, liberal, and secularly oriented people, especially union leaders and women's rights advocates, were tortured and assassinated. He was a key person in the clerical establishment, and his main base is among the Muslim clergy. The fact that he now finds himself leading the charge against the very system that he helped create is ironic to say the least.

Ahmadinejad, on the other hand, is more republican oriented, and has a stronger base among the military and the oil-sector bourgeoisie than among the clergy, despite the support from the leading ayatollah. Both represent different layers and sectors of the Iranian elite, but have different international connections and allegiances. And, if the more cynical elements of the reform movement are to be believed, both stand as obstacles to genuine democratic rights for the mass of the population, and both are happy to leave the clergy in control of social norms.

9. "Tehran voices: 'I was out in the streets 30 years ago and today I'm out again,'" *The Guardian* (December 11, 2009).
10. Ali Akbar Dareini, "Fighting intensifies in Iran over subsidy cuts," *BusinessWeek* (March 10, 2010).
11. See Iran age distribution pyramid on NationMaster (accessed May 23, 2010).
12. "Iranians send 80 million SMS per day," Payvand Iran News (November 24, 2008).
13. John Kelly and Bruce Etling, "Mapping Iran's Online Public: Politics and Culture in the Persian Blogosphere," The Berkman Center for Internet & Society, Harvard Law School (April 6, 2008).
14. See Technorati 2009 State of the Blogosphere.
15. "10 Worst Countries to Be a Blogger," Committee to Protect Journalists (April 30, 2009).
16. Bruce Etling, "The 'Freedom to Scream' in Egypt," Internet & Democracy Project, Harvard University (February 18, 2009).
17. Clark Boyd, "Arabic blogosphere begins to bloom," BBC (June 18, 2009).
18. Natasha Tynes, "G21 MidEast Arab Blogging," G21.net (2006).
19. Herbert Moller, "Youth as a Force in the Modern World," *Comparative Studies in Society and History*, vol. 10, no. 3 (April 1968).
20. Ibid.
21. WorldPublicOpinion.org undertook a poll of a nationally representative sample of Iranians. Interviewing was conducted August 27–September 10, 2009, among a national sample of 1,003 Iranian adults age eighteen and older. The margin of error for a sample of this size is no larger than +/- 3.1 percentage points.
22. Melody Moezzi, "Iranian Regime Faces Worst Enemy Yet: Itself," *The Huffington Post* (September 19, 2009).
23. Larry Jagan, "Fuel price policy explodes in Myanmar," *Asia Times* (August 24, 2007).
24. "Monks lead largest Burma protest," BBC (September 24, 2007).
25. Maung Zarni, "Burma Special: Why we must talk to the generals," *New Statesman* (August 14, 2006).
26. See United Nations Human Development Report (2009).
27. In 1995, one of those students, Maung Zarni—then a Burmese activist studying at the University of Wisconsin—founded the Free Burma Coalition (FBC). The site quickly evolved from a single Web page that aggregated several sources of information about Burma into a network of activists at hundreds of educational institutions in North America and twenty-eight other countries. It was more of a clearinghouse than a command center. Activists used the FBC site to swap ideas and intelligence about new companies and political bodies to target, along with suggested organizing and advocacy strategies.

28. Tim Johnston, "U.S. Seeks New Tack on Burma," *The Washington Post* (April 12, 2009).
29. EarthRights International, an organization of lawyers that specialize in human rights, sued U.S. energy giant Unocal to force action. Unocal eventually provided out-of-court compensation to villagers who are believed to have toiled as slave labor for the Yadana gas pipeline from southern Burma to Thailand. However, hundreds of thousands of ethnic villagers have been forced to relocate or have been conscripted into chain gangs to pave the way for similar investments from Chinese, Russian, South Korean, and Vietnamese companies. But these companies are not as susceptible to the tactics of U.S.-based NGOs as their American counterparts. Since many American and European multinationals have already divested from Burma, it makes sense for Western business leaders to put more public pressure on their foreign rivals.
30. See World Bank 2008 population data and Nielsen 2009 Internet penetration data. China has an estimated total of 300 million Internet users, representing 17 percent of the world's 1.8 billion global Internet users.
31. Jane Marshall, "CHINA: Record numbers studying abroad," University World News (October 19, 2008).
32. Evan Osnos, "Angry Youth: The new generation's neocon nationalists," *The New Yorker* (July 28, 2008).
33. See Human Rights Watch World Report (2002).
34. Mark Ward, "China clampdown on tech in Urumqi," BBC (July 6, 2009).
35. David Drummond, "A new approach to China," The Official Google Blog (January 12, 2010).
36. Maria Livanos Cattaui, "The test of practice: global progress in a world of sovereignty," openDemocracy (July 18, 2004).
37. According to organized crime experts, it's not an idle scenario. Of the almost six thousand Russian criminal syndicates operating today, more than two hundred Russian mafia structures operate internationally, with networks in twenty-nine different countries where criminal activity is closely coordinated with local syndicates. Taking virtually unlimited funds with them from Russia, these criminals buy up property from Buenos Aires to Berlin, paying $3–10 million in cash. Once settled, Russian criminals waste little time making themselves at home—getting involved in racketeering and extortion, control over prostitution, and other forms of illegal business. Gradually they consolidate their positions and begin to intervene actively in the economic life of the countries in which they reside, establishing control over industrial and financial structures with drug trafficking and gambling, while eventually displacing the local mafiosi who had first supported them. See Stanislav Lunev, "Russian organized crime spreads beyond Russia's borders, squeezing out the local competition," The Jamestown Foundation (May 30, 1997).

Chapter 18: Ground Rules for Reinvention

1. Seth Godin, *Tribes: We Need You to Lead Us* (Portfolio, 2008).
2. Sean Michael Kerner, "You Can't Control Linux," CIO Update (April 14, 2010).
3. Dennis Whittle, "GlobalGiving.com—Globalizing opportunity," *Alliance* (June 1, 2003).
4. "Ready, Set, Give," GlobalGiving.com (accessed May 21, 2010).
5. "The Tube for IDEO", IDEO.com (accessed May 21, 2010).
6. "The knowledge: Euan Semple," InsideKnowledge (June 16, 2005).

7. Jeanine Plant, "Top 5 Issues That Motivate Young Voters Today," WireTap (October 23, 2006).
8. Jackie Crosby, "Entrepreneur turned Geek Squad into a geek army," *Los Angeles Times* (April 1, 2010).

Chapter 19: Leadership for a Changing World

1. See YouTube video "Starlings on Otmoor" (February 21, 2007).
2. Dan Reed, "The 2010 Time 100: Jaron Lanier," *Time* (April 29, 2010).
3. Jaron Lanier, *You Are Not a Gadget: A Manifesto* (Knopf Doubleday, 2010).
4. Joshua Topolsky, "Live from Apple's iPhone OS 4 event!", Engadget (April 8, 2010).
5. Jean Tate, "Click on Hubble: Galaxy Zoo Now Includes HST Images," Universe Today (April 22, 2010).
6. David Morgan, "Nearly 20 percent of U.S. workers underemployed," Thomson Reuters (February 23, 2010).
7. Daniel Henninger, "Joblessness: The Kids Are Not Alright," *The Wall Street Journal* (April 12, 2010).
8. Ronald Brownstein, "Young People Seek Shelter from the Storm," *National Journal* (May 8, 2010).
9. Louisa Peacock, "IBM crowd sourcing could see employed workforce shrink by three quarters," *Personnel Today* (April 23, 2010).
10. "Where Will the Jobs Come From?" Kauffman Foundation Research Series: Firm Formation and Economic Growth (November 2009).
11. Doug Tsuruoka, "Amazon.com's Third-Party Sales Balloon," Investors.com (July 4, 2009).
12. Adam Pagnucco, "Why MPW Turned Down *The Washington Post*," Maryland Politics Watch (April 20, 2010).
13. Laura M. Holson, "Tell-All Generation Learns to Keep Things Offline," *The New York Times* (May 8, 2010).
14. Don Tapscott, *Grown Up Digital: How the Net Generation Is Changing Your World* (McGraw-Hill, 2008).
15. Ibid.
16. Nicholas Carr, "Is Google Making Us Stupid?", *The Atlantic* (July/August 2008).
17. Nicholas Carr, *The Shallows: What the Internet Is Doing to Our Brains* (Atlantic Books, 2010).
18. James Harkin, *Lost in Cyburbia: How Life on the Net Has Created a Life of Its Own* (Knopf Doubleday, 2009).
19. Evgeny Morozov, "Think Again: The Internet," *Foreign Policy* (May/June 2010).
20. Nicholas Kimbrell, "Reinventing Detroit," *The National* (February 27, 2010).
21. "OECD Environmental Outlook to 2030: Key Results," Organisation for Economic Co-operation and Development (accessed May 15, 2010) and the CIA World Factbook global population estimate of 6.8 billion people. Also, videos like http://www.youtube.com/watch?v=BoypPMbSLp4 created by the World Economic Forum and partners play into our ignorance of the impact of water systems on our existence, the scope of which could fill its own entire book.
22. See TED Video "Carolyn Steel: How food shapes our cities" (October 2009).
23. "The Ripple Effect from IDEO," IDEO (2009).
24. See the last chapter of *Wikinomics*.
25. Bradford Plumer, "Is There Enough Food Out There For Nine Billion People," *The New Republic* (February 3, 2010).

INDEX